Atomic Number	Element's Name	Chemical Symbol	Atomic Number	Element's Name	Chemical Symbol
1	Hydrogen	H	54	Xenon	Xe
2	Helium	He	55	Cesium	Cs
3	Lithium	Li	56	Barium	Ba
4	Beryllium	Be	57	Lanthanum	La
5	Boron	B	58	Cerium	Ce
6	Carbon	C	59	Praseodymium	Pr
7	Nitrogen	N	60	Neodymium	Nd
8	Oxygen	O	61	Promethium	Pm
9	Fluorine	F	62	Samarium	Sm
10	Neon	Ne	63	Europium	Eu
11	Sodium	Na	64	Gadolinium	Gd
12	Magnesium	Mg	65	Terbium	Tb
13	Aluminum	Al	66	Dysprosium	Dy
14	Silicon	Si	67	Holmium	Ho
15	Phosphorus	P	68	Erbium	Er
16	Sulfur	S	69	Thulium	Tm
17	Chlorine	Cl	70	Ytterbium	Yb
18	Argon	Ar	71	Lutetium	Lu
19	Potassium	K	72	Hafnium	Hf
20	Calcium	Ca	73	Tantalum	Ta
21	Scandium	Sc	74	Tungsten	W
22	Titanium	Ti	75	Rhenium	Re
23	Vanadium	V	76	Osmium	Os
24	Chromium	Cr	77	Iridium	Ir
25	Manganese	Mn	78	Platinum	Pt
26	Iron	Fe	79	Gold	Au
27	Cobalt	Co	80	Mercury	Hg
28	Nickel	Ni	81	Thallium	Tl
29	Copper	Cu	82	Lead	Pb
30	Zinc	Zn	83	Bismuth	Bi
31	Gallium	Ga	84	Polonium	Po
32	Germanium	Ge	85	Astatine	At
33	Arsenic	As	86	Radon	Rn
34	Selenium	Se	87	Francium	Fr
35	Bromine	Br	88	Radium	Ra
36	Krypton	Kr	89	Actinium	Ac
37	Rubidium	Rb	90	Thorium	Th
38	Strontium	Sr	91	Protactinium	Pa
39	Yttrium	Y	92	Uranium	U
40	Zirconium	Zr	93	Neptunium	Np
41	Niobium	Nb	94	Plutonium	Pu
42	Molybdenum	Mo	95	Americium	Am
43	Technetium	Tc	96	Curium	Cm
44	Ruthenium	Ru	97	Berkelium	Bk
45	Rhodium	Rh	98	Californium	Cf
46	Palladium	Pd	99	Einsteinium	Es
47	Silver	Ag	100	Fermium	Fm
48	Cadmium	Cd	101	Mendelevium	Md
49	Indium	In	102	Nobelium	No
50	Tin	Sn	103	Lawrencium	Lw
51	Antimony	Sb	104	(Kurchatovium)	Ku
52	Tellurium	Te	105	(Hahnium)	Ha
53	Iodine	I			

Chemistry, Man, and Environmental Change
An Integrated Approach

J. Calvin Giddings
The University of Utah

ILLUSTRATIONS BY ALEXIS KELNER

CANFIELD PRESS ⏀ SAN FRANCISCO
A Department of Harper & Row, Publishers, Inc.

New York   *Evanston*   *London*

To my wife Sue
and other river folk:
Roger, Jim, and J.

Sponsoring Editor: *R. Wayne Oler*
Assistant Developmental Editor: *Malvina Hindus*
Editorial and Design Supervision: *Brian Williams*
Production and Manufacturing Supervision: *Christine Schacker*
Designer: *Joseph Fay*
Copyeditor: *Edna Miller*
Cover photo: *Robert Ames, Photofind, S.F.*

Contents

Preface

The main characters in the drama of man's changing environment are chemicals. The roles are acted out by water, oxygen, mercury, lead, DDT, carbon monoxide, sulfur dioxide, nitric oxide, benzopyrene, phosphates and plutonium—as fine a cast of heroes and villains as one is likely to find anywhere. But this drama is deadly serious, with man's fate hinging on the outcome. The rules of the play are the rules of chemistry. In this book I have attempted to present these rules simply and clearly. I have used the rules to follow the ebb and flow of the cast across the landscape of earth; to chart their repercussions for man; and to predict the direction of future entanglements between these chemical actors and human society.

Very simply, then, *Chemistry, Man, and Environmental Change* was written to forge the crucial link between chemical principles and our natural environment. The book is intended for nonscience students who desire an introduction to the science of environment without an undue burden of scientific detail. No previous chemistry on the part of the student is presumed.

The text of this book is functionally divided into two parts. The first part—consisting of the initial five chapters—establishes a background in chemistry, showing at each stage how chemical fundamentals have a major bearing on current environmental questions. The close integration of chemical topics and environmental consequences is unique among texts for introductory chemistry courses. This focus is continued and expanded through the second portion of the book.

Chapters 6 through 11 comprise the second part. Here we explore, chapter by chapter, major environmental topics, emphasizing their chemical origins. These last six chapters have been written in such a way that they can be read in any sequence. If time is short, any chapter or combination of chapters may be omitted. This approach offers great flexibility to the instructor and the reader in selecting and arranging environmental topics to suit their own purposes and time budget.

The exercises in this book deserve particular attention. They are intended to stimulate the student's interest and provide him with a widened perspective, while at the same time encouraging him to become a quantitative environmental critic. The exercises of the second part are particularly directed at current environment problems and their broad implications.

The usefulness of *Chemistry, Man, and Environmental Change* has been greatly enhanced by four coordinating supplements to the text: the laboratory manual by Dorothy Barnes of the University of Massachusetts at Amherst, the student guide by Edna Miller, the instructor's manual by Jack Healey of Chabot College in Hayward, California, and *Our Chemical Environment*, a reader in environmental chemistry by Manus B. Monroe and me.

Many people were instrumental in the creation of this book. Professor David M. Grant of the University of Utah, aware of my long-standing interest in the environment, suggested in 1970 that I develop a course in environmental chemistry for nonmajors. This book is an outgrowth of that course. Going back another dozen years, an exceptional gentleman and outdoorsman named Carl Bauer first kindled my interest in conservation and environmental protection. More recently this book has been aided by the critical readings and suggestions of Professor Ronald O. Ragsdale, Professor Charles Mays, Jim Byrne, and Dr. Manus Monroe. Professor William Weir of Reed College performed two readings of the manuscript—in both its early and later drafts—as did Professor Jack Healey. Henry DePhillips of Trinity College also provided helpful comments and suggestions. My secretary, Dorothy Martin, did exceptional service in the squeeze between my schedule and the publisher's deadlines. And speaking of such deadlines, I would also like to express thanks to Wayne Oler, editor-in-chief of Canfield Press; Brian Williams, former managing editor; and Mal Hindus, assistant developmental editor, for their efforts on behalf of this project.

I believe that the environmental focus will continue to grow and develop as a basis for the teaching of chemical principles in colleges and universities. If this text is to contribute to that development, the criticism, corrections, and suggestions from those who use it will prove essential. I welcome the opportunity to hear from its readers.

J. Calvin Giddings
Salt Lake City, Utah

ALASKA'S MOUNT McKINLEY At 20,320 feet, the highest peak in North America. Alpine tarn in foreground. (*Photo by Ed Cooper/Photofind.*)

1 CHEMISTRY, ENVIRONMENT, AND POLLUTION

NATURE OF ENVIRONMENTAL THREATS

Environmental writers of the past decade have forecast an assortment of catastrophes, including:

- A hot earth, new deserts, melted ice caps, and flooded coastal cities—all due to the billions of tons of carbon dioxide released into the air each year by man's enterprise.
- An atmosphere running out of oxygen because of pollution's threat to the microscopic, ocean-dwelling phytoplankton, a primary source of oxygen on this planet.
- Oceans filling with toxic mercury, washed to sea by the cesspool rivers from cities around the world.
- A stratosphere no longer shielding man from deadly ultraviolet light because of the depletion of its protective ozone screen by nitric oxide exhaust from high-flying SST aircraft.

The key factors in these predicted disasters are chemical: carbon dioxide, oxygen, mercury, ozone, and nitric oxide. All these chemicals were present on earth long before the advent of man. Their amount in any place is determined by dynamic natural cycles that shuttle materials back and forth among earth, air, water, and living creatures. Life has become marvelously adapted to the chemical balance established by such worldwide cycles. If man does threaten the global environment, it is by his tampering with the natural cycles and his disturbance of the decisive chemical balances that these cycles create on our planet.

In this light, most critical questions about environmental degradation must be asked and answered in quantitative terms—in numbers that express absolute or relative amounts. For instance, it is not truly meaningful to ask whether man is releasing mercury to the environment: he has done so since the first land was cleared, the first fire was lit, and the first ore was smelted. However, it is significant to ask if the amount released is enough to disturb natural mercury balances and affect life. Whether or not all the threats posed above (and many others) are real must ultimately be decided in terms of "how much" of this or that chemical is added or altered. We shall confront these and related threats throughout the book, as soon as we establish some basic ideas about the chemical nature of our environment.

THE ROLE OF CHEMISTRY

To justify a chemical study of environment and of environmental pollution, we must answer two questions.

What has chemistry to do with environment and pollution? First of all, pollution is a contamination of environment that is harmful or potentially harmful to human life and other life forms in the total ecological community. Now, whether or not a contaminant is harmful depends on its chemical makeup. Furthermore, the contaminant probably was produced by some chemical process or chemical reaction in the first place, and if it is to be disposed of, it will require another chemical reaction. Thus the genesis of pollutants, the damage they cause, and their removal all involve chemical topics.

Our natural environment is made up of earth, air, water, and living things. We have refashioned it to include concrete, asphalt, plastics, automobiles, and many other synthetic objects and devices. To it we have added man-made pollution. Each of these environmental components is a form of matter.

Chemistry, in a broad sense, is the study of all matter, its composition, its properties, and its transformation from form to form. The principles of chemistry are therefore needed to understand the nature of every form of matter, whether of environmental significance or not. For example, metals, drugs, gasoline, foods, the earth's crust, water, radioactive materials, and the human brain cell all have properties determined mainly by chemical principles. Chemistry is clearly very broad. Here, we focus on a single area of chemistry: our environment. This subject is so important to the future of life on this planet that every educated citizen should have some knowledge of its scientific basis.

In studying environment and pollution, should we limit ourselves to chemical aspects? The environment and its changes at the hand of man are now being studied by biologists, engineers, meteorologists, lawyers, economists, social scientists, urban planners, politicians, and many others, but to understand all aspects of the subject would require years. Perhaps first of all we should learn the basic science of our environment and of its pollution. The laws of science ultimately dictate what we can and cannot do with the environment. Thus all studies and efforts to control our environment and environmental pollution are, in the end, bound by the limits imposed by science.

The science of chemistry is central to environmental problems, but we have no intention of limiting our discussion to chemical aspects. We shall try to show the implications of our chemical studies to health, daily living, social institutions, and the destiny of civilization. We shall emphasize the hard choices we must make if we are to stop and reverse the deterioration of our environment.

The science of environment and environmental pollution is very much a "live" science in which new discoveries are made every day. Controversy and uncertainty surround many topics, as they always do on the growing edge of science.

We shall point out regions of obscurity as we proceed so that we can better appreciate the nature of science and its advances. The growth of science, like most human endeavor, is irregular and characterized by diverse factors. In it are laboratory experiments, theoretical calculations, flashes of insight, suspense, momentary failures, puzzlement, and laborious thought. Eventually a rather clear understanding emerges from all this effort. The world usually sees the results in a concise, finished form, minus the sweat, toil, and doubt.

The science of environment is not a finished edifice, and we can view easily the very human task of its construction.

MATTER AND ATOMS

Matter is composed of *atoms*. Ninety different kinds of these small building blocks occur naturally. Another fifteen are man made. The nature of matter is determined by the atoms that make it up: their kind, their number, and their arrangement. Similarly, a building's appearance and function depend on the type, number, and arrangement of bricks. Because it is built from nearly one hundred different kinds of atomic "bricks," the matter present on earth is very diverse in its properties, as we can observe by looking about. Depending on the atomic architecture, matter can be in the form of a gas, a liquid, or a solid; it can be poisonous or nonpoisonous; explosive or inert; metallic or nonmetallic; and living or nonliving. It can vary in shape, density, flexibility and color. The possible forms are almost endless.

Atoms and Molecules. Atoms often combine with one another to form *molecules*. A molecule, then, is simply a discrete (distinct) cluster of atoms held tightly together. The linkage holding one atom to another is called a *chemical bond*. The formation of molecules from atoms via chemical bonds is the single most crucial step in the architecture of matter.

Single molecules may have anywhere from two to millions of atoms. For instance, the common air pollutant carbon monoxide has two atoms, one of carbon and one of oxygen. Two-atom molecules like this one are called *diatomic* molecules. Sulfur dioxide is a pollutant with three atoms, one of sulfur and two of oxygen, and is called *triatomic*. Some protein molecules, essential to life, are much larger, containing hundreds or even thousands of atoms.

The States of Matter. Matter may exist as a *gas*, a *liquid*, or a *crystalline solid*—the three *states of matter*. The way molecules and atoms are arranged in the three states of matter is illustrated in Figure 1-1.

In a gas, molecules (or atoms, if molecules have not formed) are widely separated from one another. They move freely around, expanding to occupy whatever space is available to them.

In a liquid, molecules or atoms cling closely to one another, but in a rather random and mobile arrangement. Liquids have a fixed volume, determined by the space occupied by the molecules. Liquids are fluid because the molecules can slip over one another. For this reason they adjust to the shape of their container.

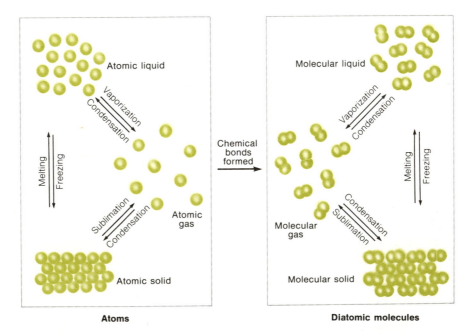

Figure 1-1. States of matter for atoms and diatomic (two-atom) molecules. Note the terms that are used for the processes of changing from one form to another.

In a crystalline solid, atoms or molecules attach to one another in a regular, evenly spaced network. Many metal atoms form crystals. Table salt is a crystal in which charged sodium and chloride atoms occupy alternate sites on the network. Sometimes atoms form molecules first, and then the molecules join together in a regular crystal network. Crystal formation, then, may occur *in place* of molecule formation, or *in addition* to it. In either case, crystalline solids are rigid because atoms and molecules do not readily slip out of their positions in the network.

A so called *pure substance,* or simply *substance*, is matter in its most uniform atomic or molecular condition. It is made up of only one kind of atomic or molecular unit. It may consist of molecules all of one kind or of atoms or sets of atoms all of one kind. It may or may not be crystalline. If the atoms are all one kind, the substance is called an *elementary substance* or simply an *element*. Since there are 90 different kinds of atoms in nature, there are 90 natural elements. For example, pure iron is an element, consisting only of iron atoms arranged in a crystal. Pure oxygen is an element, ordinarily existing as a gas made up of diatomic molecules rather than as a crystal (Figure 1-2).

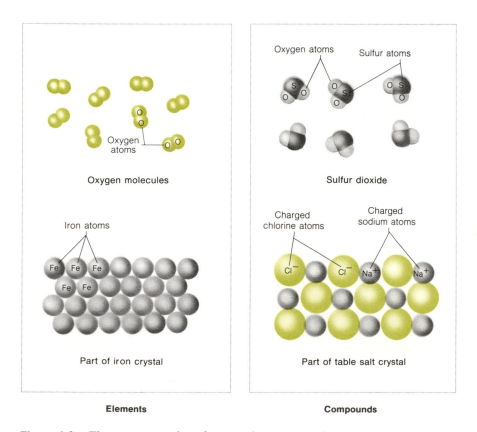

Figure 1-2. The two categories of pure substance are elements and compounds. Examples of each are illustrated in this figure.

If molecules making up a pure substance have different kinds of atoms in them, the substance is called a *compound*. Thus pure sulfur dioxide is a compound because the sulfur dioxide molecule has both sulfur atoms and oxygen atoms. Since atoms can be combined in many different ways to form molecules, there are far more compounds than elements: compounds number in the millions, while elements are limited to the number of basic atomic building blocks, about ninety.

Some crystals in which different kinds of atoms alternate regularly in the network are also called compounds. Common table salt, mentioned above, is an example. A crystal of table salt is absolutely uniform with its precise, alternating array of charged sodium and chlorine atoms. (See Figure 1-2.)

MIXTURES, MIXING, AND THE SPREAD OF POLLUTION

If we stir several pure substances together, we get a *mixture*. A mixture, then, consists of molecules and atoms of different kinds all mixed together.

Air is a good example of a mixture. Natural air has nitrogen molecules, oxygen molecules, argon atoms, carbon dioxide molecules, neon atoms, hydrogen molecules, and various other kinds of molecules and atoms mixed together. However, there are more nitrogen molecules in this mixture than all the rest combined: 78 percent of the total. Oxygen is next at 21 percent. The other substances are in much smaller amounts, each less than 1 percent.

Polluted air has small added amounts of sulfur dioxide, carbon monoxide, nitrogen dioxide, ozone, and many other substances, the exact amount of each depending on the source of the pollution and its proximity. None of these substances is ordinarily present in large amounts, but many are so toxic that one part per million[1] or less (very dilute) may be dangerous to health. Therefore we are particularly interested in the *amount* of different substances, even those present in small amounts, in the mixture known as air.

Although there are exceptions, mixtures usually are formed quickly and spontaneously from the component substances. A sugar cube dissolves in coffee and mixes throughout the cup. Cigarette smoke mixes with the air of a room. The mixing processes may be thought of as the spreading of one or several substances into another. The substance may originally be confined to some small region, but with mixing it occupies an increasing volume and becomes increasingly dilute. This tendency is illustrated in Figure 1-3 for salt mixing into a glass of water. This kind of process occurs again and

---

[1]We shall explain the meaning of this measurement and others a little later.

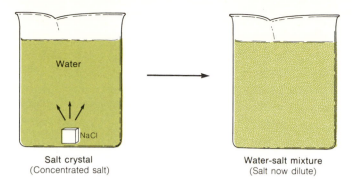

Salt crystal
(Concentrated salt)

Water-salt mixture
(Salt now dilute)

Figure 1-3. Salt dissolving and mixing with water.

again throughout nature. The inevitable mixing process governs the consequences of nearly every form of environmental pollution.

Pollutants generally originate in some relatively small region of space: a car engine, a blast furnace, or perhaps a sewer line. They emerge and spread (Figure 1-4). Smoke from a smokestack or a tailpipe mixes quickly with the surrounding air and eventually spreads into the world's atmosphere. Wastes from a sewage outfall mix with the water of rivers, lakes, and finally oceans. DDT used on a farm to control the cotton bollworm spreads throughout the environ-

Figure 1-4. A column of smoke from trees felled in clearing a ski run in the Wasatch National Forest. Notice how the column broadens as it ascends, dispersing smoke over an ever-widening region of the atmosphere.

ment; traces are found in antarctic wildlife thousands of miles from any place where a substantial amount was used.

The mixing and dilution of pollutants was a very helpful thing in earlier times. Pollutants released to the environment were diluted and carried away, causing no trouble. The environment, both water and atmosphere, seemed like a handy and almost infinite garbage dump in which all wastes could be discarded, diluted, and forgotten.

Now, with an increased population and a widened industry, the environmental reservoir no longer can dilute and dissipate all of man's wastes. We can easily choke a river or dirty the air around a city. The mixing and spreading processes, once advantageous, simply make pollutants harder to capture and control. Now we are beginning to affect the ultimate reservoirs, the world's oceans and atmosphere, through our incessant, reckless dumping.

THERMODYNAMICS, ENERGY, AND NATURAL CHANGE

A scientific discipline basic to both chemistry and environment is *thermodynamics*. Thermodynamics is the study of energy and of natural change. Energy is a key component both of our civilization and of our present environmental crisis. The science of natural change includes all natural processes that have taken place in the past and all that will take place in the future, and it states some important limitations on the degree to which man can influence environmental processes.

Thermodynamics is an extensive and far-reaching discipline. Much of it is abstract and mathematical in nature. Here we will summarize in nonmathematical form the most important conclusions of the two major laws of thermodynamics. But if thermodynamic concepts still seem less than crystal clear after one or two readings, the student should not be discouraged. Much environmental insight can be gained by simply getting a feel for the kinds of forces that underlie the fascinating, relentless changes on planet earth.

First Law of Thermodynamics: Energy.　Energy makes its appearance in many forms: as work; as a capacity for work; as heat; as electrical, gravitational, chemical, and nuclear energy; as the kinetic energy of motion; and so on. The *first law of thermodynamics* is simply a statement that energy is conserved. It says that although energy may be transformed back and forth from one of the above forms to another, its amount is never diminished. We shall refer often to transformations of energy, for they are of great environmental significance. For instance, the chemical energy of gasoline and other fuels is converted to heat energy by burning. The heat energy may then be converted in part to the energy of motion of an automobile, an airplane, or an ocean liner. Nuclear energy in a nuclear power plant is converted to heat energy and then in part to waste heat (thermal pollution) and in part to electrical energy. Important cases crop up endlessly in areas of environmental concern.

Energy may be stored in various ways. We shall call stored energy *potential energy*. It is energy that can be released to do work, to create heat, and so on. Chemical energy (such as the energy stored in fuels), nuclear energy, or simply the energy of a book held above one's head (energy which can be released as kinetic energy when the book is dropped) will all be lumped together as forms of potential energy.

Potential energy is closely related to various natural forces. A book held up in the air has potential energy only because the force of gravity, in pulling the book down, can create motion and kinetic energy. Chemical energy is a consequence of the strong forces between atoms which lead to chemical bonds. Energy and force almost always appear side by side in scientific discussions.

Later we shall refer to an energy crisis or energy shortage facing mankind. Actually, there are enormous quantities of energy around. There is enough heat energy in the oceans to drive civilization for hundreds of centuries. It need only be transformed to an equivalent amount of electrical energy, chemical energy, or some other useful form of energy. Here lies the crux of the matter. The first law of thermodynamics tells us that energy is neither created nor destroyed in its various transformations, but it does not tell us which transformations or changes are possible and which inherently impossible. It so happens that it is not possible to transform more than a very small part of the heat energy of the oceans to useful energy. The crisis is one not of energy, but of *useful* energy—energy in a form that can be utilized by man. The natural law that dictates which transformations are able to produce useful energy and that furthermore spells out the conditions for any natural change is the second law of thermodynamics.

Second Law of Thermodynamics: Natural Change. Through everyday experience we learn to expect certain processes to occur by themselves. We find that other events simply never happen. What rules govern spontaneous occurrence and nonoccurrence? Why study them? To answer the last question first, our focus here is on the environmental consequences of human activity. If we are to understand environmental repercussions, we must know something about natural processes. A simple example is the spontaneous spreading of released pollutants, mentioned earlier. Knowing that a pollutant will mix with the environment, we can judge the consequences of pollutant release (some of it will end up everywhere) and the cure (since it cannot be gathered from everywhere, it must be controlled at the source). Other environmental processes are less obvious and require more critical thought.

The *second law of thermodynamics* gives very precise criteria for the occurrence or nonoccurrence of various imaginable events. From this law we can extract two simple rules which together govern natural occurrences.

Energy (or Force) Criterion. Processes in which potential energy is decreased tend to occur spontaneously. Conversely, processes accompanied by an increase in potential energy tend not to occur. For instance, snow on a steep roof has a high potential energy because of the earth's gravity and will tend to tumble down. Water runs only one way over a waterfall, from a point of high gravitational energy to low. Some of the energy can even be tapped if the falling water is made to drive a turbine-generator. In any event, the force of gravity imparts energy to any object high up, and as gravity pulls the object down, potential energy is released as kinetic energy, which finally becomes heat. The natural trend always is to move down.

Scientifically, it makes no difference whether we think of spontaneous downward motion as due to the force of gravity or to the loss of potential energy. Force and energy, as we mentioned earlier, always go hand in hand. For most students it may be easier to think of natural changes as being pushed along by some force, but the connection with potential energy should be kept in mind.

Other forces influence natural events. Electrical forces pull unlike charges together and push like charges apart. Chemical forces push atoms around. The pressure of the atmosphere is, essentially, a force responsible for many aspects of atmospheric circulation and weather. As these forces act on our natural world, heat is released to the environment.

Heat may be thought of as a final resting place for energy, as Figure 1-5 suggests. The energy of a moving car, for instance, becomes heat energy (Figure 1-5b) because of brake friction or simply the frictional resistance of air. The energy of all moving objects except those in a vacuum is lost to heat energy as they slide, bump, or glide along. If they are to be kept moving in spite of these frictional heat losses, new energy must be supplied continuously via the burning of fuel, or by other means.

While in general all energy tends to become environmental heat, it tends specifically to heat the coldest object around (thus heat will invariably pass from a hot object, such as a stove element, to a cold object, such as a coffee pot). The change from potential energy to low-temperature heat is irreversible. The second law makes it clear that this energy path is one way only, and that energy so transformed cannot by path reversal be used to run machines, generate electricity, or perform other useful tasks.

The relentless and universal degradation of energy to low-temperature heat has led astrophysicists to suggest that our final destiny is wrapped up in the *heat death of the universe*. Whether their

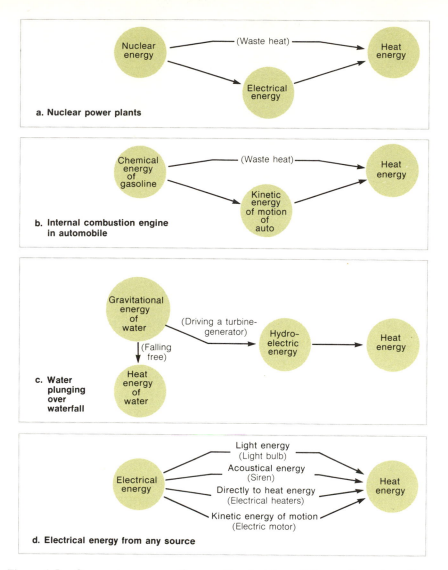

Figure 1-5. Common energy pathways. The first law of thermodynamics tells us that the total amount of energy is unchanged as the above transformations, and other like them, occur. Thus, in case *a*, 68 percent of released nuclear energy goes directly to environmental waste heat and 32 percent to electrical energy. The electrical energy may be transformed to other kinds of energy, as detailed in *d*, but it eventually ends up as heat also. Thus nuclear energy is finally converted 100 percent to heat. The figure shows clearly that energy coursing along natural pathways tends always to end up as heat energy. While this result is not required by the first law (where energy could run backward along these arrows as well, provided its amount was unchanged), it is dictated by the second law of thermodynamics, which specifies the *direction* of all natural change.

prediction is true or not, such a "death" is billions of years away. On earth, and on a much shorter time scale, we are encouraging a miniature heat death: the useful energy of fuels is being transformed with considerable haste into useless, irretrievable heat, radiated eventually into space.

The irreversibility of energy changes makes useful energy a unique resource: it is the only resource that cannot possibly be recycled. Once used, useful energy is gone forever. For this reason, energy reserves must be treated with utmost care to ensure a viable future for man.

While low-temperature heat is essentially useless, the heat energy of hot materials can be partially recovered as useful energy. Thus in both coal-burning and nuclear power plants, chemical and nuclear energies are transformed to hot steam. As the steam drives turbine blades, part of its heat energy is converted to electrical energy, but an even larger part enters the environment as low-temperature, waste heat, otherwise known as *thermal pollution.*[2]

Entropy Criterion, or Criterion of Randomness. Processes in which *entropy* is increased tend to occur spontaneously. Entropy is a thermodynamic measure of disorder and randomness. We are well aware of increasing disorder in many processes. A deck of cards or a sequence of numbered pages becomes disordered if casually rearranged. Marbles dropped on a floor tend to scatter about in disorder rather than to collect themselves in a neat pile. Molecules, too, tend to scatter about in a random fashion. As they scatter, they frequently mix with other molecules in the vicinity. Thus the mixing of air pollutants into the atmosphere, referred to earlier, is part of a larger trend toward universal disorder.

The tendency noted earlier for all energy to transform to heat energy also is explained by the entropy criterion. Heat may be considered as energy mixed thoroughly with matter. Like other mixing processes, this one occurs spontaneously. Thus

---

[2]The percentage of heat energy converted to useful energy is called the *efficiency*; it is about 40 percent for a modern coal-burning plant and 32 percent for nuclear plants. The large heat energy remaining—60 percent and 68 percent of the total, respectively—enters the environment as thermal pollution.

Why not improve engine efficiency, gaining more useful energy and reducing thermal pollution? First of all, the second law of thermodynamics sets an absolute ceiling on the efficiency of any engine that derives power from heat. Second, there are practical stumbling blocks (such as friction, poor heat transfer) which make it impossible to achieve even this limited efficiency.

The second-law ceiling on efficiency is

$$\text{efficiency} = \frac{\text{useful energy (work, electrical energy, etc.)}}{\text{heat energy supplied (by chemical or nuclear energy)}} = \frac{T_{\text{hot}} - T_{\text{cold}}}{T_{\text{hot}}}$$

where T_{hot} is the absolute temperature of the hot steam or gases driving the engine and T_{cold} is the temperature to which the steam must be lowered (usually by an environmental coolant like water) before reentering the engine cycle. This equation shows that the efficient generation of useful energy requires a considerable temperature differential; that hot and cold regions must exist side by side, neither one by itself being adequate. Therefore the enormous heat energy of the ocean or atmosphere cannot be converted practically to useful energy because temperature differences are relatively slight in any particular region. (The temperatures T_{hot} and T_{cold} are absolute temperatures, measured from absolute zero, the point where molecular motion ceases. We get absolute temperatures either by adding 273° to the centigrade scale or 460° to the Fahrenheit scale.)

heat energy is an inevitable consequence of letting energy loose to mix in the environment.

The tendency of molecules to scatter about can be stated more precisely: molecules and atoms, unless restrained by some force or barrier, will tend to occupy uniformly all the space available to them. To illustrate, Figure 1-6 shows two containers joined by a stopcock. In them are different gases, each gaseous substance having its own unique kind of molecule. With the stopcock closed, each gas is forced to remain in its own compartment. When the stopcock is opened, molecules of type A quickly spread into the right-hand compartment, and B molecules diffuse to the left. Shortly we reach the condition in which each substance is spread uniformly over the space available, as in Figure 1-7.

Equilibrium.   Spontaneous events in nature take place, change occurs, and finally the process runs its course and stops. A rock rolls downhill and comes to rest at the bottom because a counterforce from the solid earth stops its descent. Gases mix in a chamber until mixing is complete, then no further change occurs. When change ceases absolutely, the process is said to have reached *equilibrium*. We shall amplify this important concept later.

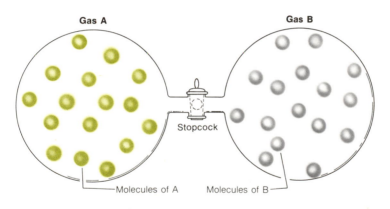

Figure 1-6.   Unmixed gases.

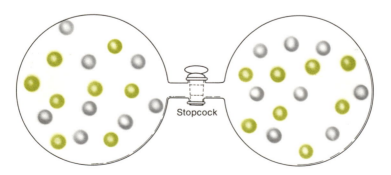

Figure 1-7.   Uniform mixture of gases.

Why Molecular Mixing Does Not Always Occur. Recall that spontaneous processes are directed by both energy (force) and entropy (disorder) criteria. If these criteria should give opposing "instructions," as they frequently do, then nature must strike a compromise. However, it is often an unequal compromise in which one factor tends to dominate. For instance, the complete mixing that the entropy criterion "wants" is often subdued almost entirely by forces between molecules.

There are attractive forces between molecules which frequently make them cling to one another. These forces are not the strong ones —the chemical bonds—that hold atoms together within a molecule, but much weaker *intermolecular forces* among all neighboring molecules (Figure 1-8). It is these forces that hold the molecules of liquids (and some solids) together. By so doing, they prevent the liquid from entering the atmosphere and, occasionally, from entering other liquids. Mixing cannot occur unless substances first dissolve in one another. Once they dissolve or penetrate, mixing is spontaneous.

Intermolecular forces, for example, keep the ocean together and prevent its total penetration of and uniform mixing with the atmosphere. However, the forces are limited, and as part of the compromise a small fraction of the water molecules do escape into the atmosphere as vapor, as in Figure 1-9. The vapor then mixes thoroughly, as the entropy criterion would suggest.

For some substances, intermolecular forces are so weak that the molecules fly apart entirely to form a gas. Gases, then, always mix uniformly because there are no forces sufficient to restrain them from penetrating into one another. Here the tendency to disorder is overriding.

When different liquids dissolve in one another, they mix. Some liquids, however, are not mutually soluble (do not dissolve in each other). For instance, oil and water will not dissolve together and mix.

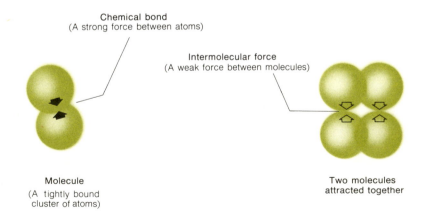

Chemical bond
(A strong force between atoms)

Intermolecular force
(A weak force between molecules)

Molecule
(A tightly bound
cluster of atoms)

Two molecules
attracted together

Figure 1-8. Forces between atoms and molecules.

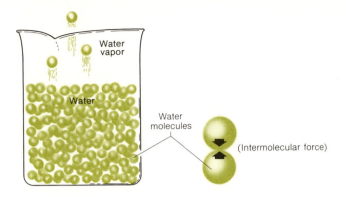

Figure 1-9. The intermolecular force between water molecules (or other liquid molecules) keeps the liquid intact, except for a few molecules that manage to break away into the atmosphere as vapor. (Molecules, as drawn here, are idealizations. The actual shape of molecules will be discussed in Chapter 4.)

There are many nonsoluble and thus nonmixing combinations in nature: various solids and liquids fail to dissolve appreciably in other liquids. This nonsolubility is usually a result of unequal intermolecular forces.

Oil will not dissolve in water because the forces between water molecules are so great that oil molecules cannot penetrate between them (Figure 1-10). Other substances, with molecules more strongly attracted to water, can penetrate and dissolve.

Because oil does not dissolve appreciably in water, spilled and dumped oil does not mix thoroughly with lake, stream, or ocean water. But even here a kind of mixing occurs—along the surface

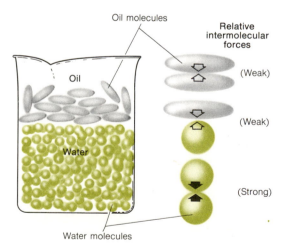

Figure 1-10. Oil will not mix with water because water molecules are held together by intermolecular forces that are relatively strong. This fact makes it difficult for oil molecules to slip between water molecules and thus enter the water.

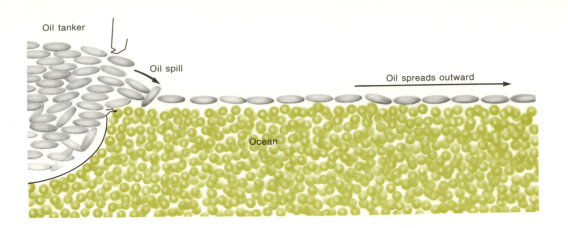

Figure 1-11. Oil spilled on water spreads along the surface.

of the water (Figure 1-11). The surface represents the only "space" available to the oil, and it becomes rapidly occupied. Oil spills from tankers and faulty oil wells soon cover many square miles of ocean surface, eventually reaching beaches and wildlife. Oil from the split-open tanker *Torrey Canyon* spread across 360 square miles of ocean before it became dilute enough to be harmless.

SEPARATION AND POLLUTION CONTROL

The reverse of mixing is *separation*. By separation, we mean the total or partial sorting of a mixture back into its component substances. Mixing and separation are contrasted in Figure 1-12.

Recall that entropy (disorder) favors mixing, not separation. In the natural processes we most commonly see, entropy is overriding and separation thus is ruled out. Smoke does not collect back into a smokestack. Sewage fails to concentrate itself at the outfall. There is no inclination for an oil spill to collect back into a massive blob where it could be collected or treated.

Nonetheless, separation occurs quite often in nature in circumstances where the energy or force criterion becomes overriding. For instance, water vapor from the ocean leaves salt behind because the intermolecular forces for salt are so great that it cannot escape the ocean as vapor. These forces clearly override the tendency for salt to mix into the atmosphere. Therefore when the vapor condenses to rain or snow, it is virtually salt free, a fortunate circumstance, since man's fresh water supply depends ultimately on ocean-nourished rainfall and snowfall.

Another separation process—settling, or *sedimentation*—depends on overriding gravitational forces. Most *particulate* (particle) matter mixed in the atmosphere is eventually pulled to earth as a rain of fine

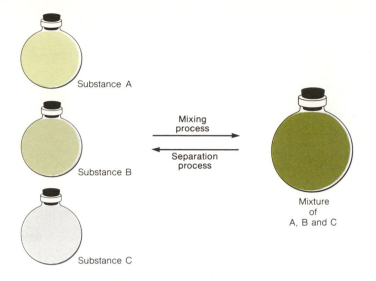

Figure 1-12. Mixing and separation.

dust. Also, silt will separate from water by settling, and thus clear water is produced.

Separation processes on a human scale are of great importance. Many industrial operations depend on some separation process to rid their products of undesirable components or to divide raw materials into their useful components. For instance, petroleum, a mixture of hundreds of substances, is separated, into many fractions, some being gasoline and others used as fuel oil, roofing tar, lubricating oil, diesel fuel, and so on. Petroleum fractionation is done by selective vaporization—called *distillation*—in imitation of ocean water vaporizing and freeing itself of salt. The petroleum is heated, and the smaller petroleum molecules, with weaker intermolecular forces, vaporize. The vapor is condensed and collected, giving one fraction. This process is repeated, with application of more and more heat, so that one obtains successive fractions that are graded according to their ease of vaporization. (See Figure 1-13b on page 20.)

Separation processes are extremely important in pollution and environmental work. First of all, separation is essential if we are to collect the pollution load of various "streams" entering the environment so as to avoid the wide-scale, irreversible environmental mixing mentioned earlier. These streams include sewage, smokestack gases, and various plant effluents. Each is itself a mixture, usually consisting mainly of water or air, but having a rather high pollutant content. The pollutants are easiest to remove while they are still concentrated in the efflux stream.

Second, unfit water can be purified by different techniques for removing contaminants. Most exciting is the prospect of removing salt from ocean water economically and on a large scale, thus pro-

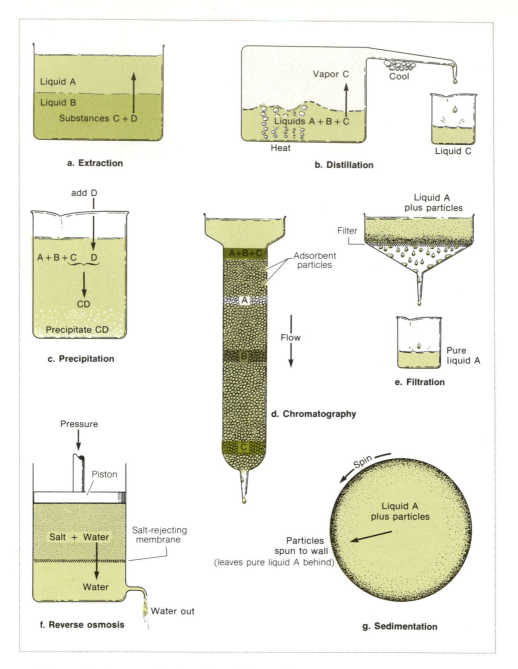

Figure 1-13. Schematic illustrations of how various separation methods work. (See text opposite.)

viding a virtually unlimited source of water to quench the world's thirst. Unfortunately, a redistribution of water over the earth's surface would be likely to cause serious ecological disturbances and would perhaps lead to disastrous changes in climate. Few plans for the controlling and harnessing of nature on a massive scale can escape serious repercussions. The attempt to subjugate insects by

DDT has illustrated this problem most forcefully: but more about DDT later.

Third, separation processes are frequently used in the *chemical analysis* of pollutants. In chemical analysis the atomic and molecular makeup of matter is unraveled. After polluting substances are separated from small air and water samples, they can be identified and their amounts determined. In this way we can find out how grave a given pollution problem is, and we can start to use specific remedies to combat those contaminants that we found by analysis to be most damaging.

Several important separation methods are described below. The description is keyed to Figure 1-13, where various hypothetical substances (A, B, C, etc.) are moved around to achieve separation. We put emphasis on the forces which cause these substances to move from place to place, because these and similar forces are used whenever matter is manipulated by man. Foremost among these forces are intermolecular forces, but also included are hydrostatic (pressure) forces, needed to make liquids flow; centrifugal forces; and so on. The descriptions are brief, but they cover the salient features of how these forces are utilized to achieve separation.

a. Extraction. In extraction, two nonmixing liquids, such as oil and water (A and B in Figure 1-13a), are brought into contact. One liquid may contain several substances (C and D) to be separated. Because of differences in intermolecular forces, one substance (D) but not the other (C) may be strongly attracted to one of the liquids (A). It moves into that liquid, leaving the other substance behind. Thus separation occurs.

b. Distillation. We have mentioned distillation as the selective vaporization by heat of molecules (C) with the weakest intermolecular forces. After vaporization, the vapor is cooled and condensed to a pure or relatively pure liquid, as in Figure 1-13b.

c. Precipitation. In precipitation, a substance (D) added to a mixture may form a chemical bond with just one substance (C) in the mixture, forming a molecule (or crystal) that is no longer soluble in the mixture because of altered intermolecular forces. The newly formed molecules (CD) form a solid *precipitate* at the bottom of the container, thus removing (separating) the single reacting substance (C) from the mixture.

d. Chromatography. In chromatography, a mixture of substances is washed through a tube containing an *adsorbent*, a material that attracts and holds certain substances on its surface. One substance (A) may be adsorbed more strongly than are others because of greater intermolecular forces with the adsorbent. Hence A may be held back as the other substances (B and C) wash through the tube. Because of variable intermolecular forces, dozens of substances can be separated in one run. The method is particularly useful when a gas is used to drive substances through the tube. This variation is called *gas chromatography*.

e. *Filtration*.   A filter is simply a barrier with many small holes. Particles cannot pass through such holes, and so they are separated from liquids that are forced through the filter (Figure 1-13e). A special filter with holes in the molecular size range will allow the passage of small molecules but not large, thus separating molecules of different sizes. This filtration of molecules is known as *ultrafiltration* because of the use of ultrasmall molecule-size holes.

f. *Reverse Osmosis*.   Here water or some other liquid is forced through a thin membrane, such as a cellophane sheet. Salt will not dissolve in some membranes because of unlike intermolecular forces. Hence water will pass through the membrane and separate from salt and other impurities. This method is a prominent one for producing pure water from brackish water and seawater.

g. *Sedimentation*.   We have mentioned that sedimentation is the settling of particles from liquids or gases. With the aid of a spinning centrifuge, very small particles and even some large molecules can be spun out of liquids or gases.

ZERO POLLUTION?

Some people say that pollution standards should be reduced to zero. This statement means that no pollution at all is to be tolerated. Although pollutant levels ought to be reduced immensely, a zero standard is scientifically absurd and totally unenforceable. We shall explain why.

Atoms and molecules are tiny beyond imagination. They are too small to see and too numerous to keep track of. In a single drop of water there are 1,500,000,000,000,000,000,000 molecules. If we had counted them steadily, one per second, since the earth began, we would have barely scratched the surface: only 1/10,000 of the drop's molecules would be counted in this vast interval. There are more water molecules in this drop than snowflakes in a storm covering the entire land area of the earth to a depth of one foot. Zero pollution implies that not a single one of these molecules is to be a pollutant molecule. Separation and purification methods are utterly incapable of the demanding task of such ultimate purification. Separating out the last pollutant molecule would be like finding and removing a single red snowflake in an earth-covering snowfall. Such purity does not occur even in nature. A drop of her purest water or cleanest air contains billions of molecules that we would classify as pollutant. It appears pure because these billions are lost in the vast sea of molecules, as easily as a billion red snowflakes could disappear from sight in a massive snowcover of the earth.

Nature pollutes constantly because her ever-present "bad" molecules tend to mix with the "good." Deadly mercury erodes from rock into rivers, mixes, and washes into the seas: each drop of natural seawater contains about 15 billion mercury atoms. Decaying vegetation loses sulfur compounds, which become air-borne sulfur dioxide. Plants and animals emit hydrocarbons and various other organic compounds, which mix into water and air. Some sea plants produce carbon monoxide. Volcanoes spew particulate matter (volcanic dust) into the atmosphere. Lightning produces oxides of nitrogen. All animals, including humans, have radioactive breath.

The virtue of nature is not that pollution is avoided, but that pollution is so light. Man has stepped this level up enormously. Where a billion or a trillion "bad" molecules in a drop may go unnoticed, a billion billion will be suffocating. The difference is primarily one of degree and not of kind. This fact emphasizes again the quantitative nature of problems in environmental pollution.

Finally, if by some super-separation method we did attempt to rid our air and water of all pollutant molecules, there would be no chemical instrument or technique for checking our effectiveness. Just as we cannot pluck a few molecules out of the vast molecular sea, neither can we sense their presence by any method of chemical analysis. With rare exceptions, well in excess of a billion molecules are needed to affect a measuring device and thus show us that they are indeed there.

Pollution standards should be rigorously low, but they should be set at finite levels that show man's understanding of the workings of the real world of molecules.

MAN'S CHEMICALS VERSUS NATURE'S CHEMICALS

Let us dispel one more somewhat related myth. The thesis of many people, particularly some of those favoring natural foods or opposing water fluoridation, is that the chemicals produced by man are somehow more insidious than those produced by nature. Even identical chemicals, like sodium fluoride, are purported to be safe if occurring naturally, unsafe if added by man. There are valid arguments favoring some of these causes, but this is not one of them. Man does not have the power to change the fluoride ion (a charged form of the fluorine atom) if he wished: they all come ultimately from nature, having been on earth long before man. They come to us down through the ages, unchanged and unchangeable.

It is true that man has assembled some of nature's atoms into new kinds of synthetic molecules, never before present on earth. Many of these molecules are harmful or toxic. However, nature is not always benevolent either. Peace Creek in Florida naturally carries twice the fluoride level that will damage teeth. Also, witness the toxic alkaloids, common to the poisonous amanita mushrooms and other plants, which in minute doses will kill a human being. Deadly curare, used to tip arrowheads in South America, is a natural substance. Many insects are engaged in a virtual state of chemical warfare using natural toxins. There are literally hundreds of toxic substances in the natural landscape.

What is truly different about man's chemicals is that he has allowed them to spread copiously throughout the earth's air and water. Few of us consume any significant amount of curare, but we all feast on DDT and other pesticides every day. We breathe little of nature's ozone, abundant in the upper atmosphere, but we can be made ill by exactly the same substance in photochemical smog down here on earth. Small amounts of radioactivity from naturally occurring substances are present everywhere, but man can raise its level of intensity significantly.

Man's chemicals are not uniquely insidious (although many are far from harmless): they are simply too abundant. When vast quantities are released to the environment, we invite trouble. The trouble is with us now, and it is called pollution.

MEASUREMENT AND ENVIRONMENTAL POLLUTION

We have suggested that the study of pollution is almost meaningless without some measure of the *amount* of pollutant. Large quantities are dangerous. Very small quantities are often acceptable and may even be generated by natural sources. To judge the threat in each case, we must know how much there is. We must learn to express this amount in numbers and to interpret the numbers used by other people in describing the intensity of pollution.

Any expression of amount contains a number and a *unit*. Both are necessary. Thus "12 tons" represents a very specific amount. It would not be specific if we omitted either the number, 12, or the unit, tons. If we omitted tons, it would tell us nothing about quantity; likewise, if we replaced tons by pounds, the meaning would be altogether different. The units that go with a number must always be expressed, whether they be tons, gallons, tablespoons, quarts, minutes, inches, or miles per hour.

METRIC SYSTEM

The *metric system* of units is used over most of the world. It was developed in France in 1790 to provide a logical and simple framework for measurement. The metric system is a decimal system, based on multiples of ten. It is much easier to work with than is the English system, presently used in the United States. Where we convert from one unit to another with awkward numbers like 12 inches to a foot or 5280 feet in a mile, the metric system has a simple 100 centimeters in a meter. We have 16 ounces per pound and 2000 pounds per ton, instead of an elementary 1000 grams per kilogram. Because conversion factors between metric units are numbers like 10, 100, 1000, and so on, mathematical calculations are much easier in metric units than in the units we commonly use. Conversion becomes a simple matter of moving the decimal point.

We shall use metric units frequently in dealing with the chemistry of the environment. However, learning metric units is worthwhile for several additional reasons: (1) it is probable that the United States will eventually convert to the metric system; (2) it is the most common system worldwide; (3) nearly all scientific work is reported in metric units; (4) most federal and state pollution standards utilize metric units, and these units are being reported increasingly by newspaper and magazine articles about pollution; and (5) once learned, the metric system is very easy to work with.

Five basic units provide a foundation for the metric system. Three of these units are *grams* to express mass, *meters* to express length, and *liters* to express volume. Time is expressed in seconds, already common practice. Temperature is expressed in *degrees centigrade*.

The standard abbreviation and the English equivalents of the first three metric units are given in Table 1-1. It is well to fix these units in mind and to try to visualize their size. A gram (g), for instance, is truly quite small—1/454 of a pound or only 0.0352 of an ounce. It is roughly the weight of three aspirin tablets or twenty water droplets. A penny weighs just over 3 grams; a dime just over 2.

A liter (ℓ) is slightly larger than a quart. A meter (m) is just over a yard.

Table 1-1. BASIC METRIC-ENGLISH EQUIVALENTS AND CONVERSION FACTORS

Metric Unit and Abbreviation		Equivalent English Unit	Conversion Factor
1 gram (g)	=	1/454 pound (lb)	1/454 lb/g
1 meter (m)	=	39.4 inches (in.)	39.4 in/m
1 liter (ℓ)	=	1.06 quarts (qt)	1.06 qt/ℓ

In the last column of Table 1-1, metric-English equivalents are shown by a conversion factor. The conversion factor is a shorthand method for showing the relationship between units of all kinds; its full utility will be explained later. For now we note only that an equivalence like 60 minutes per hour can be expressed simply as 60 min/hr; the metric-English conversion of 39.4 inches in one meter (m) can be stated as 39.4 in./m, and so on.

Very often it is convenient to work in a metric unit that is different from the basic unit. Most of the other metric units have names logically connected to the basic three. For instance, the prefix "kilo" always means "1000," so we know that a kilogram is 1000 grams,

Table 1-2. COMMON METRIC PREFIXES

	Prefix	Abbreviation	Multiplication Factor
	Mega	M	1,000,000
	Kilo	k	1000
Most used	Centi	c	1/100 or 0.01
	Milli	m	1/1000 or 0.001
	Micro	μ	1/1,000,000 or 0.000001
	Nano	n	1/1,000,000,000 or 0.000000001
	Pico	p	1/1,000,000,000,000 or 0.000000000001

a kilometer is 1000 meters, and so on. The common prefixes, with abbreviations, are shown in Table 1-2.

With the last two tables we can interpret most metric measures. For instance, a mass shown as "1 mg" is 1 milligram, since m = milli and g = gram. From the last table, "milli" is seen to be one thousandth: "1 milligram" is thus one thousandth of a gram. From Table 1-1, we find a conversion to pounds: 1 g = (1/454) lb and thus 1 mg = (1/1000) × (1/454) lb = 1/(454,000) lb. This quantity is indeed small, a pound divided by 454,000, equivalent to 1/50 of a drop of water; but some pollutants are so toxic that a few milligrams (mg) are dangerous to health. Hence the milligram (mg) is a handy unit, and we shall use it often.

Units of mass are used more than are units of length, volume, time, and temperature in environmental chemistry. Mass is the most direct measure of amount. Mass and weight are treated as equivalent here; but technically weight is a measure of mass only under the pull of gravity, and it would be zero in outer space. On the surface of the earth, mass and weight amount to the same thing and will be so treated.

To familiarize the reader with the most important units of each kind, common units of mass, length, and volume are summarized in the following sections.

Mass. The *kilogram* (kg), equal to 1000 grams (g) or 2.20 pounds (lb), is often used to represent large masses, such as body weight. Very large amounts are occasionally measured in *metric tons*, equal to 1000 kg or 2200 lb. The metric ton is an exception to our logical way of naming units: a metric ton would more properly be termed a "megagram." Despite the curious name, multiples of 10 still relate it to other metric units.

Units smaller than the gram in common use are *milligram* (mg), 0.001 g, already discussed, and the *microgram* (μg), 0.000001 g. Table 1-3 summarizes the major metric mass units and their equivalents in grams and pounds. The ratio of the size of each unit to the one below it in the table is shown at the left. This ratio is simply the conversion factor between neighboring units, and it is expressed as such.

Table 1-3. SUMMARY OF COMMON METRIC MASS UNITS AND CONVERSION FACTORS

Ratio of Unit Size	Unit and Abbreviation	Gram Equivalent	Pound Equivalent
1000 kg/met. ton	Metric ton	1,000,000 g/met. ton	2200 lb/metric ton
1000 g/kg	Kilogram (kg)	1000 g/kg	2.20 lb/kg
	Gram (g)	1 g/g	0.00220 lb/g, or (1/454) lb/g
1000 mg/g			
1000 μg/mg	Milligram (mg)	0.001 g/mg	0.00000220 lb/mg
	Microgram (μg)	0.000001 g/μg	0.00000000220 lb/μg

Length. The *kilometer* (km), equal in length to 1000 meters (m), is most often used to express large distances, such as those between cities. Small distance units in common use are the *centimeter* (cm), 0.01 m, and the *millimeter* (mm), 0.001 m. The really small units in regular use have names which are not in strict accord with our prefix rules. The *micron* (μm or μ), which should be the "micrometer," is 0.000001 m. This unit is used often to measure particle size in particulate air pollution. The *angstrom* (A) is 10,000 times smaller, 0.0000000001 m, and is about the size of an atom. (See Table 1-4.)

Table 1-4. SUMMARY OF COMMON METRIC LENGTH UNITS AND CONVERSION FACTORS

Ratio of Unit Size	Unit and Abbreviation	Meter Equivalent	English Equivalent
1000 m/km	Kilometer (km)	1000 m/km	3280 ft/km, or 0.621 mile/km
1000 m/m	Meter (m)	1 m/m	39.4 in./m, or 1.09 yards/m
10 mm/cm	Centimeter (cm)	0.01 m/cm	0.394 in./cm
1000 μm/mm	Millimeter (mm)	0.001 m/mm	0.0394 in./mm
10,000 A/μm	Micron (μm)	0.000001 m/μm	0.0000394 in./μm
	Angstrom (A)	0.0000000001 m/A	0.00000000394 in./A

Volume. The standard unit, the liter (ℓ), is the volume of a cube 10 cm along each edge. A *cubic meter*, since it is a cube with all three edges 10 times longer (1 m or 100 cm) than the edges of a liter-sized cube, has a volume 10 × 10 × 10 = 1000 times greater. Thus the cubic meter (m³), a unit commonly used in stating the concentration of air pollutants, equals 1000 ℓ. It is equivalent to a "kiloliter."

The *milliliter* (ml) is, of course, a volume 1000 times smaller than the liter, 0.001 ℓ. One *cubic centimeter* (cc or cm³) is also 0.001 ℓ since it is a cube with each edge 10 times smaller than those of the cube containing a liter. Hence 1 ml = 1 cc = 1 cm³. (Volume units are shown in Table 1-5 on page 28.)

It might seem that the great diversity of units above is unnecessary. However, the diversity is actually a relatively modest one, and it assures us that there is a convenient unit to express most measured quantities, large and small. A much greater diversity exists in present

**Table 1-5. SUMMARY OF COMMON METRIC VOLUME UNITS
AND CONVERSION FACTORS**

Ratio of Unit Size	Unit and Abbreviation	Liter Equivalent	English Equivalent
1000 ℓ/m^3	{ Cubic meter (m^3)	1000 ℓ/m^3	1.31 cu yards/m^3
	{ Liter (ℓ)	1 ℓ/ℓ	1.06 qt/ℓ
1000 ml/ℓ	{ Milliliter (ml)	0.001 ℓ/ml	0.00106 qt/ml
	{ Cubic centimeter (cc, cm^3)		

usage (mile, furlong, rod, yard, foot, inch, etc.), with over 85 units actively used in the United States. However, familiarity makes the present system seem less formidable, despite its greater complexity. With repeated use the metric system becomes not only familiar but also extremely convenient.

We shall now introduce two important quantities whose units are combinations of those already discussed.

Density.   *Density* is mass per unit volume. It is thus a derived quantity, being a combination of basic mass and volume units. It is often expressed in g/ml or the equivalent g/cm^3. Cold water (at 4° on the centigrade scale) has a density of exactly 1.00 g/ml and is thus a convenient reference for comparing densities. Metallic mercury has a density of 13.6 g/ml; a given volume of mercury therefore weighs 13.6 times as much as an equal volume of water. Air has a much lower density than either of these substances, roughly one thousand times less than water: if it is dry, its density is 0.00120 g/ml at room temperature and at the normal pressure of the atmosphere.

Concentration.   While density measures the total mass of all the matter found in a unit volume, *concentration* measures only the amount of a specified substance in a unit volume (or mixed with a fixed amount of other substance). It is a unit used to describe mixtures. It does so by giving the relative amount of one or more component substances in the mixture. It is the most important unit in pollution studies and pollution control, for it reflects the relative abundance of pollutants in air or water that we may breathe or drink.

The concentration of a single substance can be expressed in two somewhat different ways. First, it is an amount per unit volume. This form may have units of milligrams per liter (mg/ℓ), commonly used for water pollutants, or milligrams per cubic meter (mg/m^3), an air-pollutant standard. Other units are used occasionally. These units are like those of density, but here the mass is that of only one substance in a mixture.

Second, concentration is the relative amount of one substance in a mixture of substances. *Percentage* is an example. Percentage is parts per hundred. Five percent means 5 parts in 100 parts total. If it is *weight percent*, it means 5 g in 100 g of mixture, or 5 kg out of 100 kg, and so on. The units used, g or kg, are unimportant since the amount is relative. Hence we use the general term, "parts": substitute *parts by weight* for grams or kilograms above.

A *volume percent* is much like weight percent. If we mix 5 volume units (ml or ℓ) of carbon dioxide into 95 volume units of air, forming 100 volume units total, we have a mixture that is 5 volume percent carbon dioxide. It is not, however, 5 weight percent, because carbon dioxide is denser than air, and the 5 volume parts contain over 5 percent of the weight. (The weight percentage turns out to be 7.4 percent.)

Many pollutants, such as mercury in food, are a threat at concentrations far below one part per hundred. For these substances, the more convenient units *parts per million* (ppm), or sometimes *parts per billion* (ppb), are used. The logic is the same as before. We can have ppm by weight, used commonly with water and food contaminants, or ppm by volume, used sometimes in air-pollution work. As an example of how this system works, the upper limit on mercury in seafood, established by the Food and Drug Administration (FDA), is 0.5 ppm by weight. This rule means that fish is unacceptable if it contains over 0.5 weight units of mercury (say mg or g) per million weight units of food.

One part per million is such a small concentration that it is hard to visualize. It is equivalent to 1 apple in 10,000 crates of oranges, or 1 lb of oats in 500 tons of wheat. Figure 1-14 (p. 30) shows 1 dot among 10,000, a ratio of dots tiny in appearance, but still 100 times as large as the miniscule ratio of one part per million. It is remarkable that a few parts per million of any chemical can influence our environment, but as we go along we shall find that such low concentrations often have profound consequences.

EXPONENTIAL NUMBERS

The numbers that express environmental pollution are commonly very large or very small. Numbers like 500,000,000, the tons of carbon monoxide in the earth's atmosphere, are awkwardly large. A small concentration of phosphorus, 0.00000001 g per ml of lake water, can cause the lake to begin to choke on algal growth. Extreme numbers like these crop up again and again, the big ones often for expressing some total amount of pollutant and the small ones as a measure of pollutant concentration. Small numbers also surface when we express the size of atoms, molecules, and the particles of particulate air pollution.

Big and small numbers, written as above, have too many zeroes to keep track of easily. They are clumsy to write and difficult to multiply and divide. It is best to handle them by using exponential notation.

A typical *exponential number* is shown below:

Exponent

$$10^3$$

Figure 1-14. One white dot among 10,000 is equivalent to 100 parts per million (ppm). A ratio of 1 ppm, then, would be only one white dot in 100 pages of dots.

CHEMISTRY, ENVIRONMENT, AND POLLUTION

The *exponent* is shown as a superscript. This number, 10^3, is equal to 1000, a one followed by three zeroes. In general, the exponent tells us the number of zeroes following a one. One billion (1,000,000,000), with nine zeroes, is simply 10^9.

Essentially, all exponents do is shift the decimal place. Think of 10^9 as 1.0×10^9, the multiplication by 1.0 leaving the value unchanged. When we multiply 1.0 by 10^9, the decimal place of the 1.0 goes 9 places right, to give one billion.

added zeroes

$$1.000000000 \times 10^9 = 1,000,000,000$$

decimal to here

The "added zeroes" necessary to indicate where the decimal point must go are shown.

If the exponent is negative, the decimal goes left. Thus we can express extremely small numbers in exponential notation. For instance, one millionth equals 10^{-6}, as seen below:

added zeroes

$$0.000001.0 \times 10^{-6} = 0.000001$$

one millionth

Large and small numbers of greater complexity can be shown in the same way. If we multiply 412.0356 by 10^3, the decimal point must move right three places:

$$412.035\,6 \times 10^3 = 412,035.6$$

If the exponent is a minus number, the decimal point moves left:

added zeroes

$$0.00412.0356 \times 10^{-5} = 0.004120356$$

In simplest terms, then, an exponent tells us how to move decimal points back and forth. We can also consider the exponent of an exponential number as representing the number of times that 10 is multiplied by itself. For instance, in 10^2 the exponent 2 tells us to multiply 10 by itself 2 times: $10 \times 10 = 10^2$. Similarly, $10^3 = 10 \times 10 \times 10 = 1000$. Likewise a negative exponent tells the number of times we divide by 10. For instance,

$$10^{-3} = \frac{1}{10} \times \frac{1}{10} \times \frac{1}{10} = 0.001.$$

Table 1-6 (p. 32) shows how positive and negative exponential

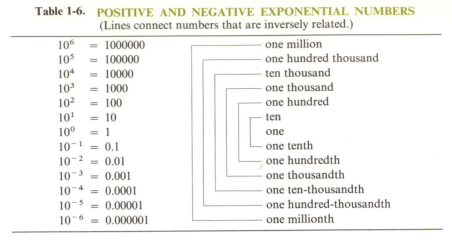

Table 1-6. POSITIVE AND NEGATIVE EXPONENTIAL NUMBERS
(Lines connect numbers that are inversely related.)

10^6	$= 1000000$	one million
10^5	$= 100000$	one hundred thousand
10^4	$= 10000$	ten thousand
10^3	$= 1000$	one thousand
10^2	$= 100$	one hundred
10^1	$= 10$	ten
10^0	$= 1$	one
10^{-1}	$= 0.1$	one tenth
10^{-2}	$= 0.01$	one hundredth
10^{-3}	$= 0.001$	one thousandth
10^{-4}	$= 0.0001$	one ten-thousandth
10^{-5}	$= 0.00001$	one hundred-thousandth
10^{-6}	$= 0.000001$	one millionth

numbers are inversely related to one another. For instance, where 10^6 is one million, 10^{-6} is one divided by one million. The first is a million times larger than one; the second a million times smaller.

Suppose we run across a very long, cumbersome number such as 1,500,000,000,000,000,000,000, given earlier as the number of molecules in a drop of water. How do we convert this number into convenient exponential notation? We can start by writing it as

$$1.5 \times 10^? = 1,500,000,000,000,000,000,000$$

21 places

We ask, what should the exponent be to move the decimal point of 1.5 twenty-one places to the right in order to obtain the long number, 1,500,000,000,000,000,000,000? It should, of course, be 21:

$$1.5 \times 10^{21} = 1,500,000,000,000,000,000,000$$

Usually, as in the case above, the decimal point of a number expressed exponentially is just to the right of the first nonzero number:

(first nonzero number)

1,500,000,000,000,000,000,000

↑

(decimal to be placed here)

Occasionally, however, the decimal is placed somewhat to the left or right of this favored spot, giving such equivalent forms as

$$1.5 \times 10^{21} = 15 \times 10^{20} = 0.15 \times 10^{22}$$

You can verify that these numbers are equal by writing them out in full length. Each converts to the original 1,500,000,000,000,000,000,000.

Multiplying Exponentials. Suppose we wish to multiply $10^1 \times 10^2$. This expression is $10 \times 100 = 1000 = 10^3$. Notice that the exponent in the answer, 3, is the sum of the exponents of the multiplied numbers, $1 + 2$. This rule is general. Another example:

Sum these two to get

$$10^3 \times 10^2 = 10^5$$

This simple scheme works as well for the multiplication of a big group of numbers,

Multiply numbers: $\quad 10^4 \times 10^2 \times 10^3 \times 10^5 = 10^{14}$

(Add exponents: $\quad 4 + 2 + 3 + 5 = 14$)

and it works for negative exponentials:

Multiply Numbers: $\quad 10^8 \times 10^{-2} \times 10^{-3} = 10^3$

(Add exponents: $\quad 8 + (-2) + (-3) = 3$)

If the multiplication involves numbers such as

$$(1.5 \times 10^3) \times (4.0 \times 10^{12})$$

it is simplest to multiply first the nonexponential terms together, then the exponential terms:

nonexponential terms		exponential terms
$\underbrace{1.5 \times 4.0}$	$\times$	$\underbrace{10^3 \times 10^{12}}$
6.0	$\times$	10^{15}

which provides the answer, 6.0×10^{15}.

Dividing Exponents. We know that $10^3/10^1 = 1000/10 = 100 = 10^2$. In this case, the final exponent, 2, is obtained by subtracting the denominator, or lower exponent, from the numerator, or upper one: $3 - 1 = 2$.

In general, any exponent appearing in a denominator must be subtracted:

Divide numbers: $\qquad \dfrac{10^9}{10^3} = 10^6$

(Subtract exponents: $\quad 9 - 3 = 6$)

This rule holds for large collections of numbers

Divide numbers: $\qquad \dfrac{10^4 \times 10^{21}}{10^{10} \times 10^{11}} = 10^4$

(Subtract exponents: $\quad 4 + 21 - (10 + 11) = 4$)

and for negative numbers

Divide numbers: $\qquad \dfrac{10^4}{10^{-3}} = 10^7$

(Subtract exponents: $\quad 4 - (-3) = 7$)

As with multiplication, division of more complicated numbers is achieved by dividing first the nonexponential numbers, then the exponential terms.

$$\frac{6.0 \times 10^{15}}{1.5 \times 10^3} = \underbrace{\frac{6.0}{1.5}} \times \underbrace{\frac{10^{15}}{10^3}}$$
$$4.0 \times 10^{12}$$

In this way, the answer to the above example is found to be 4.0×10^{12}.

CONVERSION OF UNITS AND ENVIRONMENTAL PROBLEM SOLVING

It often is necessary to convert quantities from one unit to another. We may need to convert inches to feet, cents to dollars, minutes to hours, or milligrams to kilograms. Conversion know-how is essential in dealing with and understanding all quantities, including environmental quantities.

From simple conversions, we shall proceed to multiple conversions and to the solving of problems that deal with environmental quantities. Why should the nonscience student learn to work environmental problems? There are several important reasons. We must know how to deal with our quantitative world, whether it be dollars and cents, the ratios in cooking recipes, or miles per gallon of gasoline. More importantly, we wish to take the black magic out of environmental concern. We have noted that environmental problems always have a quantitative basis (although qualitative factors such as the quality of life are far from unimportant). Yet many environmental discussions are largely emotional, with little quantitative foundation. There are enough serious environmental problems on earth that we cannot afford to worry about "false" problems that are not backed up by quantitative facts. We must learn to be critical in assessing other people's statements about our environment. People concerned about their environment must make a credible study with logical and often quantitative foundations, or else the environment of earth is truly in danger.

Conversion Factors. There are 1000 grams (g) in a kilogram (kg). As we noted earlier this equivalence is expressed in the *conversion factor* 1000 g/kg. By adding "g/kg", we specify that this is the gram-kilogram conversion factor. Other conversion factors from the tables given earlier are 1000 mg/g, 2.20 lb/kg, 100 cm/m, 1000 ℓ/m^3, and so on. The "ratio of unit size" given at the left of Tables 1-3, 1-4, and 1-5 is the conversion factor between a unit and the next higher unit. Thus from the length table, the angstrom-micron conversion factor is 10,000 A/μm, or the equivalent 10^4 A/μm.

A conversion factor of 1000 g/kg specifies grams per kilogram, but how do we express the inverse case, kilograms per gram? Here we take the reciprocal of the first conversion factor:

$$1000 \ \frac{g}{kg} \rightarrow \frac{1}{1000} \ \frac{kg}{g}$$

That is, we turn everything upside down, both units (g and kg) as well as the number 1000. (We treat them all as simple mathematical quantities that can be multiplied, divided, and otherwise manipulated.) In this way, we find that 1/1000 or 0.001 of a kilogram makes one gram, as we would logically expect. This method works for all conversion factors. Some worth remembering are (1/454) lb/g, which becomes 454 g/lb; 0.394 in./cm, which becomes 2.54 cm/in.; and 1.06 qt/ℓ, which is 0.946 ℓ/qt.

This way of writing conversion factors has great advantages. It leads to a method of making conversions of all kinds, and it gives us a way to check the correctness of our work.

Conversion Method. The method of making simple conversions, and for working a host of environmental problems as well, is the *factor-unit method*.[3] Suppose we have 0.5 kilograms (kg) of some substance ("0.5" is the factor and "kg" is the unit). How many grams (g) is this? Let us write down 0.5 kg, multiply it by the conversion factor 1000 g/kg, and see what happens.

$$0.5 \ kg \ \times \ 1000 \ \frac{g}{kg}$$

Multiplying the two numbers together, 0.5 × 1000, provides the right answer, 500. More than that, we can cancel out the kg units

$$0.5 \ \cancel{kg} \ \times \ 1000 \ \frac{g}{\cancel{kg}} \ = \ 500 \ g$$

and bring the g unit through into the answer. Recall that no answer is complete without units: this method provides both the number (factor) and the units.

By supplying units, the factor-unit method provides a check on our calculation. Suppose that we had incorrectly multiplied the 0.5

---

[3]The algebra of conversion, which is the basis of the factor-unit method, is quite straightforward and can be illustrated by the conversion of inches to feet. The equivalence of one foot and 12 inches can be expressed mathematically as 1 ft = 12 in. Both sides can be divided by the unit "ft," giving 1 = 12 in/ft. The term 12 in/ft (= 12 in/1 ft) is a conversion factor. It is equal to one. Any quantity can be multiplied by one without altering its value; so any quantity can be multiplied by a conversion factor like 12 in/ft because it is equal to one.

A distance of 5 ft is a fixed distance that will not be changed if multiplied by 12 in/1 ft:

$$5 \ ft = 5 \ ft \ \times \ \frac{12 \ in}{1 \ ft} = \frac{5 \ ft}{1 \ ft} \ \times \ 12 \ in = 5 \ \times \ 12 \ in = 60 \ in$$

The key step is canceling the "foot" units to find the factor 5: 5 ft/1 ft = 5.

kg by the reciprocal conversion factor, 0.001 kg/g. We would have

$$0.5 \text{ kg} \times 0.001 \frac{\text{kg}}{\text{g}} = 0.0005 \frac{\text{kg}^2}{\text{g}}$$

In this case no units cancel, and the answer comes out in the peculiar units "kg^2/g." Since we wanted the answer in grams (g), we know immediately that this answer, and hence our approach, is wrong.

Suppose now that we wish to convert from cubic meters (m^3) to milliliters (ml). The table of common volume units (Table 1-5) does not contain this conversion factor, although it gives the factor for going from m^3 to liters, ℓ, and then for converting from ℓ to ml. We need merely use two steps of our method. Our logic is that we have m^3, and from this unit we go to ℓ, thence to ml. Symbolically, the *logic chain* of steps is m$^3 \rightarrow \ell \rightarrow$ ml. It is often helpful to write this chain out to be sure that we can get from the starting point to the desired end using only those steps with known conversion factors.

If, then, we have 5 m^3, we write

$$5 \text{ m}^3 \times 1000 \frac{\ell}{\text{m}^3} \times 1000 \frac{\text{ml}}{\ell} = 5 \times 10^6 \text{ ml}$$

$$(\text{m}^3 \quad \rightarrow \quad \ell \quad \rightarrow \quad \text{ml})$$

showing that there are 5×10^6 (five million) ml in 5 m^3.

Notice that for the step m$^3 \rightarrow \ell$, we use the conversion factor 1000 ℓ/m^3, not its reciprocal, 0.001 m^3/ℓ. The ℓ, to which we are going, must appear on top. This rule is a general one for each step of a conversion problem.

Another example will be useful. In one average day, 1.2 trillion or 1.2×10^{12} gallons of fresh water enter and flow along streams and rivers in the United States. How many liters flow per day? How do we get from gallons to liters? We know that 1 gallon contains 4 quarts. We have given several times the conversion factor between quarts and liters. Thus logically, we follow the path gal $\rightarrow$ qt $\rightarrow \ell$. Our rules suggest the following setup:

$$1.2 \times 10^{12} \text{ gal} \times 4.0 \frac{\text{qt}}{\text{gal}} \times 0.95 \frac{\ell}{\text{qt}} = 4.6 \times 10^{12} \ell$$

All units cancel out except the one desired in the final answer, ℓ. This result shows that we did the right thing.

Conversion to ppm. Conversion to parts per million (ppm) or percentage involves similar principles. Here we are expressing the ratio of one component to the entire mixture. It is helpful to write down this ratio if conversions are to be made.

The FDA standard for mercury in fish, 0.5 ppm by weight, is an expression of the allowed weight ratio of mercury to fish. It means that the limit is 0.5 parts by weight (perhaps mg) in 10^6 parts of fish.

In ratio form,

$$0.5 \text{ ppm (mercury in fish)} = \frac{0.5 \text{ parts mercury}}{10^6 \text{ parts food}}$$

$$= \frac{0.5 \text{ mg mercury}}{10^6 \text{ mg food}}$$

If the ratio is not in this form, with 10^6 (parts, mg, etc.) in the denominator, it should be so converted. The conversion to a ratio of this form is done as shown in the following example.

In 2000 mg of cold water, 0.02 mg of life-supporting oxygen are found. What are the ppm by weight of oxygen in such water? The answer is important, for a result below 5 ppm will indicate a serious pollution problem, as we shall discuss in a later chapter. To answer this question, then, we first form the expressed ratio

$$\frac{0.02 \text{ mg oxygen}}{2000 \text{ mg water}}, \quad \text{or} \quad \frac{0.02 \text{ parts oxygen}}{2000 \text{ parts water}}$$

which answers nothing yet, because 10^6 does not appear in the denominator. This lack can be remedied if we multiply by the factor

$$\frac{10^6 \text{ parts water}}{10^6 \text{ parts water}}$$

This factor equals one, and thus it does not change the value of the ratio. We now have

$$\frac{0.02 \text{ parts oxygen}}{2000 \text{ parts water}} \times \frac{10^6 \text{ parts water}}{10^6 \text{ parts water}}$$

The "parts water" can be canceled like any other unit. Cancel as follows:

$$\frac{0.02 \text{ parts oxygen}}{2000 \text{ parts water}} \times \frac{10^6 \text{ parts water}}{10^6 \text{ parts water}}$$

As a result, we have

$$\frac{\dfrac{0.02 \times 10^6}{2000} \text{ parts oxygen}}{10^6 \text{ parts water}}$$

where $(0.02 \times 10^6/2000)$ came from collecting the various numbers together, all except the 10^6 in the denominator, which must remain there. Using our rules of exponential multiplication and division, we find that $(0.02 \times 10^6/2000) = 10$. Our ratio is thus

$$\frac{10 \text{ parts oxygen}}{10^6 \text{ parts water}} = 10 \text{ ppm (oxygen in water)}$$

This amount is well in excess of 5 ppm, which suggests that the water is relatively nonpolluted and is capable of supporting an abundance of fish life, and so on.

If we work in the above way, any ratio can be converted to ppm, or with slight modification, to parts per hundred (percent), parts per thousand, or parts per billion.

The only other matter which may require attention is the need for the terms in original ratio to be expressed in like units. In our example, we had mg oxygen in mg water, the two "mg" canceling to form a true ratio.

Suppose that 1 kg of water were found to contain 10 mg of oxygen. The ratio

$$\frac{10 \text{ mg oxygen}}{1 \text{ kg water}}$$

is not in suitable form, because mg and kg are unequal. We thus convert the kg to mg, or vice versa. There are 10^6 mg in 1 kg, as we can see by going through the conversion sequence kg → g → mg:

$$1 \text{ kg} \times \frac{10^3 \text{ g}}{\text{kg}} \times \frac{10^3 \text{ mg}}{\text{g}} = 10^6 \text{ mg}$$

Then the above ratio becomes

$$\frac{10 \text{ mg oxygen}}{10^6 \text{ mg water}}$$

which is now a proper ratio, having mg both above and below.

This particular ratio needs no further work to get ppm, since 10^6 is in the denominator. Our answer, again, is 10 ppm by weight of oxygen in water.

Since, as we just saw, a kilogram contains 1 million milligrams, any expression of concentration in mg/kg is the same as mg/10^6 mg and automatically comes out in ppm. Thus 7 mg/kg equals 7 ppm by weight, and so on. Both units are used frequently in expressing food and water contamination, and it is important to remember that they are the same.

SIGNIFICANT FIGURES

No measurement is exact. Uncertainty exists in all measured quantities, although in some cases the uncertainty can be reduced to exceedingly low levels.

By convention, no more digits are reported for any number than are warranted by knowledge of its accuracy. Sometimes (very frequently in this text) we use fewer digits than we could because we do not need ultimate accuracy, but we never use more digits.

If length measured on a ruler has an uncertainty of a few tenths of a millimeter (mm), we can report lengths such as 37.2 mm. The

Table 1-7. SIGNIFICANT FIGURES

Number	Number of Significant Figures	Comments on Zeroes
7	1	
18.4	3	
18.40	4	zero *does not* fix decimal pt.
1008	4	zeroes *do not* fix decimal pt.
1.04452	6	zero *does not* fix decimal pt.
1.3	2	
0.13	2	zero *does* fix decimal pt.
0.013	2	zeroes *do* fix decimal pt.
2.70×10^4	3	zero *does not* fix decimal pt.
27,000	?	? ? ? ? ?

last digit, 2, may carry some slight uncertainty, but the digits preceding it do not. It would be incorrect to report this number as 37.24 mm because the 4, representing hundredths of a mm, must be completely uncertain, as our accuracy is limited to a few tenths of a mm. If a more accurate means of measurement were used, with uncertainty reduced to a few hundredths, we could present quantities like 37.24 mm with justification.

The three digits 3, 7, and 2 constitute the *significant figures* for this particular measurement. There are three of them. No more are warranted. If we use more, we incorrectly imply an accuracy that does not exist.

Usually the number of significant figures is simply the number of digits appearing in a number. Zeroes are excluded only if they serve solely to fix the decimal point. Some examples are shown in Table 1-7. Only the last example is uncertain, for we cannot tell by inspection whether the last three zeroes are merely fixing the decimal place, which by implication is right after the last zero, or whether they represent a significant part of the measurement. For this reason, such numbers are best expressed in exponential form, such as the 2.70×10^4 shown in the next to last entry. Here three significant figures are clearly intended.

When we combine numbers by multiplication, division, addition, or subtraction, accuracy is limited by the least accurate number. Its entire uncertainty carries through into the answer.

In multiplication and division, we cannot retain more significant figures than the minimum possessed by the entering numbers. For instance, we must report 4.0×1.4111 as 6.6, not as 6.6444.

When we add or subtract small groups of numbers, digits can be retained only as far right as the least accurate number.

$$
\begin{array}{r}
4.11 \\
+\ 21.3662 \\
\hline
25.4762
\end{array}
\quad \xrightarrow[\text{off to}]{\text{round}} \quad 25.48
$$

Note that the answer has more significant figures, four, than does the entry 4.11, with three. This result can happen in addition and subtraction, but not in multiplication and division. In addition and subtraction, we must mainly check to see that we have not gone more places beyond the decimal point than is warranted, two being allowed in the last case.

Some digits must generally be dropped, as above, when we are rounding off to the correctly expressed answer. If the first digit to be dropped is equal to or greater than 5, we round up. We did so in the foregoing addition example, where $25.4762 \rightarrow 25.48$. If the first dropped digit lies below 5, it is simply dropped: $25.326 \rightarrow 25.3$.

One case exists in which a number has no uncertainty at all. This case occurs when a direct count is possible. If we count four smokestacks, it is exactly four smokestacks and could be expressed with an infinite number of significant figures, $4.00000 \ldots$. While zeroes beyond the decimal are rarely shown, they should be regarded as there when such numbers are used in calculations.

Exercises

1. Pick one of the four suggested catastrophes used to introduce this chapter and list the environmental quantities (with appropriate units) that you feel you would need to know in order to decide whether the threat was significant.

2. Newspaper, magazine, and television accounts have suggested many environmental catastrophes other than the four listed at the beginning of the chapter. Name one. Can you show how this suggested catastrophe relates to chemistry? Is the quantity of some chemical substance involved in any important way?

3. Give five specific examples in which you have observed the spontaneous mixing of pollutants with the environment. (Hint: You probably detect the mixing and spreading of pollutants as often using your sense of smell as your sense of sight.)

4. A large power plant releases about 2×10^{11} BTU's (British thermal units) of chemical energy stored in coal each day. If the plant is 40 percent efficient, how many BTU's will be released to the environment as thermal pollution?

5. (Refer to Problem 4.) A BTU is a quantity of energy that will heat 1 lb of water $1°$ Fahrenheit (F). How many pounds of water would be heated $1°F$ by waste heat from the plant described in the foregoing problem? How many tons? (Water is usually heated more than $1°F$ so that less is needed to carry away the waste heat. However, even if the amount of water is 10 times less than calculated in this problem, it is still an enormous quantity of water, reflecting the large magnitude of the thermal pollution problem.)

6. How many water molecules are there for each atom of mercury in seawater? Use approximate figures from the text giving the number of water molecules and mercury atoms in a drop of water. If you could reduce yourself to molecular size, do you think it would be easy to find a mercury atom in seawater? Explain.

7. Write out the names of the units having the following abbreviations: kg, ml, Ml, cm, cg, mg, mm, μg, ng, pg, μl, nl, and μm.

8. Express the following numbers in exponential notation: one trillion, one billion, one million, one thousand, one hundred, one hundredth, one thousandth, one millionth, one billionth, and one trillionth.

9. Express the following quantities in exponential notation:

   a. The area of the earth, 197 million mi^2.
   b. The water used daily by industry, 400 billion kg.
   c. Carbon monoxide in the atmosphere, 500,000,000 tons.
   d. Carbon dioxide in a liter of air, 0.00032 ℓ.
   e. Cadmium in a ml of seawater, 0.00000008 g.

10. Write out in full length the following numbers. Give each number its proper name (for instance, 10^6 is one million):

    a. The water used to produce a ton of steel, 5×10^4 gal.
    b. Mining wastes produced per year in U. S., 10^{12} kg.
    c. The *Torrey Canyon* oil spill, 1.19×10^5 tons.
    d. Yearly U. S. weight of scrapped autos, 1.5×10^9 kg.
    e. Carbon monoxide in a m^3 of rural air, 1.0×10^{-4} g.

11. Multiply the following exponential numbers:

    a. $10^4 \times 10^2$
    b. $10^{12} \times 10^5$
    c. $10^7 \times 10^{-5}$
    d. $10^{12} \times 10^{-3}$
    e. $(2 \times 10^3) \times (5 \times 10^6)$

12. Express the following in exponential notation:

    a. $10^8/10^2$
    b. $10^{15}/10^9$
    c. $(10^3 \times 10^7)/10^5$
    d. $(10^5 \times 10^{14})/(10^2 \times 10^6)$
    e. $(8 \times 10^7)/(2 \times 10^4)$

13. What is the equivalent in ounces of 1 μg, 1 mg, and 1 kg, given that 1 g equals 0.0352 ounce?

14. The size of a small atom is about 1 A, or 0.0000000001 m. Express this diameter in terms of μm, mm, and cm. How many atoms, lined up end to end, would it take to reach 1 in.? (Hint: Decide whether this number of atoms is simply equal to the number of angstroms in 1 in.)

15. Make the following conversions:

    a. Express 220 lb in kg.
    b. Express 6.6 lb in mg.
    c. Express 4.54 kg in lb.
    d. Express 908 g in lb.
    e. Express 3280 ft in cm.

16. Make the following conversions:

    a. Express 394 in. in mm.
    b. Express 762 mm in in.
    c. Express 50.8 cm in in.
    d. Express 106 qt in ml.
    e. Express 0.131 yd^3 in ℓ.

17. The density of carbon dioxide is roughly 2 g/ℓ. What is the mass of carbon dioxide occupying a volume of 12 ℓ.

18. It is common to find that a m^3 sample of city air contains 5 ml of carbon monoxide. What are the ppm by volume of carbon monoxide in air under these conditions?

19. The fat of herring gulls has been found to contain up to 3 mg of DDT per g of fat. Express this concentration in ppm by weight.

20. Mercury present at 300 parts per trillion by weight in Minamata Bay, Japan, led to the death and severe injury of over 100 people. What is this level in ppm by weight of mercury?

21. If a m^3 of air contains .0011 ℓ of sulfur dioxide, it may cause adverse health effects in several days. What are the ppm by volume of sulfur dioxide in this air?

22. Air contains 3.2 ℓ of carbon dioxide for each 10^4 liters of air. Express the concentration of carbon dioxide in air in ppm by volume.

23. A 5 kg sample of polluted water was found by chemical analysis to contain 20 mg of oxygen. What are the ppm by weight of oxygen in this water?

24. How many significant figures are in each of the following numbers?

    a. 42   b. 39.50
    c. 30.95   d. 2.5×10^9
    e. 2.500×10^9 f. 2.005×10^9
    g. 0.0025   h. 0.25
    i. 1,000,001   j. 1.000001×10^6

Glossary

Adsorbent   a material that attracts and holds another material on its surface.

Atoms   minute building blocks from which matter is made.

Centrifuge   a machine that spins a solution, causing heavier components to settle.

Chemical analysis   the unraveling of the atomic and molecular makeup of matter.

Chemical bond   the linkage holding one atom to another.

Chromatography   a separation method in which components of a mixture are adsorbed at different rates on another material in a tube.

Compound   a substance whose molecules are made up of different kinds of atoms.

Concentration   measurement of the amount of a specified substance in a unit volume or mixed with a fixed amount of other substance.

Condensation   process of changing from a gas to a liquid.

Density   mass per unit volume.

Diatomic   describes a molecule that has two atoms.

Distillation   separation of a mixture by the selective vaporization by heat of the molecules that have the weakest intermolecular forces.

Element   a substance whose atoms are all of one kind.

Equilibrium   state that is reached when spontaneous change ceases.

Exponential notation   a method of expressing very large and very small numbers by use of an *exponent* (superscript) that shows the position of the decimal point.

Extraction   a separation method in which one component is attracted out of a mixture and into a nonmixing liquid that is brought into contact with the mixture.

Filtration   a separation method in which one component is left behind when the mixture is passed through a *filter*, a barrier with many small holes.

Kinetic energy   energy associated with motion.

Mass   for our purposes, equivalent to weight.

Metric system   a decimal system of units of measurement used in scientific work and, in many countries, in everyday matters as well.

Molecule   a discrete cluster of atoms held tightly together by a chemical bond.

Percentage   parts per hundred. Percentage may be *weight percent* (for example, grams/100 grams) or *volume percent* (for example, liters/100 liters).

Potential energy   stored energy that can be released to do work, create heat, and so on.

Precipitation   a separation method in which one component of a mixture is caused to become a solid that falls (precipitates) out of the mixture. The solid is called a *precipitate*.

Pure substance   matter made up of only one kind of atomic or molecular unit.

Reverse osmosis   a separation method in which, for example, water is separated from impurities by being forced through a thin membrane.

Sedimentation   a separation method in which particles settle out of liquids or gases.

Separation   the sorting of a mixture back into its component substances.

Significant figures   those digits of a number that are known exactly with a certain precision.

Soluble   capable of being dissolved.

Sublimation   process by which a solid becomes a gas without an intermediate liquid stage.

Thermodynamics   study of energy and natural change.

Three states of matter   gas, liquid, and crystalline solid.

Triatomic   describes a molecule that has three atoms.

Vaporization   process by which a liquid becomes a gas.

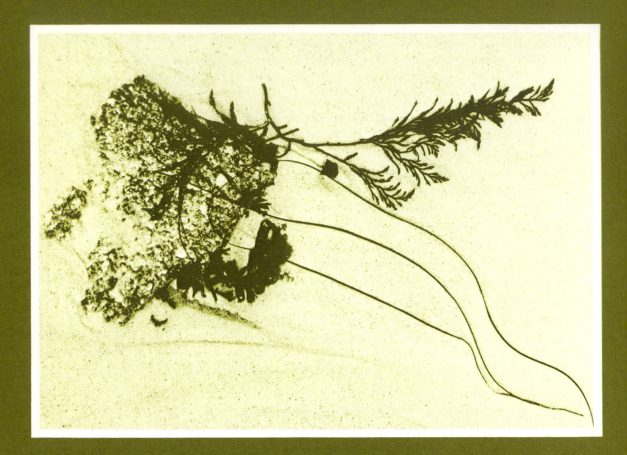

SEAWEED AND ROCK
(*Photo by Betty Berenson/Photofind.*)

2 ATOMS

BUILDING BLOCKS

This chapter is devoted to understanding atoms. Atoms, we have learned, are the unbelievably small building blocks of matter. Each kind of atom has unique properties, and these ultimately determine the diverse characteristics of all matter on earth.

Atoms, hidden by their extreme smallness, eluded detection by man until the last century. The Greeks had suggested their existence as a matter of pure speculation. These unseen building blocks remained objects of doubtful existence for centuries thereafter. Finally, long after stars and galaxies a million remote light-years away became known to man, the atoms that make up our earth and ourselves were discovered and many of their secrets unraveled. We shall probe these former secrets, so important because they underlie the ebb and flow of the environment of this planet.

THE NUCLEAR ATOM

Even a century ago, little was known about the atom and its electrons, protons, and neutrons. Atoms were presumed to exist and to be the small building blocks of matter, a point that became increasingly

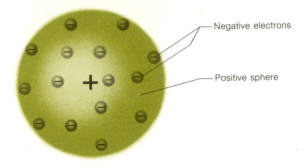

Figure 2-1. The Thomson model of the atom (1898).

certain following the 1808 atomic hypothesis of John Dalton. But what were atoms like? In 1898, J. J. Thomson suggested that the atom was a sphere of positive electricity with negative electrons imbedded in the surface (Figure 2-1). Electrically charged particles and thus electricity were produced when an electron was detached from the atom. This model of the atom seemed therefore to provide a satisfactory explanation of electricity and other experimental facts until the famed experiment of Ernest Rutherford and co-workers early in this century. The Rutherford experiment was one of the most important in modern science. As with many crucial experiments in the history of scientific discovery, it made necessary the revision of an old concept, Thomson's model of the atom, and led to a new model, the so called nuclear atom. The nuclear atom is still regarded as valid today.

Rutherford and his students, Hans Geiger and Ernest Marsden, bombarded a thin gold foil with high-velocity nuclear particles known as alpha particles (Figure 2-2). (Alpha particles are products of radioactive decay and will be described in Chapter 11.) These minute and energetic projectiles, emitted in the radioactive disintegration of the element polonium, were expected to pass directly through the gold foil as a bullet would go through a paper sheet. Most of them did. However, a few were deflected and some, about one in ten thousand, bounced back from the foil entirely. This result astonished

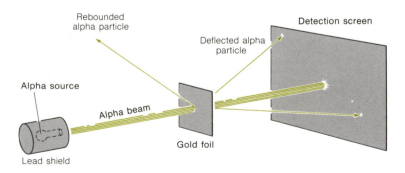

Figure 2-2. The Rutherford experiment.

and perplexed Rutherford. He later said, "It was almost as incredible as if you fired a 15-inch shell at a piece of tissue paper and it came back and hit you." Nothing in the Thomson model would suggest this result, for there were no massive lumps of matter to rebound from. The electrons were too light and the positive sphere was penetrable, as was proved by the transmitted majority of alpha particles. Yet rebounding meant that the alpha particle had collided with a particle more massive than itself, for rebounding, whether of nuclear particles, basketballs, or marbles, occurs only when the object struck is of greater mass than is the striking object. Not only must the alpha particles be somehow colliding with more massive objects in the atom, but also those objects must be so small that the great majority of alpha particles could slip through without touching them. By this reasoning Rutherford concluded that the mass of the gold atom was concentrated in extremely small blobs which were too little to be struck often, but which led to the deflection and rebounding of those alpha particles which did chance to strike them.

Rutherford's new evidence led him to propose in 1911 the *nuclear atom* (Figure 2-3). The mass of this atom was concentrated in a small, positively charged *nucleus* at the center of the atom. The diameter of the nucleus was calculated from the number of deflected and rebounded alpha particles to be about 100,000 times smaller than that of the atom itself. Since atoms were extremely minute, the nucleus must be tiny beyond imagination!

Electrons occupied the outer spaces of the atom, according to this picture. They were too light to contribute much to the mass of the atom, and too light to deflect the passing alpha particles.

The nuclear-atom model stands today, but much more is known about the nucleus of atoms and particularly about the all-important

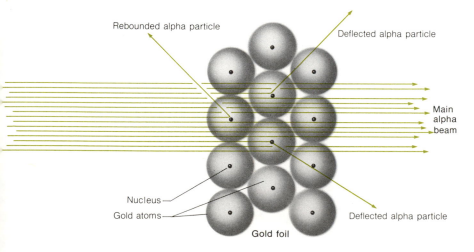

Figure 2-3. Rutherford's nuclear atoms, here assembled in gold foil. Most alpha particles in the beam miss the tiny gold nuclei. Rutherford's foil was actually several thousand atoms thick.

electrons existing in the outer atomic reaches. Our present wealth of knowledge has been gained through thousands of chemical and physical experiments and the efforts of many theoreticians. We shall now describe the salient features of atoms as they are presently understood, leaving out the great bulk of experimental and theoretical evidence which has made our present atomic picture so rich in detail.

NATURE OF THE ATOM

First of all, we reemphasize the point made in Chapter 1 that atoms are incredibly small, from one to several angstroms in diameter. If 100 million were lined up end to end, they would extend slightly beyond a centimeter, reaching about an inch. Atoms are far too small to observe in a regular light microscope. Only recently has the powerful electron microscope been barely able to detect their existence. Since they cannot be seen directly, atoms must be studied by indirect means (such as the Rutherford experiment). That a complete and consistent atomic picture has been pieced together although no atom ever has been seen is a triumph of modern science. Certainly, with such ingenuity, most environmental problems could be solved. The difficulty is that human institutions may not agree with the costly and perhaps harsh remedies necessary.

The nucleus of the atom, as we have said, is much smaller still, about 10^{-13} cm in diameter. It is composed of *protons* and *neutrons*. These two elementary particles have nearly equal mass, the neutron being only 0.1 percent heavier than the proton. The proton carries a single unit of positive electrical charge, often indicated in chemistry by a single plus, $+$. It is the smallest unit of electrical charge existing in nature. The neutron, as its name implies, is electrically neutral.

It has long been known that like charges repel each other and unlike charges attract one another. The forces of repulsion and attraction between charged bodies are called *electrostatic forces*. These forces are universal, working on the tiny scale of the nucleus as well as across miles of space. Nuclear protons, each with a like positive charge, repel one another strongly because of these forces.

It is remarkable, then, that protons assemble together in a nucleus (accompanied by neutrons) where the repulsion of their positive charges for each other would tend to make the proton-rich nucleus fly apart. They can assemble only because there are powerful counterbalancing forces holding the protons and neutrons together—the so called *nuclear forces*, which act as a strong nuclear glue. These forces successfully bind the nucleus together in nearly all atoms, but they are occasionally inadequate, and then nuclear breakup (fission) and radioactivity occur. We shall discuss these matters in a later chapter dealing with nuclear energy and radioactivity.

The atom's electrons are light by comparison to the nucleus. Each electron has a mass about 1840 times smaller than that of an individual proton or neutron in the nucleus. However, its electrical charge is as great as that on the proton, but is negative, $-$, rather than positive, $+$. The number of negative electrons in a given atom equals exactly the number of positive protons in that atom's nucleus; thus the electrical charges precisely cancel one another and the atom itself is electrically neutral.

THE PERIODIC SYSTEM OF ELEMENTS

The *atomic number*, Z, is the number of protons in the nucleus of an atom. The simplest atom, hydrogen, possesses one proton, $Z = 1$. It has also one electron to balance the proton's charge. The helium atom, next simplest, has two protons and two electrons, $Z = 2$. It also has two neutrons in its nucleus, but since neutrons do not affect electron behavior and thus chemical properties, we shall not discuss their presence in the nucleus at this time.

Other kinds of atoms possess more matched proton-electron units. Thus we proceed up the atomic scale of familiar-sounding elements: hydrogen, $Z = 1$; helium, $Z = 2$; lithium, $Z = 3$, beryllium, $Z = 4$; boron, $Z = 5$; carbon, $Z = 6$; nitrogen, $Z = 7$; oxygen, $Z = 8$; on up to $Z = 105$, an unstable, man-made element tentatively named Hahnium. These elements are shown in Table 2-1, along with their chemical symbols. The chemical symbol, a shorthand abbreviation for the element, is used frequently to simplify and condense the language of chemistry. We shall use them often hereafter. It is well to familiarize yourself with the chemical symbols of the common elements.

The one hundred or so well-studied elements show a variety of chemical properties. They vary in metallic character, in their tendency to form chemical bonds, in the number and type of bonds formed, and so on. These properties are of major concern in all fields of chemistry, including environmental pollution. If the properties of elements varied haphazardly with increase in atomic number, the study of chemistry would be enormously complicated by unconnected facts about all the hundred or so separate elements. Fortunately, many elements have closely related properties, and these *similar properties recur in a periodic way with increase in atomic number (Z)*. This rule is a statement of the *periodic law*, a cornerstone of chemistry. Here is an example of the periodic law. The density of elements in solid form varies with Z. As Z increases, the density has periodically recurring highs and lows, as illustrated in Figure 2-4. Other properties, such as chemical reactivity, atomic size, and melting point, vary in the same way, with maxima and minima similarly spaced.

Table 2-1. THE ELEMENTS

Atomic Number	Element's Name	Chemical Symbol	Atomic Number	Element's Name	Chemical Symbol
1	Hydrogen	H	54	Xenon	Xe
2	Helium	He	55	Cesium	Cs
3	Lithium	Li	56	Barium	Ba
4	Beryllium	Be	57	Lanthanum	La
5	Boron	B	58	Cerium	Ce
6	Carbon	C	59	Praseodymium	Pr
7	Nitrogen	N	60	Neodymium	Nd
8	Oxygen	O	61	Promethium	Pm
9	Fluorine	F	62	Samarium	Sm
10	Neon	Ne	63	Europium	Eu
11	Sodium	Na	64	Gadolinium	Gd
12	Magnesium	Mg	65	Terbium	Tb
13	Aluminum	Al	66	Dysprosium	Dy
14	Silicon	Si	67	Holmium	Ho
15	Phosphorus	P	68	Erbium	Er
16	Sulfur	S	69	Thulium	Tm
17	Chlorine	Cl	70	Ytterbium	Yb
18	Argon	Ar	71	Lutetium	Lu
19	Potassium	K	72	Hafnium	Hf
20	Calcium	Ca	73	Tantalum	Ta
21	Scandium	Sc	74	Tungsten	W
22	Titanium	Ti	75	Rhenium	Re
23	Vanadium	V	76	Osmium	Os
24	Chromium	Cr	77	Iridium	Ir
25	Manganese	Mn	78	Platinum	Pt
26	Iron	Fe	79	Gold	Au
27	Cobalt	Co	80	Mercury	Hg
28	Nickel	Ni	81	Thallium	Tl
29	Copper	Cu	82	Lead	Pb
30	Zinc	Zn	83	Bismuth	Bi
31	Gallium	Ga	84	Polonium	Po
32	Germanium	Ge	85	Astatine	At
33	Arsenic	As	86	Radon	Rn
34	Selenium	Se	87	Francium	Fr
35	Bromine	Br	88	Radium	Ra
36	Krypton	Kr	89	Actinium	Ac
37	Rubidium	Rb	90	Thorium	Th
38	Strontium	Sr	91	Protactinium	Pa
39	Yttrium	Y	92	Uranium	U
40	Zirconium	Zr	93	Neptunium	Np
41	Niobium	Nb	94	Plutonium	Pu
42	Molybdenum	Mo	95	Americium	Am
43	Technetium	Tc	96	Curium	Cm
44	Ruthenium	Ru	97	Berkelium	Bk
45	Rhodium	Rh	98	Californium	Cf
46	Palladium	Pd	99	Einsteinium	Es
47	Silver	Ag	100	Fermium	Fm
48	Cadmium	Cd	101	Mendelevium	Md
49	Indium	In	102	Nobelium	No
50	Tin	Sn	103	Lawrencium	Lw
51	Antimony	Sb	104	(Kurchatovium)	Ku
52	Tellurium	Te	105	(Hahnium)	Ha
53	Iodine	I			

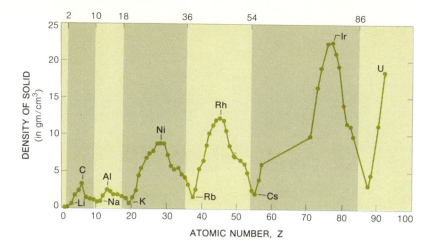

Figure 2-4. The periodic variation of the density of elements with increasing atomic number.

The periodic repetition of properties has been known for over a century. As early as 1863, John A. Newlands observed that for some elements, similar properties recurred with every eighth element. This eight-step recurrence was named the *law of octaves*. Others studied the subject, but an accurate picture was missing until 1869–1871, when the brilliant Russian chemist, Dmitri Mendeleev, established a rather complete chart of elements arranged periodically to show the relationship of their properties. There were several blanks in Mendeleev's chart, corresponding to elements unknown at that time. He boldly predicted the properties of these missing elements, basing his ideas on the periodic concept. With the help of his predictions, a number of new elements were soon discovered, three within fifteen years. Their properties were very much as Mendeleev said they would be.

The verification of Mendeleev's predictions showed the great utility of the periodic law in chemical discovery, but nowhere is the periodic law more useful than in helping the student of chemistry understand the matter making up his world.

Mendeleev's chart is called the *periodic table*. A modern version is shown in Figure 2-5. The sequence of atomic numbers is left to right, top to bottom, as you would read a book. *The key to the arrangement of elements is that those in a column have similar properties.* These properties recur periodically because as we read the chart, we return again and again to the same column in a cyclical manner.

The elements in a column are called a *family* or *group*. These groups are identified by the numbers IA, IIA, and so on, shown at the top of the periodic table.

As an example of family relationships, the group IA elements are very reactive. They all form ions (charged atoms resulting from the loss/gain of electrons) with a single positive charge. They all

IA	IIA	IIIB	IVB	VB	VIB	VIIB	VIII			IB	IIB	IIIA	IVA	VA	VIA	VIIA	O
1 H																	2 He
3 Li	4 Be											5 B	6 C	7 N	8 O	9 F	10 Ne
11 Na	12 Mg											13 Al	14 Si	15 P	16 S	17 Cl	18 Ar
19 K	20 Ca	21 Sc	22 Ti	23 V	24 Cr	25 Mn	26 Fe	27 Co	28 Ni	29 Cu	30 Zn	31 Ga	32 Ge	33 As	34 Se	35 Br	36 Kr
37 Rb	38 Sr	39 Y	40 Zr	41 Nb	42 Mo	43 Tc	44 Ru	45 Rh	46 Pd	47 Ag	48 Cd	49 In	50 Sn	51 Sb	52 Te	53 I	54 Xe
55 Cs	56 Ba	57 La	72 Hf	73 Ta	74 W	75 Re	76 Os	77 Ir	78 Pt	79 Au	80 Hg	81 Tl	82 Pb	83 Bi	84 Po	85 At	86 Rn
87 Fr	88 Ra	89 Ac	104 Ku	105 Ha	106	107											
			58 Ce	59 Pr	60 Nd	61 Pm	62 Sm	63 Eu	64 Gd	65 Tb	66 Dy	67 Ho	68 Er	69 Tm	70 Yb	71 Lu	
			90 Th	91 Pa	92 U	93 Np	94 Pu	95 Am	96 Cm	97 Bk	98 Cf	99 Es	100 Fm	101 Md	102 No	103 Lr	

Figure 2-5. The periodic table of elements.

combine with group VIIA elements in a one-to-one ratio: one atom from group IA to one atom of group VIIA. For other families, the ratios are different: two to one, three to one, and so on. Though there are some exceptions to it, this uniformity of behavior makes the study of all chemical and physical properties of the elements much easier. We shall go further into this study after we discuss *why* atomic properties are periodic. The discoveries of Newland and Mendeleev were empirical—based on observation, not theory. The necessary theoretical vehicle, quantum mechanics, was not available in the last century to assist them. With its help, today we can make much more sense out of the periodic law and the periodic table.

Just as the periodic law is founded on quantum mechanics, so quantum mechanics is based on the laws of wave motion. In the next section we shall discuss waves. Understanding something about waves will help us form a quantum concept of atoms. Even more, waves by themselves are integral to the environment. Without waves, the earth would receive no energy from the sun. From this fact alone, it becomes immediately clear that wave motion has far-reaching environmental consequences.

WAVES AND ELECTROMAGNETIC RADIATION

Most people have seen waves rolling in from the ocean. Although waves may travel endless miles across water, the water itself moves only locally as it rocks with the rhythm of the passing waves. How can the waves move along while their watery medium stays put? The answer is that a wave is nothing but a disturbance—a rocking of the water—passed from one parcel of water to the next. However, it is a universal kind of disturbance, occurring widely in nature. Most significantly, it is a disturbance that carries energy, sometimes through great distances.

The distance between the crests (tops) of successive waves is called the *wavelength*. The concept of wavelength is illustrated in Figure 2-6. Every train of waves has a characteristic wavelength. Wavelength is often indicated by the Greek letter lambda, λ.

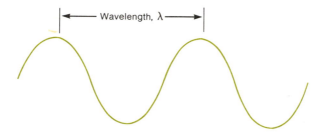

Figure 2-6. The wavelength, λ, of a train of waves.

The number of waves arriving at a fixed object each second is the *frequency*, shown by the Greek letter v. Waves of short wavelength are crowded close together, and therefore a great many of them arrive per second: the frequency is high. By this logic, wavelength and frequency are inversely related—when one is small, the other is large.[1]

---

[1] Frequency also depends on the wave velocity, c, the distance traveled by a wave in one unit of time (typical units for velocity are cm/sec, mi/hr, and so on). Note that in time t, a wavecrest will move a distance ct. In this time, all the crests out to distance ct will arrive, a number equal to

$$\text{number of crests} = \frac{\text{total distance, } ct}{\text{distance occupied by one wave, } \lambda} = \frac{ct}{\lambda}$$

Since frequency is the number of crests going by per unit of time (no. crests/t), division of each side of the above equation by t yields

$$\text{frequency, } v = \frac{\text{no. crests}}{t} = \frac{c}{\lambda}$$

This equation, $v = c/\lambda$, shows that frequency (v) goes up in direct proportion to wave velocity (c) but is still inversely related to wavelength (λ), as suggested earlier.

Electromagnetic Waves. A rich assortment of waves exists in nature, including waves on water (just discussed), sound waves, and electromagnetic waves. Of all these, electromagnetic waves are most significant, and they will occupy our attention here.

Trains of electromagnetic waves are called *electromagnetic radiation*. Ordinary light is a form of electromagnetic radiation, as are x rays, gamma (γ) rays (a form of radioactivity, discussed in a later chapter), ultraviolet (UV) radiation (black light), infrared (IR) radiation, radar, and radio waves. All of these travel at the same enormous velocity, the velocity of light (3×10^{10} cm/sec or 186,000 mi/sec in the vacuum of space). They differ from one another only in wavelength and frequency. The wavelengths cover an astounding range, having dimensions from the minute size of a nucleus to the vast distances of space. Part of this wavelength range (or spectrum), which has many roles in the environment, is shown in Figure 2-7.

Recall that waves on water are a disturbance propagated along by the periodic tossing of the water itself. Sound waves are propagated by the to-and-fro motion of air (or other matter). Similarly, electromagnetic waves are propagated by pulsing electric and magnetic fields (hence the term electromagnetic). These fields can exist in empty space, and therefore electromagnetic radiation is singular in its ability to traverse the vastness of space. By this route a part of the sun's enormous energy is transmitted to earth, causing green plant photosynthesis and warming our environment in protection against the terrible cold of outer space.

Waves of all kinds carry energy, but electromagnetic radiation is peculiar in carrying energy in small distinct parcels called *quanta*. The energy of a quantum of electromagnetic energy is proportional to the frequency of the radiation.[2] High-frequency radiation (that is, radiation of short wavelength) can release large units (quanta) of energy, large enough to strip electrons from atoms. The loss of electrons profoundly alters the chemical properties of atoms. If this stripping happens to key atoms in living systems, great biological damage can be done. For this reason, all high-frequency radiation is a threat to living systems; low-frequency radiation is not. (Low-frequency radiation will cause indirect damage only if it is intense enough to heat tissue to biologically unacceptable levels.) The dividing line between damaging and nondamaging radiation is somewhat blurred because some atoms are more easily affected than others. The approximate location of this line is shown in Figure 2-7: it occurs at a wavelength of roughly 3000 A, in the ultraviolet range of the spectrum.

By what means can electromagnetic radiation pass its quantum of energy along to an atom? Electromagnetic radiation, you recall,

---

[2]Mathematically, the energy of a quantum of electromagnetic energy is expressed by $E = h\nu$, where h is a constant of proportionality known as Planck's constant.

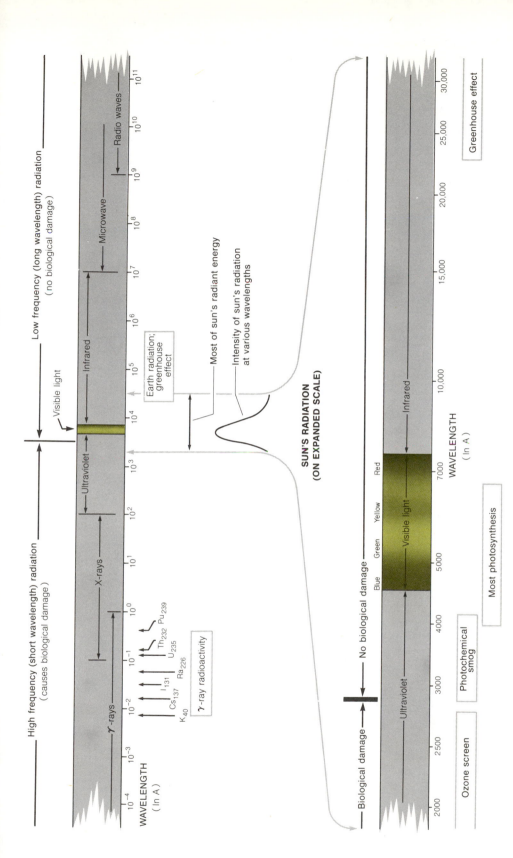

Figure 2-7. Wavelength spectrum of electromagnetic radiation, in angstroms, A. *Upper part:* A broad range of spectrum showing the location of important classes of electromagnetic radiation, visible light, gamma rays, and so on. In total, they span the range from 10^{-4} A to 10^{10} A (100 cm), a 100-trillion-fold range of wavelengths. *Lower part:* The limited wavelength range of the sun's radiant energy, expanded. The location of various parts of the spectrum having environmental impact are shown, including the wavelength range of the ozone screen, photosynthesis, photochemical smog, and the carbon dioxide (CO_2) greenhouse. Each is discussed in later chapters.

has a pulsing electrical field that is able to tug on all electrical charges. The electron is electrically charged and can therefore be jerked out of place by impinging radiation of the proper frequency. More than this, electrons moving in atoms are a source of pulsing electrical fields, and they can in some circumstances generate electromagnetic radiation.

By both emitting and absorbing electromagnetic radiation, atoms exchange energy with the environment. This exchange takes place in fixed parcels of energy (therefore at fixed frequencies), the magnitude of which depends very specifically on the atom involved. By virtue of this specificity, measurement of the frequencies of absorbed and emitted radiation provides a "fingerprint" to identify the elements in complex materials.

The measurement of the frequencies (or wavelengths) of electromagnetic radiation is called *spectroscopy*. It is responsible for much of our present knowledge about the nature of matter, and it is a prime tool for ferreting out environmental contaminants like mercury (Hg). The spectrum of mercury is shown in Figure 2-8. This unique "fingerprint" will appear whenever matter containing mercury is subjected to spectroscopic analysis.

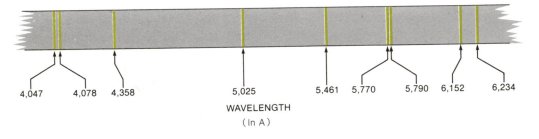

Figure 2-8. The spectrum of mercury (Hg). This spectrum is a unique "fingerprint" for the detection of Hg.

Standing Waves. Ocean waves are called *traveling waves* because they move constantly onward. Electromagnetic waves from the sun are also traveling waves, reaching us after a long journey through space. By contrast, waves that are physically confined to a limited region with no possibility of moving onward develop into *standing waves*. The waves on a violin string are standing waves, vibrating again and again in the same place.

Standing waves are all limited to certain discrete wavelengths that have definite frequencies associated with them. With a violin string, each frequency (wavelength) emits a distinct musical note: these are the fundamental note and various overtone notes. The shape of the waves corresponding to several notes is shown in Figure 2-9. (Discrete standing waves are found in wind instruments also, but these waves are acoustical waves in a confined chamber rather than the waves on a confined string.) In this figure, waves have wavelengths

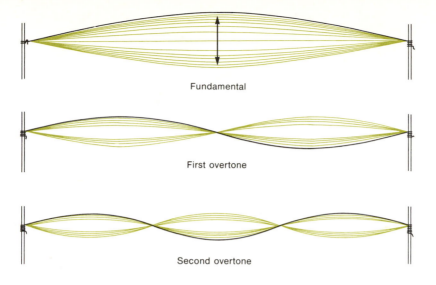

Fundamental

First overtone

Second overtone

Figure 2-9. The discrete waves on a violin string (or any vibrating string or wire) leading to the musical notes shown.

equal respectively to 2/1, 2/2, 2/3, . . . times the length of the string. Waves of intermediate wavelength never occur. Such discreteness in standing waves is responsible for the music which enriches our lives, and it is responsible for the order of the universe because it fixes the structure of atoms. We shall explain this idea starting with the simplest atom in existence—that of ordinary hydrogen (H).

THE HYDROGEN ATOM

In 1913 Niels Bohr, a Danish physicist, proposed that the hydrogen atom's electron was spinning around the proton, held in orbit by the attraction of negative to positive charge. In one way his model resembled a small solar system, with electrons imitating the planets circling the sun. In another way his atomic model was very different, for the electrons were limited to certain discrete orbits at certain fixed distances from the proton, and they could not revolve in orbits between (Figure 2-10). This idea was a revolutionary picture of orbital motion and all of the basic physics from which it derived, because, as it was then conceived, a planet could occupy any orbit around the sun if it just started out right. This classical picture of orbits has been verified in recent years by space capsules, which can be put in any desired orbit around the earth or moon. But electrons are different. They clearly play by other rules, the quantum rules of discreteness. Quantum behavior has worked itself into everyday language: the term "quantum jump" is used to describe anything that makes a sudden and discrete change from one level (orbit) to another, not stopping between.

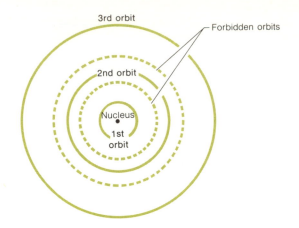

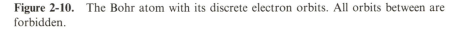

Figure 2-10. The Bohr atom with its discrete electron orbits. All orbits between are forbidden.

How did Bohr presume to know about these discrete electron orbits if he could not see the electrons in motion? He deduced his picture from the light emitted from hydrogen atoms—that is, from the spectroscopy of hydrogen. If an electron were to fall from the second orbit to the first, it would give up energy as a quantum unit of light, the frequency of which would indicate the magnitude of the energy. By observing this emitted light, Bohr could deduce the energy of different electron orbits, along with their distance from the nucleus.

Just as spectroscopy led Bohr to his picture of discrete orbital energies, it has since become the chief tool in deducing atomic and molecular structure. It is an indirect tool, like all the rest, but it measures with great accuracy the many energy changes that are associated with electron, nuclear, and atomic motions. Thus it provides a rigorous check on our picture of atoms and molecules. Spectroscopy verified certain features of Bohr's atom, but a more detailed spectroscopic analysis showed that Bohr's model was incomplete and inaccurate. However, the main fault of the Bohr atom was its failure to explain how atoms bond together into molecules, a process fundamental to all chemistry. This failure led in 1926 to the second revolutionary model of an atom's electrons, the quantum mechanical model.

According to quantum mechanics, electrons are no longer to be considered as revolving in definite orbits about the nucleus. Instead the electron forms a "fuzzy" electron cloud extending over much of the atom. The cloud is held together by its attraction to the positive nucleus. The exact position and momentum (motion) of the electron within the cloud at any moment are uncertain, no longer describable by physical law, nor in accord with our usual rules of common sense. Hence arises the famous *uncertainty principle* of quantum mechanics.

The strange existence of electron clouds is due to the fact that electrons in atoms act like standing waves. Different clouds may be identified with different standing waves in the atom. This fact explains several important features of atoms. First, waves always extend over a finite region of space: for this reason, their location cannot be pinpointed like the location of a small particle. Thus the uncertainty principle is explained.

Second, we have noted that standing waves exist only in certain discrete waveforms. This fact explains the discrete "orbits" thought to exist by Bohr. In actual fact, there are no orbits: instead there are only the discrete electron clouds, each having a precisely defined energy. The unique energy associated with each cloud replaces the Bohr picture of a unique energy associated with each orbit. A quantum of light is emitted when the electron changes from one cloud form to another rather than from one orbit to another.

The concept of electrons acting as waves was perplexing at first because particles such as electrons were considered to be entirely unrelated to waves. Yet here was a particle (or presumed particle) acting as a wave would. We now think of the electron and other elementary "particles" as having *dual wave-particle characteristics*. The "particle" picture, by itself, simply cannot explain the observed characteristics of matter. Since wave motion is involved, the theory of electron behavior is sometimes called *wave mechanics*, identical in meaning to quantum mechanics.

According to the new rules of quantum mechanics, electron clouds (standing waves) possess definite shapes, sizes, and energies. The shapes of several electron clouds are shown in Figure 2-11. There are many other cloud configurations, mostly more complicated.

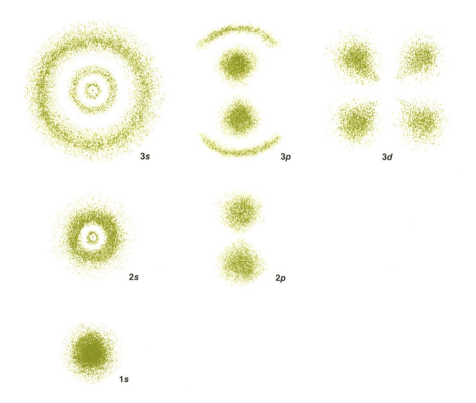

3s   3p   3d

2s   2p

1s

Figure 2-11. Typical electron clouds.

The seeming strangeness of quantum mechanics is beautifully illustrated by the 2p electron cloud. The cloud completely vanishes in the center plane between its two lobes. This means electrons are never to be found there. However, they pass back and forth from the left lobe to the right lobe rather frequently. They do so without, apparently, ever occupying the corridor between. It is like going in and out of a garden without ever being found at the gate. This idea is maddeningly incomprehensible until we remember that electrons travel in the form of waves. Waves can indeed be transmitted through regions of zero amplitude: thus we explain the electron dilemma. But such action is very unparticlelike, and it warns us that other unexpected and fascinating phenomena are to be found as part of the quantum world of atoms.

THE SIMPLIFIED SHELL MODEL

For a rigorous study of atoms and molecules, we would need to examine the shape and size of electron clouds in detail. This task would be arduous: we would need to understand the difficult mathematics of quantum mechanics. For our purposes, we need only to extract and use those major features of quantum mechanics that give direction to our thinking about atoms. Two features are most important.

First, while electron clouds have varied shapes, they fall into distinct size groups. (The size of an electron cloud refers to the average distance of the cloud from the nucleus.) Electron clouds in a given size group are said to occupy a *shell*. A shell, then, is simply a collection of like-sized electron clouds. For convenience we can imagine that the shell of electrons encircles the nucleus uniformly, as shown in Figure 2-12. The distance from the nucleus out to the shell is equal to the mean distance of that shell's electron clouds from the nucleus, so that the shell idea shows very well the size and distance from the nucleus of the underlying electron clouds. In a way these shells resemble the old Bohr orbits, but we must regard them as more diffuse and cloudlike so that they are in accord with quantum mechanical realities.

Soon we shall see that the successive shells of electron clouds which cloak the nucleus give rise to the layers of elements that constitute the periodic table.

Second, quantum mechanics specifies that each electron shell has a very definite limit to its capacity for electrons. After a fixed number of electrons enter a shell, no more can be accommodated. Electrons beyond a shell's capacity must enter another, unfilled shell. This, too, is basic to the periodic table, for the filling of one shell and the beginning of another amounts to the completion of one row or period of the periodic table and the start of another.

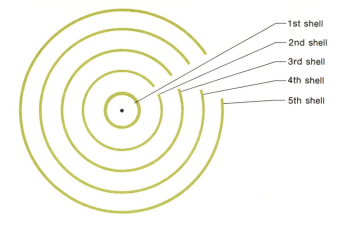

Figure 2-12. Shell model of the atom.

When examined in detail, electron shells are found to have an orderly substructure (corresponding to different shape features among electron clouds of comparable size). The subunits of this substructure also have a fixed capacity. The rules that govern these capacities, combined with the rules that fix in what order the various subunits may be filled by electrons, are all that we need to construct the periodic table in all its important detail.

SHELL-FILLING RULES

Quantum mechanics is very specific about the nature of shell subunits. Thus the various shells are composed of *subshells*, and the subshells are further divided into so called *orbitals*. Each orbital can hold, at most, exactly two electrons. Does it sound complicated? It is no more so than this apartment complex: the building is divided into floors (the shells); each floor may have several wings (the subshells); and each wing contains rooms (the orbitals), each one capable of housing at most two people (the electrons). Here are the essential features of this comparison:

Atoms: shells subshells orbitals: electrons

Apartment complex: floors   wings   rooms : people

The unique distribution of electrons within their atomic apartment complex is responsible for the structure of the periodic table of elements. Electron distribution, as with the distribution of people in a city dwelling, hinges on two factors. First, it depends on the electron capacity of the various parts—the orbitals, subshells, and shells—within the total complex. Second, distribution is determined by the electrons' tendency to occupy and fill completely some parts of the complex in distinct preference to other parts. Definite rules exist both for capacity and the order of occupancy. These rules follow.

Electron Capacity

Rule I. The first capacity rule was stated above: only two electrons are permitted per orbital (room).

Rule II. The second capacity rule says that there are several types of subshells (wings), identified by the letters *s*, *p*, *d*, and *f*. These subshells contain 1, 3, 5, and 7 orbitals, respectively. Thus at a capacity of two electrons per orbital, the *s*, *p*, *d*, and *f* subshells hold a maximum of 2, 6, 10, and 14 electrons.

Rule III. The first shell has only a single subshell, the *s* subshell; the second shell has both *s* and *p* subshells; the third shell has *s*, *p*, and *d* subshells; the fourth and succeeding shells can use all four (*s*, *p*, *d*, and *f*) subshells.

The three capacity rules are summarized below:

Rule I
Electrons per orbital | 2 electrons/orbital

Rule II
Orbitals per subshell | 1 orbital/*s* subshell
3 orbitals/*p* subshell
5 orbitals/*d* subshell
7 orbitals/*f* subshell

Rule III
Subshells per shell | 1 subshell (*s*)   /1st shell
2 subshells (*s* and *p*)   /2nd shell
3 subshells (*s*, *p*, and *d*)   /3rd shell
4 subshells (*s*, *p*, *d*, and *f*)/4th shell
4 subshells (*s*, *p*, *d*, and *f*)/5th shell
4 subshells (*s*, *p*, *d*, and *f*)/6th shell
4 subshells (*s*, *p*, *d*, and *f*)/7th shell

Figure 2-13 illustrates rules II and III. It diagrams the subshells in each shell and the total number of orbitals both per shell and per subshell. This figure may also be regarded as a diagram of the apartment building analog, showing the correspondence between shells and floors, and so on.

If we apply rule I and allow only two electrons in each of the orbitals of Figure 2-13, an electron count—subshell by subshell—gives Figure 2-14, the electron capacity of the various subshells. Note that all *s* subshells, no matter what shell they are in, hold at most two electrons. Likewise, all *p*, *d*, and *f* subshells hold a maximum of 6, 10, and 14 electrons, respectively, the same numbers we noted before.

The standard designation for subshells, shown in Figure 2-13, is 1*s*, 3*p*, 4*d*, and so on. The number refers to the shell; the letter to the subshell.

Order of Occupancy. As we feed electrons into this shell-subshell-orbital structure in order to build up the atoms, we must

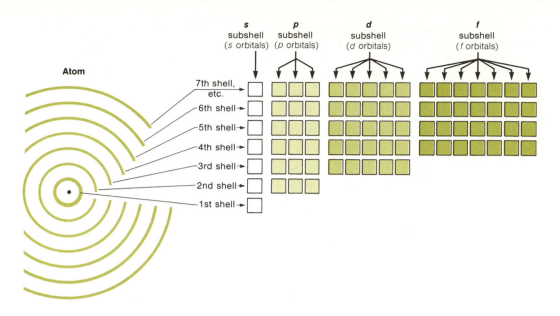

Figure 2-13. The capacity of the atomic shells for subshells and orbitals. Each little box is an orbital which can hold two electrons.

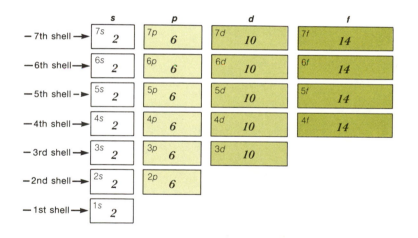

Figure 2-14. Electron capacity of subshells.

determine which orbitals they occupy. The one basic rule here is that the electrons will seek the lowest possible potential energy level, as dictated by the second law of thermodynamics. In other words, they prefer being as close to the nucleus as possible, pulled in by the positive charge of the protons. Clearly, then, the electrons will tend to occupy innermost shells first. However, because of subtleties in the shape of the underlying electron clouds, the electrons also have a subshell preference: *s*, *p*, *d*, and *f*, in that order.

The results of the tug-of-war between the two preferences—one for certain shells and one for certain subshells—is not easy to predict beforehand for higher elements, as it is for the first 18. However, the actual order of filling has a great deal of regularity which, we shall discover, is the basis of the periodic table.

Because the preference is for the lower shells and for the left-hand subshells (the s subshell is furthest left) in Figure 2-13, we can envision the electrons "gravitating" to the lower left-hand corner. The "preferred housing" is near this corner, and electrons will fill these subshells first, then those further out. The precise order of subshell filling is shown in Figure 2-15. The order of filling is seen to be $1s \rightarrow 2s \rightarrow 2p \rightarrow 3s \rightarrow 3p \rightarrow 4s \rightarrow 3d \rightarrow 4p \rightarrow 5s \rightarrow 4d \rightarrow 5p \rightarrow 6s \rightarrow [4f\text{-}5d] \rightarrow 6p \rightarrow 7s \rightarrow [5f\text{-}6d]$.

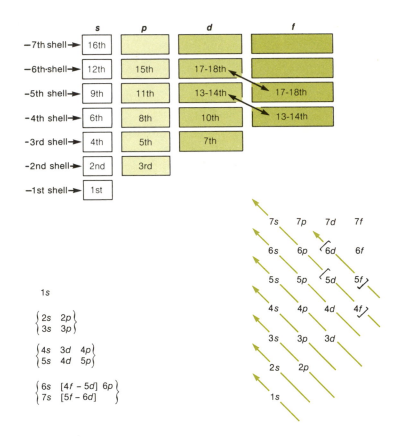

Figure 2-15. The order in which subshells are filled by electrons. Two subshells share 13–14th places and 17–18th places, which means, for example, that 13 and 14 fill (in part) at the same time. The lower right-hand figure is a simple diagram to aid in remembering the filling order. Follow the arrows, the bottom one first. The lower left-hand diagram gives the filling order arranged to show the origin of the periodic table (to be explained shortly). When it is read from left to right and top to bottom, like the periodic table itself, a regular arrangement becomes apparent, with all s subshells on the left, all p subshells on the right, all d subshells one shell number lower than the preceding s subshell, and so on.

Reread the rules of this section. They are essential to all chemistry.

We shall now discuss the filling of shells, showing the successive chemical elements that are created as electrons (along with the complementary protons) are added to the atom as dictated by the foregoing rules.

SHELLS AND ELEMENTS

The first and second electrons go into the single $1s$ orbital of the first shell. By the rules, these two electrons fill the shell completely. The atoms constructed are hydrogen and helium, chemical symbols H and He. The shell is represented by

The third and fourth electrons go into the second shell, the $2s$ orbital. The $2p$ subshell is filled next. There are three $2p$ orbitals in this subshell, able to accommodate six electrons, the fifth through the tenth. These eight successive electrons complete another shell, giving the atoms lithium (Li), beryllium (Be), boron (B), carbon (C), nitrogen (N), oxygen (O), fluorine (F), and neon (Ne):

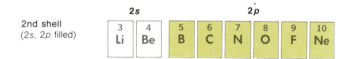

Similarly the next eight electrons fill the s and p orbitals of the third shell. In theory the third shell could accommodate 10 more electrons in its d subshell, but recall that electrons prefer the s subshell above all other subshells. The next electron chooses to fill the $4s$ rather than the $3d$ orbital, even though it is one shell further out. Thus the third shell is temporarily rounded out with only eight elements:

These elements are sodium (Na), magnesium (Mg), aluminum (Al), silicon (Si), phosphorus (P), sulfur (S), chlorine (Cl), and argon (Ar).

The filling of the $4s$ orbital starts the fourth shell. Next we might anticipate that either the 4*p* subshell or the missed 3*d* subshell will fill. As it turns out (Figure 2-15), the 3*d* fills first with its 10 electrons, followed by the 4*p*. Thus the filling of the fourth shell, interrupted by the completion of the 3*d* subshell, is represented by

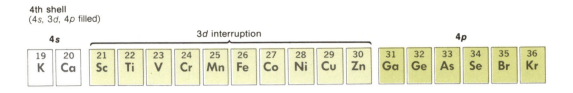

the elements being potassium (K) and calcium (Ca) on through to krypton (Kr), element 36. (The names of the other elements appear in Table 2-1.)

The same thing, exactly, happens with the fifth shell: the 5*s* subshell fills, then we go back to fill the 4*d* subshell, and finally on with the 4*p* subshell. We have

5th shell
(5*s*, 4*d*, 5*p* filled)

5*s*		4*d* interruption										5*p*					
37 Rb	38 Sr	39 Y	40 Zr	41 Nb	42 Mo	43 Tc	44 Ru	45 Rh	46 Pd	47 Ag	48 Cd	49 In	50 Sn	51 Sb	52 Te	53 I	54 Xe

In this series are elements 37 through 54, rubidium (Rb) to xenon (Xe).

The sixth shell begins in much the same way, with the 6*s* subshell filling and the next lower *d* subshell, 5*d*, also beginning to fill. No sooner has one electron gone into the 5*d* subshell, however, than seemingly as an afterthought, the next electron strikes out for the long neglected 4*f* subshell. Beginning here, the added electrons appear vaguely confused in their preference, not always filling one subshell completely before starting another. Furthermore, the electrons already present occasionally change between *d* subshells and *f* subshells as atomic number increases. The reason for all this confusion is that a *d* subshell and the next lower *f* subshell have nearly identical energies. Each electron added affects the energy somewhat because of its repulsive force on other electrons. Thus these two subshells can shift above and below one another in energy as electrons are added, the electrons as usual choosing the subshell that is lowest in energy.

With this single complication, the filling of the next level is represented by

6th shell
(6s, 5d-4f, 6p filled)

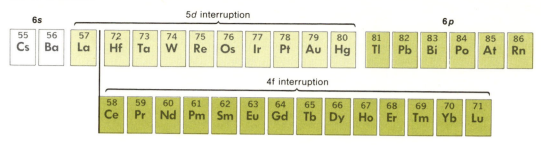

This diagram shows how the filling of the 4f subshell (capacity: 14 elements) breaks into the filling of the 5d subshell. Previously, subshells did not break into other subshells; they sandwiched nicely between adjacent subshells. This change reflects the complication we mentioned earlier.

The seventh shell is similarly interrupted by both 6d and 5f subshells:

7th shell
(7s, 5f filled)
(6d, partly filled)

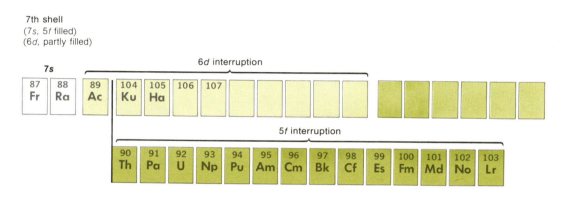

However, we see that the 6d subshell is never completed: we have run out of elements. The last, Z = 105, is Hahnium (Ha). Credit for its initial synthesis is presently in dispute between American and Soviet scientists.

BASIS OF THE PERIODIC TABLE

If we stack the rows of elements from the past few pages on top of one another, we get the periodic table (Figure 2-16). The way in which the various parts of the periodic table are related to the filling of subshells is clear.

We noted before that columns of elements are called "groups" or "families" because their properties are so similar. *The similarity*

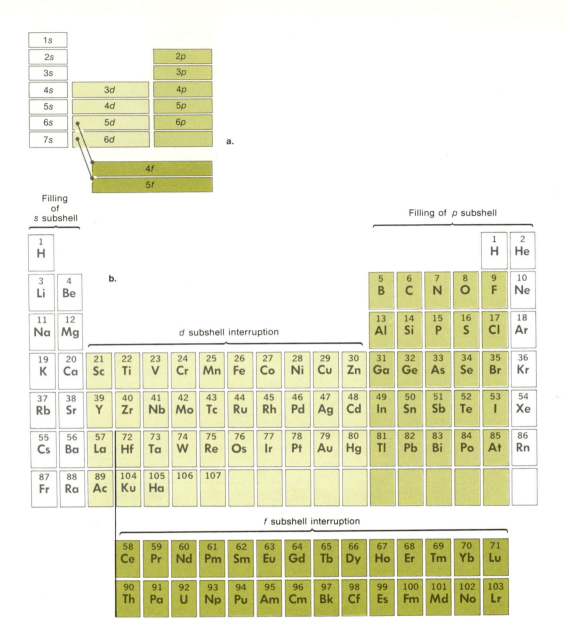

Figure 2-16. The periodic table and its relationship to subshell filling. (*a*) Basic subshell structure of the periodic table. (*b*) Subshell structure with elements filled in, yielding the full periodic table.

occurs because we are at the same stage of filling subshells of the outer shell. The outer-shell electrons are all important for chemical properties. They are at the "surface" of the atom, where contact with other atoms is made, and they are relatively mobile because of their large distance from and weakened attraction to the nucleus. These electrons are the ones responsible for forming chemical bonds between atoms. Therefore all atoms with, for instance, one electron in the outermost shell are expected to behave similarly, as indeed they do. Such atoms are found in the first column (group IA) of the periodic table. A like similarity occurs in other columns of the periodic table because they too contain atoms that all have the same number of "surface" (outer-shell) electrons.

The first row of the periodic table is unusual. It holds only two elements, hydrogen (H) and helium (He), because the first quantum mechanical shell holds only two electrons. The other shells first fill out with eight electrons, although more may be added subsequently. With which families of the eight do hydrogen and helium belong? On first inclination we would place them in the first two columns, since they have one and two outer-shell electrons, respectively. However, helium belongs more correctly in the last column, because its shell is complete, a property shared with neon (Ne) and the elements below neon. Chemical studies verify that helium belongs there: it is an inert, relatively unreactive gas like others in the final column.

Hydrogen is more difficult. It can be grouped with the first-column elements because of its lone outer-shell electron, or it can be shifted, following helium, to the next to last column, where all elements are one electron short of a completed shell. Both positions have justification, and both are used in different versions of the periodic table. Hydrogen belongs in both places and shares the properties of both groups. It is not entirely like either one because of this compromising role. It is the most exceptional element in the periodic table.

CLASSIFICATION OF ELEMENTS

Elements in certain regions of the periodic table are given collective names because of certain comparable properties. We have mentioned the designation of "families" of elements as group IA, group IVA, and so on. Some of these families are given special names—for example:

group IA (below H):   *alkali metals*
group IIA: *alkaline earth metals*
group VIIA: *halogens*
group 0:   *noble gases* (or *inert gases*)

These and other collective names are shown in the periodic table of Figure 2-17.

The horizontal rows of the periodic table are called *periods*. One cannot call them "shells," because the *d* electrons along a row are an inner shell, interrupting the filling of the principal shell.

Elements of the large central region of the periodic table where the inner *d* and *f* subshells are being filled are called *transition metals*.

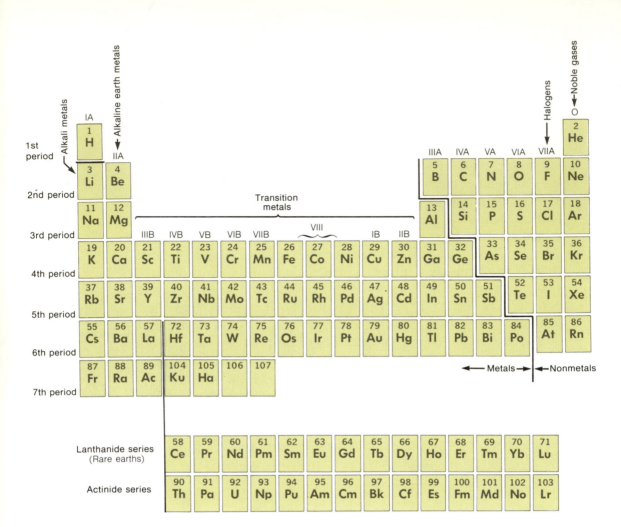

Figure 2-17. Classification of the elements in the periodic table.

The *lanthanide series*, or *rare earth* metals, and the *actinide* series (both "inner" transition metals) represent those elements in which 4*f* and 5*f* subshells are being filled, respectively. Elements within a series have very similar properties. We expect this similarity because electrons are being added to inner shells, leaving essentially a constant number of electrons in the outer shell.

There are far more metals than nonmetals among the elements. The heavy diagonal line shown in Figure 2-17 divides metals from nonmetals. The elements on the left and in the central portion of the periodic table are all metals, with the exception of hydrogen. Only 22 elements on the right-hand side, out of 105 elements total, are nonmetals. Some elements (metalloids or semimetals) touching the line may share metallic and nonmetallic properties. Arsenic (As) is an example.

Nonmetals appear uniformly on the right-hand side of the periodic table, and metals on the left. This metal-nonmetal division again illustrates the periodic law and the rational ordering of significant atomic properties by the periodic table.

ELECTRONIC STRUCTURE OF ELEMENTS

Each element has a unique number of electrons in its shells and subshells. The distribution of electrons in subshells is referred to as the *electronic structure*. The number of electrons in each subshell is shown by a superscript to the subshell symbol. For instance, two electrons in the $3s$ subshell are indicated by

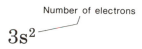

For the whole atom, electron occupancy is indicated one subshell at a time, in the order of filling. Oxygen (O), with two electrons in the $1s$ subshell, two in $2s$, and four in $2p$, is represented by

$$\text{oxygen (O)} \qquad 1s^2 2s^2 2p^4$$

The sum of the superscripts, $2 + 2 + 4 = 8$, equals the total number of electrons. When the electronic structure is given correctly, this sum must equal the atomic number (element number) Z, a number we can always find from the periodic table. (Although the superscripts must add up to Z, the fact that they do does not guarantee that an electronic structure is correct, since one can err by scrambling the numbers without changing their sum.)

The electronic structure of lead (Pb), element number 82, is more complicated. From our knowledge of the periodic table and the position of lead therein, it is clear that the $1s$, $2s$, $2p$, $3s$, $3p$, $4s$, $3d$, $4p$, $5s$, $4d$, $5p$, $6s$, $5d$, and $4f$ subshells are filled. Two electrons have entered the $6p$ subshell. Therefore the electronic structure of lead is

$$\text{lead (Pb)} \qquad 1s^2 2s^2 2p^6 3s^2 3p^6 4s^2 3d^{10} 4p^6 5s^2 4d^{10} 5p^6 6s^2 5d^{10} 4f^{14} 6p^2$$

The superscripts add to 82, the atomic number of lead.

The electronic structures of other elements can be obtained in a similar way. To get them, we must know three things: the atomic number of the element, the order in which subshells are filled, and subshell capacities.

ELECTRON DOT FORMULAS

Because outer electrons are of dominant importance, it is useful to have a way of representing these electrons by themselves. We can do so by using the electron dot formula. Here one simply shows by dots the outer electrons surrounding an atom. For example, sodium (Na) has one outer electron and thus one dot:

$$Na \cdot$$

The next element, magnesium (Mg), has two electrons in the outer shell:

$$\overset{\displaystyle\cdot}{Mg} \cdot$$

Magnesium is followed by elements 13 through 18, aluminum (Al) through argon (Ar), each with one more electron than the last:

$$\cdot \overset{\displaystyle\cdot}{Al} \cdot \quad \cdot \overset{\displaystyle\cdot}{\underset{\displaystyle\cdot}{Si}} \cdot \quad \cdot \overset{\displaystyle\cdot}{\underset{\displaystyle\cdot}{P}} \colon \quad \cdot \overset{\displaystyle\cdot}{\underset{\displaystyle\cdot}{S}} \colon \quad \colon \overset{\displaystyle\cdot}{\underset{\displaystyle\cdot}{Cl}} \colon \quad \colon \overset{\displaystyle\cdot\cdot}{\underset{\displaystyle\cdot\cdot}{Ar}} \colon$$

The number of dots, of course, is the same for each element of a family.

With the electron dot formula, it is easy to picture an atom's outer electrons and how, by the addition and subtraction of electrons, charged ions are formed from these atoms. However, the greatest use of the electron dot formula is in understanding chemical bonding, which we shall discuss in the next chapter.

IONS, METALS, AND THE PERIODIC TABLE

We have noted that electrons in the outer shell determine the properties of elements. Much depends on how tightly these electrons are bound, which in turn depends on the element's position in the periodic table.

The group IA elements, the alkali metals, have one electron in the outer shell. The electron's attachment to the atom is relatively weak because (1) it is furthest removed from the positive nucleus and (2) it is pushed outward even more by repulsion from electrons in the inner shells, and this repulsion offsets much of the nuclear attraction that does exist. This offsetting effect is called *shielding*. The inner electrons effectively shield the nucleus so that the outer electron is less attracted to it.

Ions. Because the single outer electron of alkali metal atoms is loosely held, it is easily ripped off, leaving behind a charged atom or *ion*. Since a single negative ($-$) charge is removed, a single positive ($+$) charge is left on the ion. (That is, removal of an electron gives an atom having one fewer electrons than protons, thus having a net positive charge of one.) Positive ions are sometimes called *cations*. The alkali metal cations are designated by the symbols Li^+, Na^+, K^+, Rb^+, Cs^+, and Fr^+. They are called the lithium ion (or cation), the sodium ion, the potassium ion, and so on.

The alkaline earth metals, group IIA, have two electrons in the outermost atomic shell. These two can be pulled away to form the doubly charged ions Be^{2+}, Mg^{2+}, Ca^{2+}, Sr^{2+}, Ba^{2+}, and Ra^{2+}. (Another, older, notation for doubly charged ions is Be^{++}, Mg^{++}, and so on. This notation is still commonly used.) However, these electrons are not stripped off as easily as are those attached to the alkali metals. There are two reasons for this difference. First the nuclear charge has increased by one $(+)$ unit in going from group IA to IIA, offering a greater attractive force. Second, the shielding is not as complete, since electrons in the outer shell (now two in number) do not appreciably shield the nucleus from one another. Therefore alkaline earth metals form cations more reluctantly than do alkali metals, a greater pull being necessary to dislodge the electrons.

The group IIIA elements have three electrons in the outer shell and occasionally lose all three to form the triply charged ions B^{3+}, Al^{3+} (sometimes written as B^{+++} and Al^{+++}), and so on. This ionization is still more difficult for the same reasons: increased nuclear charge and decreased shielding. The attraction for electrons and thus the difficulty of forming positive ions continues to increase as we proceed left to right in the periodic table. Group IVA elements and beyond rarely form positive ions. Quite the opposite happens: the elements to the right of the periodic table, with their increased electron attraction, form ions with a negative charge, *anions*.

Group VIIA elements, the halogens, have seven electrons in the outer shell, two in the outer *s* subshell and five in the outer *p* subshell. The outer shell by now is strongly attracted to the nucleus, because nuclear charge has increased still further and shielding has decreased. There is one empty position in this shell, and it is very attractive for electrons. Halogens therefore tend to acquire an extra electron to fill this empty slot. Negative ions, or anions, are thus formed, with symbols F^-, Cl^-, Br^-, I^-, and At^-. They are called the fluoride ion, chloride ion, bromide ion, and so on.

Group VIA elements have two empty slots in the outer shell. They may therefore acquire two electrons, forming the ions O^{2-}, S^{2-}, and so on. However, they have less affinity for electrons than do the halogens, a part of the general trend. Likewise, group VA elements form triply charged negative ions, N^{3-}, P^{3-}, and As^{3-}, but with even greater reluctance.

The major trend, then, is one in which elements to the right, except the inert gases, are electron hungry and negative-ion formers, and elements to the left are electron donors and thus positive-ion formers. The change is gradual across the periodic table.

Ion formation among the transition metals is complicated because electrons from the inner shells that are being filled may or may not be pulled away with one or two outer electrons. Therefore these elements can form ions having different charges, depending on conditions. Iron may be Fe^{2+} or Fe^{3+}, called iron (II) and iron (III) ions. The relative amount of these two is important to the chemical balance of natural waters. Likewise, copper may form Cu^+ and Cu^{2+}, and so on.

Group 0 elements, the noble gases, have completed outer shells of eight electrons each. The electrons of this shell are held very tightly, so that positive ions cannot form, and there are no unfilled slots which electrons could enter to form negative ions. These atoms are therefore relatively resistant to change, and they are often called the inert gases. Shells of eight are generally stable, as we shall emphasize further on.

Electronic Structure of Ions. When positive ions are formed, the outer electrons are stripped away, exposing the completed shell that would be characteristic of the preceding inert gas. The electronic configuration of ion and inert gas are identical; only the nuclear charge is different. For instance, sodium has the electronic configuration

$$\text{sodium (Na)} \qquad 1s^2 2s^2 2p^6 3s^1$$

Loss of the single $3s$ electron to form the sodium ion (Na^+) can be represented by

$$\text{Na} \ (1s^2 2s^2 2p^6 3s^1)$$

$$\Downarrow \qquad \searrow \text{(this electron lost)}$$

$$\text{Na}^+ (1s^2 2s^2 2p^6) + \text{one electron}$$

Therefore the electronic configuration of Na^+ is

$$\text{sodium ion (Na}^+\text{)} \qquad 1s^2 2s^2 2p^6$$

which is identical to the electronic configuration of the inert gas neon (Ne). The same electronic configuration describes the magnesium ion (Mg^{2+}) and the aluminum ion (Al^{3+}).

Negative ions form by the completion of a shell, so that they, too, generally end up with an inert-gas electron structure. For instance, the fluorine atom is represented by

$$\text{fluorine (F)} \qquad 1s^2 2s^2 2p^5$$

One electron added to the $2p$ shell gives the anion

$$\text{fluoride ion (F}^-\text{)} \qquad 1s^2 2s^2 2p^6$$

also identical to neon in electronic configuration. The anions O^{2-} and N^{3-} have the same electron structure. Ions further down the periodic table have electronic configurations identical with the noble gases argon (Ar), krypton (Kr), and so on.

Metals. Before, we noted that metals occupy the left side of the periodic table. We have also noted that elements to the left of the periodic table have loosely attached electrons in the outer shells.

Are these two characteristics at all connected? The answer is yes, for the mobile electrons are responsible for metallic properties, such as shiny surfaces, high electrical conductivity, and high heat conductivity. Outer-shell electrons, being loosely attached to metal atoms, can move around with relative freedom in metals. Because electrons are charged, they carry electric current as they move. Hence metallic conductance occurs. Electrons also carry heat energy, giving the characteristic high heat conductivity, and at the surface they can respond to impinging light, reflecting that light and making the metallic surface seem shiny.

We see in total that metallic properties, loosely attached electrons, and the tendency to form positive ions are all related. Likewise, nonmetallic properties, tightly bound electrons, and the tendency to form negative ions are connected. All these phenomena originate in the electronic structure of atoms and are correlated by the periodic table of elements.

ATOMIC AND IONIC SIZE

As we proceed from left to right in the periodic table, the atoms become smaller because the outer shell is drawn in more and more tightly. As we proceed from top to bottom within a family, the atoms become larger because new shells are added. Both trends are shown in Figure 2-18. This figure shows the radius of atoms as they are chemically bonded to other atoms (see p. 76).

A cation is smaller than the parent atom because one shell is lost in ion formation. Cations typically are half as large as their parents. An anion, by contrast, is larger than the parent atom because the mutual repulsion of electrons increases as new electrons are added and this repulsion expands the outer electron shell. The size typically doubles as a consequence.

ABUNDANCE OF ELEMENTS

In this vast universe, hydrogen (H) and helium (He) are the dominant elements. Neither is pivotal to life as we know it. The core of our earth, by contrast, is mainly iron (Fe) and nickel (Ni), heavier elements which also fail to provide a foundation for living systems, although they are necessary in trace amounts to many life forms. On the surface of the earth's massive metallic core and its similarly dense mantle layer floats a thin film of slag, the crust of the earth. The crust, along with even thinner films of ocean and atmosphere, is rich in intermediate elements 6 to 20, those most central to the machinery of life. Oxygen (O) is most abundant, but carbon (C), nitrogen (N), phosphorus (P), and sulfur (S) are all elements integral

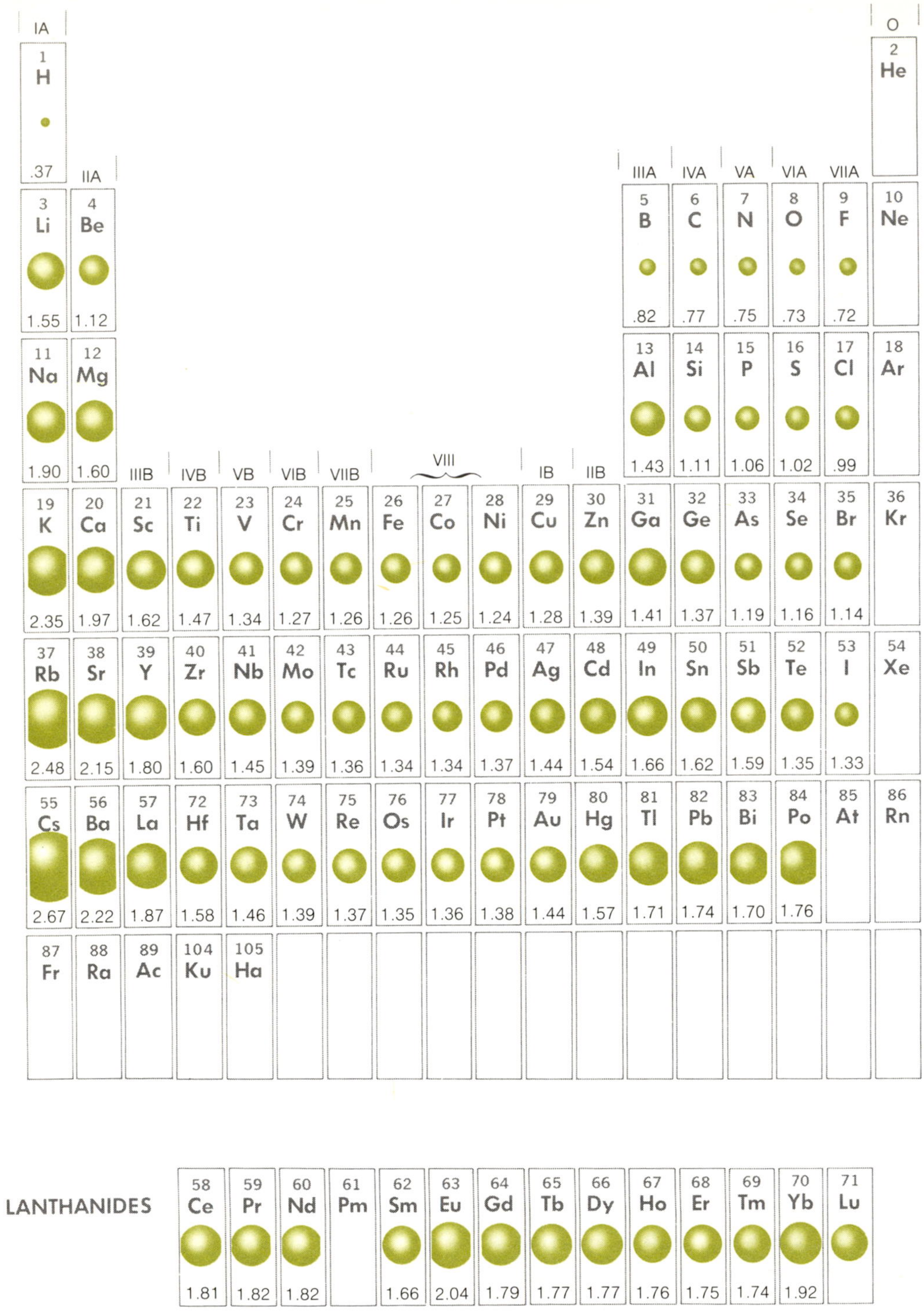

Figure 2-18. The radii of atoms, in angstrom units.

to life that are among the most abundant twenty. They are supplemented by hydrogen (H), sodium (Na), potassium (K), calcium (Ca), magnesium (Mg), and iron (Fe), all necessary but somewhat less vital to most life forms. Some other elements are needed by living systems in trace amounts, but there is little reason to believe that life would have developed along much different lines if they had been absent from the environment.

The abundance in weight percent of elements of the earth's crust, including ocean and atmosphere, is given in Table 2-2.

Table 2-2. ABUNDANCE OF TWENTY MOST COMMON ELEMENTS IN EARTH'S CRUST, INCLUDING OCEANS AND ATMOSPHERE

Atomic Number	Element	Weight Percent	Atomic Number	Element	Weight Percent
8	"Oxygen (O)	49.5	17	Chlorine (Cl)	0.2
14	Silicon (Si)	25.7	15	"Phosphorus (P)	0.1
13	Aluminum (Al)	7.5	25	Manganese (Mn)	0.09
26	"Iron (Fe)	4.7	6	"Carbon (C)	0.08
20	"Calcium (Ca)	3.4	12	"Sulfur (S)	0.06
11	"Sodium (Na)	2.6	56	Barium (Ba)	0.04
19	"Potassium (K)	2.4	7	"Nitrogen (N)	0.03
12	"Magnesium (Mg)	1.9	9	Fluorine (F)	0.03
1	"Hydrogen (H)	0.9	28	Nickel (Ni)	0.02
22	Titanium (Ti)	0.6	38	Strontium (Sr)	0.02
				All others	0.07

"Indicates elements of prime importance to the chemistry of life.

POLLUTION AND THE PERIODIC TABLE

In Figure 2-19 are spotlighted sixteen elements that are prominent in pollution. It would be too categorical to say that they are the most important pollutant elements, because importance cannot be weighed precisely. There are certainly other valid candidates for the list. Nonetheless, we call them the "sordid sixteen" for convenience, a group to whose members we shall come back often.

We are taking considerable liberty in using the adjective "sordid" to describe these elements. One of them, carbon (C), is the basic element of life. Oxygen (O) supplies the driving energy for most forms of life, and phosphorus (P) is an integral part of the energy machinery. Nitrogen (N) is a cornerstone of all protein molecules and sulfur (S) holds the protein molecules in shape so they can function. These elements are absolutely vital. Like most elements, they are a threat only when misused.

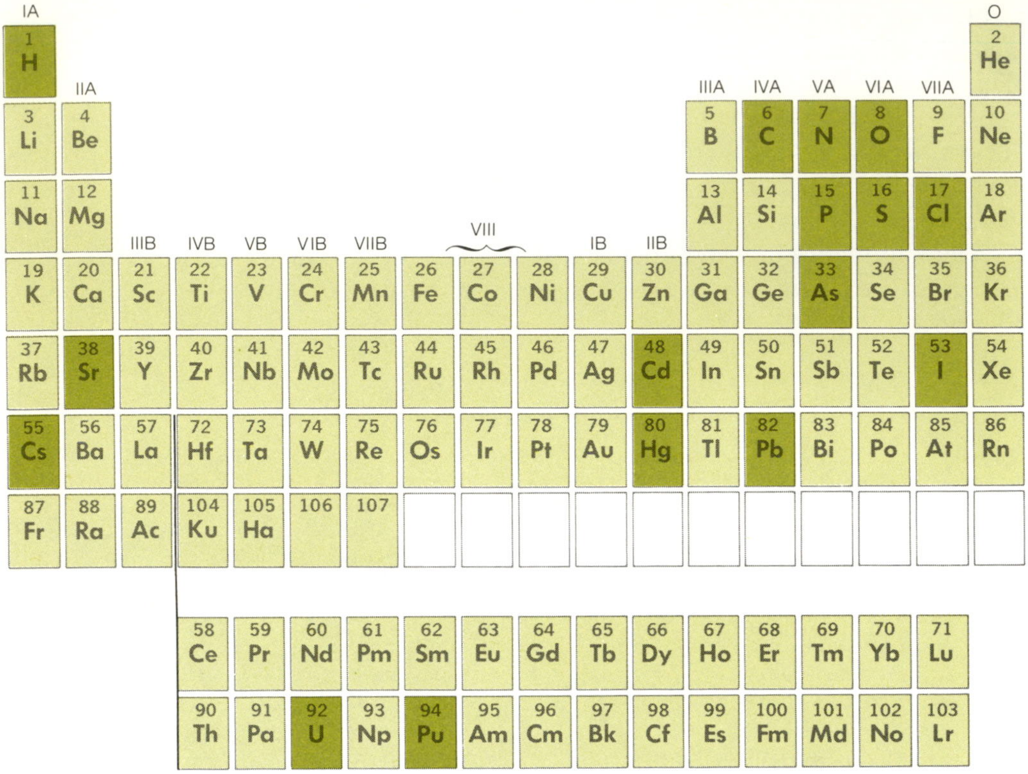

Figure 2-19. The "sordid sixteen," a group of elements playing a prominent part in environmental pollution.

Brief highlights of pollution by the "sordid sixteen" is given below in the order of increasing atomic number.

1. *Hydrogen (H), element 1.* A constituent of many air pollutants, water pollutants, and pesticides.
2. *Carbon (C), element 6.* An element integral to pollution by carbon monoxide, hydrocarbons, and pesticides.
3. *Nitrogen (N), element 7.* Oxides of nitrogen and photochemical smog, from automobiles and industry.
4. *Oxygen (O), element 8.* Has a crucial role in forming carbon monoxide, oxides of nitrogen, sulfur dioxide, ozone, and others.
5. *Phosphorus (P), element 15.* Causes water eutrophication and excessive algae growth.
6. *Sulfur (S), element 16.* Leads to sulfur dioxide pollution from fossil-fuel power plants and smelters.
7. *Chlorine (Cl), element 17.* A major ingredient of persistent pesticides.
8. *Arsenic (As), element 33.* A deadly water and air pollutant; also in many pesticides.
9. *Strontium (Sr), element 38.* One form (isotope) is radioactive; collects in bones.

10. *Cadmium (Cd), element 48.* A heavy metal water pollutant.

11. *Iodine (I), element 53.* Radioactive isotope, collects in thyroid gland.

12. *Cesium (Cs), element 55.* Radioactive isotope.

13. *Mercury (Hg), element 80.* A toxic water pollutant.

14. *Lead (Pb), element 82.* A toxic metal in water and air. Becomes airborne through the burning of leaded gasolines and through some industrial sources.

15. *Uranium (U), element 92.* The "mother" element of artificial radioactivity.

16. *Plutonium (Pu), element 94.* A synthetic, highly toxic radioactive element; produces other radioactive elements by nuclear fission.

Exercises

1. Electrostatic forces are forces between charged particles, as noted on page 48. State whether a force of attraction, a force of repulsion, or no electrostatic force at all exists between the following pairs of particles:

   a. proton-proton
   b. proton-neutron
   c. proton-electron
   d. neutron-neutron
   e. neutron-electron
   f. electron-electron
   g. positive ion-positive ion
   h. positive ion-negative ion
   i. proton-positive ion
   j. proton-negative ion

2. Small atoms are about 1 A in diameter, and nuclei are about 100,000 times smaller. How many nuclei side by side would reach 1 cm?

3. Join together with a line those elements having similar chemical properties according to the periodic law.

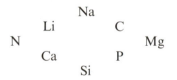

4. Platinum (Pt) is a heavy metal having a high density, while cesium (Cs) is a metal of low density. Use the periodic law and the periodic table to predict whether iron (Fe) or potassium (K) would have the higher density. Use Figure 2-4 to confirm your deduction.

5. By reference to Figure 2-7, would you predict that the following environmentally significant classes of electromagnetic radiation would cause direct biological damage?

   a. Radiation absorbed by the stratosphere's ozone screen.
   b. Radiation that causes photosynthesis.
   c. γ ray radioactivity.
   d. The radiation that causes photochemical smog.
   e. Radiation from earth that participates in the greenhouse effect.
   f. Radiation near the short-wavelength end of the sun's spectrum.
   g. Radiation near the long-wavelength end of the sun's spectrum.

6. Calculate the frequency of radiation at the threshold of the range causing biological damage. (Hint: Use the formula in footnote 1 and the velocity of light given in the text. Check your units.)

7. Name all the subshells filled in atoms of the following elements: He, Be, Mg, Ca, Zn, Kr, Cd, Xe, Ba, Hg.

8. Name the subshell that is only partially filled in atoms of the following elements: lithium, carbon, nitrogen, sulfur, potassium, iron, arsenic, iodine, cesium, and lead.

9. Name the elements associated with the following electronic structures:
$1s^2 2s^2$
$1s^2 2s^2 2p^1$
$1s^2 2s^2 2p^6$
$1s^2 2s^2 2p^6 3s^2$
$1s^2 2s^2 2p^6 3s^2 3p^1$
$1s^2 2s^2 2p^6 3s^2 3p^6$
$1s^2 2s^2 2p^6 3s^2 3p^6 4s^1$
$1s^2 2s^2 2p^6 3s^2 3p^6 4s^2 3d^1$
$1s^2 2s^2 2p^6 3s^2 3p^6 4s^2 3d^{10} 4p^5$
$1s^2 2s^2 2p^6 3s^2 3p^6 4s^2 3d^{10} 4p^6 5s^2 4d^{10} 5p^2$

10. Write down the electronic structure for the following elements: helium (He), lithium (Li), silicon (Si), calcium (Ca), and krypton (Kr).

11. Give in standard notation the electronic structure of the first ten (through strontium) of the "sordid sixteen" elements of environmental pollution.

12. Show the electron dot formulas for the elements that make up the first two rows of the periodic table: H, He, Li, Be, B, C, N, O, F, Ne.

13. Show the electron dot formulas for potassium (K), cesium (Cs), calcium (Ca), strontium (Sr), radium (Ra), arsenic (As), bismuth (Bi), selenium (Se), bromine (Br), and iodine (I).

14. The atom of a certain element has 11 electrons. Write the name and symbol of the element. How many electrons are in the $1s$ subshell? The $2p$ subshell? The $3s$ subshell? How many electrons does the singly charged positive ion of this atom have?

15. Give the electronic structure of the following ions, and note for each the inert gas having the same electronic structure: Li^+, Be^{2+}, O^{2-}, F^-, Na^+, Mg^{2+}, Al^{3+}, P^{3-}, S^{2-}, Cl^-, Ca^{2+}, Sr^{2+}, Cs^+, Ba^{2+}, Fr^+.

16. The electron configuration, $1s^2 2s^2 2p^6 3s^2 3p^6$, exists for a neutral atom; for a singly charged positive ion; for a doubly charged positive ion; for a singly charged negative ion; and for a doubly charged negative ion. Identify these five chemical species using the proper chemical symbols and the appropriate $(+)$ and $(-)$ signs.

Glossary

Alkali metals   group IA of the periodic table.
Alkaline earth metals   group IIA of the periodic table.
Anion   a negative ion.
Atomic number (Z)   the number of protons in the nucleus of an atom.
Cation   a positive ion.
Electron   an atomic component having a negative charge of 1 and a mass 1/1840 times that of the proton.

Electron dot formula   a picture that shows the outer-shell electrons of an atom as dots surrounding the atom's symbol.

Electromagnetic radiation   electromagnetic waves; for example, light waves, radio waves, and x rays.

Electronic structure   the distribution of an atom's electrons in shells, subshells, and orbitals.

Electrostatic forces   the forces of repulsion and attraction between charged bodies.

Family   the elements in one column of the periodic table. Also called a *group*.

Frequency   the number of waves arriving at a fixed point each second.

Halogens   Group VIIA of the periodic table.

Heavy metals   lower right-hand metals in the periodic table.

Inert gases   group 0 of the periodic table.

Ion   a charged atom that has lost or gained electron(s).

Law of octaves   eight-step recurrence of properties among elements.

Metals   elements to the left of the periodic table, including the transition (central) elements.

Neutron   an elementary particle of the nucleus, having the same mass as that of the proton and no charge.

Noble gases   inert gases, Group 0 of the periodic table.

Nonmetals   elements to the right of the periodic table.

Nuclear atom   model of the atom in which a small, massive, positive nucleus is surrounded by negative electrons having little mass.

Nuclear forces   nuclear "glue"—forces that keep the protons and neutrons of the nucleus together.

Nucleus   massive, positively charged center of the atom, containing protons and neutrons.

Orbital   a subdivision of a subshell, capable of holding a maximum of exactly two electrons.

Period   a horizontal row of the periodic table.

Periodic law   similar properties of the elements recur in a periodic way with increase in atomic number (Z).

Periodic table   a table in which the elements are arranged according to the periodic recurrence of their properties and outer-shell electronic structure.

Proton   elementary nuclear particle that has a charge of $+1$ and a mass nearly equal to that of the neutron.

Quantum   a small, distinct parcel, usually of energy. (*pl.:* quanta)

Quantum mechanics   the theoretical explanation of the structure of the atom and the periodic law. Also called *wave mechanics*.

Shell   a collection of like-sized electron clouds in an atom.

Shielding   offsetting of the nuclear attraction for an electron by the repulsion between it and the other electrons in inner shells.

Spectroscopy   the measurement of the frequencies (or wavelengths) of electromagnetic radiation absorbed or emitted by an element.

Subshell   a subdivision of a shell. The four types of subshells are denoted by the letters s, p, d, f.

Transition metals   elements of the central region of the periodic table, where the inner d and f subshells are being filled. Groups IIIB–VIIB, VIII, IB, IIB.

Uncertainty principle   states that the exact position and momentum of an electron within the cloud at a given time is uncertain.

Wave   a disturbance that carries energy through a distance. A *traveling wave* moves constantly onward. A *standing wave* is confined to a limited region and vibrates over and over in the same place.

Wavelength   the distance between the crests of successive waves.

CAPE HATTERAS Here the ocean is always turbulent. (*Photo by Bruce Roberts/Rapho Guillumette.*)

3 CHEMICAL BONDS AND INTERMOLECULAR FORCES

THE ARCHITECTS OF MATTER

Earlier we alluded to the strong links between atoms which lead to molecules, tightly bound atomic clusters. The binding links are chemical bonds. They arise when the outer-shell electrons of two atoms come together and interact. We noted in Chapter 1 that the formation of molecules via chemical bonds is the single most crucial step in the architecture of matter. The strength of the bonds, together with their number per atom, their direction, and their electron distribution, fixes the properties and importance of molecules; it determines whether in aggregate they form matter useful in commerce, essential to life, or a threat to health. Chemical bonds are like the mortar between bricks, holding atoms together and giving shape to the entire substructure of matter.

We found also that molecules are "sticky," adhering to one another by intermolecular forces, an attraction considerably weaker than the chemical bonds between atoms. Because these forces are weak, the union between molecules is far less lasting than that between atoms within a molecule. Only occasionally are molecules held rigidly enough in place by intermolecular forces to form a crystalline solid. More often they slip around in the molecular crowd, attracted in turn to various neighbors, maintaining the mobility of a liquid. Sometimes intermolecular forces are so weak that molecules break away entirely from one another, becoming a gas or a vapor. In this way the strength of intermolecular ties determines whether molecular matter is solid, liquid, or gas; thus these ties give form to our oceans and atmosphere. They also determine the extent to which one substance is soluble in another, and consequently the extent to which environmental pollutants enter water, air, and living organisms. Therefore these forces determine not only the form of our oceans and atmosphere, but also their composition and habitability for life.

Most world matter acquires its all-important embryonic design, as made clear above, by chemical bonds between incredibly small atoms; it acquires form and location as the molecules are forced into prescribed roles by intermolecular forces; and it gains visibility and effect as trillions upon trillions of such molecules provide the bulk that affects every aspect of our existence.

In this chapter the principal kinds of chemical bonds will be explained, as will the origin of intermolecular forces. The way in which molecules and matter are bound together by them will be discussed, with an emphasis on those substances having greatest impact on the environment of earth.

INERT-GAS ELECTRONIC STRUCTURES

In the last chapter we learned two important facts about inert gases and the completed outer shells that characterize them. First, inert gases are stable against change, with little tendency to lose or gain electrons. Second, other families of elements lose and gain electrons in such a way that they tend to achieve the closed-shell structure of inert gases. These two facts taken together suggest that *the closed-shell structure is generally preferred in nature, and that atoms not having a closed shell are inclined to achieve one.* This closed shell is attained by the shifting of outer-shell electrons back and forth between two or more atoms. In the process a chemical bond is formed. We shall describe in the next five sections of this chapter the different ways of rearranging electrons to form chemical bonds.

The natural inclination of atoms to form closed outer shells is called the *octet rule.* "Octet" derives from the eight stable electrons in the outer shell of most inert gases.[1] Only helium, in the first period, achieves shell completion with two electrons rather than eight.

---

[1] Although we refer to shells of eight as "complete" or "closed," most shells technically can accommodate additional electrons in *d* and *f* subshells. We found in the last chapter that this accommodation did not readily occur, that before adding to *d* and *f* subshells, the electrons would enter a higher shell. The explanation is found in subshell energies: subshells of the lowest energy fill first, as explained in Chapter 2. The energy of *d* and *f* subshells is normally so much higher than the energy of *s* and *p* subshells of the same shell that they behave as if they are

IONIC BONDS

Ionic bonds are the simplest kind of chemical bonds. They are the bonds between positive and negative ions. They result from the strong electrostatic attraction between opposite charges. Positive and negative ions are pulled tightly together by these strong forces, thus forming ionic bonds.

To understand ionic bonding, we must look again at the way ions are formed. Recall that positive ions (cations) are formed as electrons are ripped off atoms. Negative ions (anions) are created as electrons add to atoms. These two events are always coupled. We simply do not have free electrons around that we can add to or subtract from as desired in forming cations and anions. Any electron pulled off an atom to form a cation will be immediately drawn to another atom to form an anion. In fact, the anion-former strips the electron away from the cation-forming atom in the first place because of its great affinity for electrons. Thus for each positive (cationic) charge there exists a nearby negative (anionic) charge. Total positive and negative charge must exactly balance, because an equal amount of each is created in the transfer of electrons. The principle of *charge balance* is essential in determining how ions combine with one another.

The charge on individual ions is determined by the number of electrons they must gain or lose to form closed shells (Chapter 2). Thus the octet rule gets involved in the formation of ions and ionic bonds.

Consider the formation of common table salt, a compound known chemically as sodium chloride (NaCl). The shell structures of the ingredient atoms sodium (Na) and chlorine (Cl) are shown in Figure 3-1. First the chlorine atom plunders an electron from a nearby sodium atom, giving, as shown in the figure, a sodium ion (Na^+) and a chlorine ion (Cl^-), both with closed shells. One ion of each kind is formed: the ratio is always one to one to maintain the necessary charge balance.

The sodium and chlorine ions, once formed, draw together because of their opposite charges. The resulting ionic bonding of the pair is made explicit when we show ionic charges: Na^+Cl^-. Other sodium and chlorine ions join the growing cluster. It becomes a crystal of table salt, a regular array of sodium ions and chlorine ions in a one-to-one ratio. Each sodium ion is surrounded by six chloride ions and each chloride ion has six sodium-ion neighbors. The lattice

---

part of a higher shell. The *d* and *f* subshells are even aligned with the higher shells in the horizontal rows of the periodic table because these subshells fill at the same time that the higher shells fill. Therefore in terms of energy and ultimate chemical effects, the *d* and *f* electrons of a shell are most frequently divorced from the rest of the shell. We are usually justified in calling the shell "complete" when only *s* and *p* subshells are filled with their combined eight electrons.

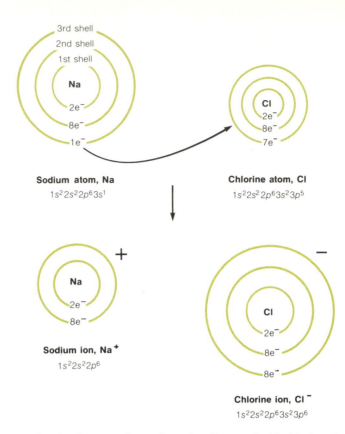

Figure 3-1. The simultaneous formation of sodium and chloride ions by electron transfer between outer shells.

arrangement of the sodium chloride (NaCl) crystal is shown in Figure 3-2.

In the same way that NaCl forms, any of the alkali metals (Li, Na, K, Rb, Cs) can combine with any of the halogens (F, Cl, Br, I, At) by electron transfer. Since all alkali metals give up one and only one

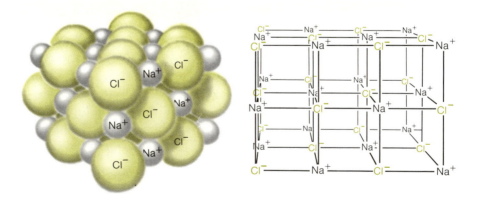

Figure 3-2. The regular arrangement of Na^+ and Cl^- in a crystal of NaCl. *Left*, the relative size of ions; it is a space-filling model. *Right*, the open framework, allowing us to visualize the regularity in three dimensions.

electron and all halogens accept only one electron, the ratio of cations to anions is always one to one. Some of the combinations are shown in Figure 3-3.

The alkaline earth metals, group IIA, lose two electrons to form the doubly charged cations Be^{2+}, Mg^{2+}, Ca^{2+}, Sr^{2+}, and Ba^{2+}. If halogens are present, they will receive these electrons. Each halogen atom can accept one electron, so two halogen atoms are needed to accept the two electrons from a single alkaline earth metal atom and thus balance its double charge. The ratio is one atom of the alkaline earth metal to two halogen atoms.

The electron dot representation below, showing only the outer electrons, pictures the electron transfer and ion formation for magnesium (Mg) and chlorine. The same electron dot representation works, of course, for all the other combinations involving alkaline earth metals and halogens, because outer shells are identical.

$$:\overset{..}{\underset{..}{Cl}}\cdot \leftarrow \cdot Mg\cdot \rightarrow \cdot \overset{..}{\underset{..}{Cl}}:$$

The transfer of the two electrons indicated by the arrows gives

$$:\overset{..}{\underset{..}{Cl}}:^{-} \qquad Mg^{2+} \qquad :\overset{..}{\underset{..}{Cl}}:^{-}$$

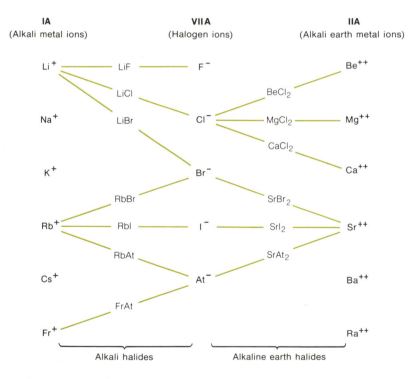

Figure 3-3. Examples of alkali halides (one-to-one ionic ratio) and alkaline earth halides (one-to-two ionic ratio) formed from the combination of specific ions. Note other possible combinations not specifically shown.

The two-to-one ratio is apparent. The charges balance, as they must: the two positive charges on Mg^{2+} balance the two negative charges on the two chlorine ions, Cl^-.

The Mg^{2+} and Cl^- ions are again attracted together and ionically bonded by electrostatic forces. They assemble together to form a magnesium chloride ($MgCl_2$) crystal whose ionic bonding is represented by $Mg^{2+}Cl_2^-$.

The chemical formula, $MgCl_2$, signifies by its subscripts that the ratio of chlorine ions to magnesium ions is two to one. The number

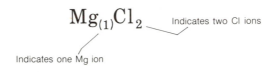

$$Mg_{(1)}Cl_2$$

Indicates two Cl ions

Indicates one Mg ion

"1" is not actually used; if no number is subscripted to the element, "1" is implied. This is the case for magnesium in $MgCl_2$. It is the case for both sodium and chlorine in sodium chloride (NaCl), a compound with a one-to-one ratio.

Other alkaline earth metals combine with other halogens to form similar compounds (see Figure 3-3). All have the same two-to-one ratio of alkaline earth metal ion to halogen ion by virtue of charge balance. The ions of one family behave consistently because of the periodic law.

Elements in other families also combine in ratios fixed by charge balance. For instance, aluminum oxide (Al_2O_3), with a two-to-three ratio, has six positive charges on the two aluminum ions (Al^{3+}) balanced by six negative charges on the three oxygen ions (O^{2-}). The way electron transfer occurs is shown below:

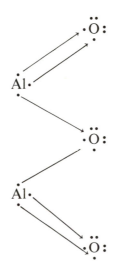

Through this kind of reasoning, we can arrive at the formula for any ionic compound if we know ionic charges. The ion's charge can generally be deduced from the periodic table, as we have emphasized all along.

Imagine the metal ions A^+, A^{2+}, and A^{3+}, where A is a general symbol for a metallic element. They combine with the general nonmetallic anions X^-, X^{2-}, and X^{3-}. The chemical formulas for various combinations of these ions are shown in Table 3-1. The families contributing ions of each charge are shown, and typical ions are given for that charge.

Table 3-1. ION RATIOS IN VARIOUS IONIC COMPOUNDS AS A FUNCTION OF IONIC CHARGE

	X^- (Halogen ions: F^-, Cl^-, Br^-, etc.)	X^{2-} (Group VI A ions: O^{2-}, S^{2-}, etc.)	X^{3-} (Group VA ions: N^{3-}, P^{3-}, etc.)
A^+ (Alkali metal ions: Li^+, Na^+, K^+, Rb^+, Cs^+, Fr^+)	AX	A_2X	A_3X
A^{2+} (Alkaline earth metal ions: Be^{2+}, Mg^{2+}, Ca^{2+}, etc.)	AX_2	AX	A_3X_2
A^{3+} (Group IIIA ions: Al^{3+}, Ga^{3+}, etc.)	AX_3	A_2X_3	AX

Transition metals, not specifically shown in Table 3-1, are unique in that several different ionic charges are possible for the same element. For instance, iron (Fe) may lose two electrons to form the iron(II) or "ferrous" ion, Fe^{2+}, or three electrons may be lost to form the iron(III) or "ferric" ion, Fe^{3+}. Table 3-2 shows the principal ions formed from transition metals in the fourth row of the periodic table. These cations, like those discussed earlier, combine with anions in a ratio determined by charge balance.

Table 3-2. SOME MAJOR TRANSITION-METAL IONS OF FOURTH-ROW ELEMENTS
(Note variable ionic charge for most of the elements shown.)

21 Sc	22 Ti	23 V	24 Cr	25 Mn	26 Fe	27 Co	28 Ni	29 Cu	30 Zn
Sc^{3+}	Ti^{2+}	V^{2+}	Cr^{2+}	Mn^{2+}	Fe^{2+}	Co^{2+}	Ni^{2+}	Cu^+	Zn^+
	Ti^{3+}	V^{3+}	Cr^{3+}	Mn^{3+}	Fe^{3+}			Cu^{2+}	
	Ti^{4+}	V^{5+}	Cr^{6+}	Mn^{4+}					

Ionic compounds are named systematically. As in the "sodium chloride" example used above, the metal's name appears first. The nonmetal's name appears last, with the ending *-ide*. Transition metals with a variable ionic charge must have that charge specified, as in iron(II), iron(III), copper(I), and so on. (In verbal use, these ions are called "iron two," "iron three," "copper one," and so on.) Examples are

sodium chloride	$NaCl$
magnesium oxide	MgO
aluminum oxide	Al_2O_3
strontium fluoride	SrF_2
iron(II) chloride	$FeCl_2$
iron(III) chloride	$FeCl_3$
copper(I) chloride	$CuCl$
copper(II) chloride	$CuCl_2$
copper(I) sulfide	Cu_2S

A number of trivial names persist for the cations, some common examples being "ferrous" for iron(II), "ferric" for iron(III), "cuprous" for copper(I), and "cupric" for copper(II).

COVALENT BONDS

With ionic bonds, closed shells are formed by a transfer of electrons from one atom to another. These bonds rarely exist unless both metal and nonmetal atoms are available to form cations and anions, respectively. When only nonmetal atoms are present, *covalent bonds* are normally formed. With covalent bonds, closed shells are formed by the sharing of electrons between atoms. Much experimental evidence, and also quantum mechanical reasoning, show that a single electron shared by two atoms counts as a whole electron to each atom's shell and subshell quota. That is, an electron, if shared, can simultaneously satisfy the electron needs of two atoms.

As an example, two atoms of hydrogen (H) may combine, sharing the two available electrons. Each atom gets a complete shell because each now effectively owns two electrons, as would the inert gas helium. The joining of two hydrogen atoms (to form a diatomic hydrogen molecule) is represented below:

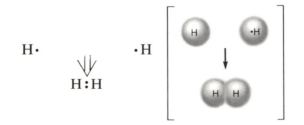

The shared electron pair is shown as dots between the participating atoms. The relative shape and size of the molecule and the atoms

from which it is formed are shown at the right. The two hydrogen atoms clustered together this way are very stable, because each now has a closed shell. The shared electrons are said to form a covalent bond or *shared electron-pair bond*, and they are responsible for holding this stable cluster tightly together.

The shared electron pair of a covalent bond may be represented in one of two ways: as a pair of electron "dots," or as a line between atoms, giving either the *electron dot formula* or the *line formula* of the hydrogen molecule:

$$H:H \qquad \text{electron dot formula}$$

$$H—H \qquad \text{line formula}$$

The two representations H—H and H:H are equivalent. Both represent the *bonding structure* of a hydrogen molecule, chemical symbol H_2.

In molecules, as opposed to ionically bonded crystals, the chemical formula expresses more than the ratio of atoms entering the combination: it expresses the actual number of atoms in a molecular unit. Thus the subscript 2 in H_2 shows that exactly two hydrogen atoms cluster together to form each hydrogen molecule.

Consider now the covalent bond in hydrogen chloride (HCl), an occasional air pollutant stemming in part from the burning of chlorine-containing plastics. The mixture of HCl and water is known as hydrochloric acid, a common laboratory acid and one used by the body to aid in food digestion. The formation of the HCl molecule from atoms is represented by

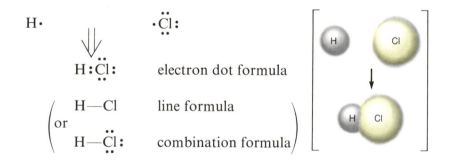

By sharing electrons, both the hydrogen atom and the chlorine atom have achieved closed shells. The hydrogen atom has the two-electron closed shell of helium, while the chlorine atom has the eight-electron closed shell (the octet) of argon.

Both hydrogen and chlorine, in the form of atoms, are one electron short of a closed shell. By sharing one electron pair, each effectively gains one electron, which results in shell completion. They can share no more than one pair, or they would exceed the electron quota of the shell. Each shared pair adds one electron to the shell quota of each atom. Therefore an atom will normally share at most the number of pairs equal to its electron shortage in the outer shell.

Covalent Bonding of the Carbon Atom.   Consider carbon (C), in family IVA. Its atom has four electrons in the outer shell, $\cdot\overset{\cdot}{\underset{\cdot}{C}}\cdot$.

By pairing with an electron from another atom, each carbon electron can become a covalent bond. The carbon atom can form four bonds, since it is four electrons short of shell completion. Compare with the chlorine atom discussed earlier, which can form at most one covalent bond.

A carbon and four hydrogen atoms can unite to form a molecule of methane, CH_4.

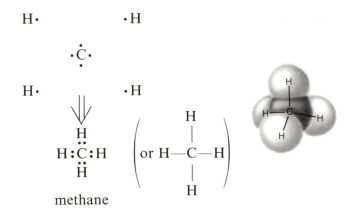

methane

Methane is the simplest organic (carbon-containing) substance. Natural gas is mostly methane.

Carbon atoms bond readily to one another as well as to hydrogen atoms, and in this way they form complex chain molecules that are integral to the machinery of life. The simplest example of carbon-carbon bonding is in ethane, C_2H_6.

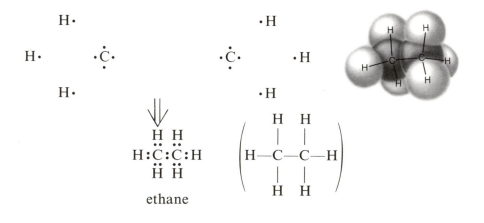

ethane

On occasion carbon atoms (and others) form *multiple bonds*, more than one bond between two atoms, stemming from the sharing of more than one electron pair. Ethylene, C_2H_4, provides an example of *double bonding*. It has the bonding structure

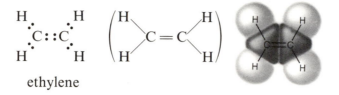

ethylene

The double bond between carbon atoms is represented by two shared electron pairs or two lines, as shown.

In acetylene, C_2H_2, a *triple bond* exists, represented by three electron pairs or three lines

$$H\!:\!C\!:\!:\!C\!:\!H \qquad (H\!-\!C\equiv C\!-\!H)$$

acetylene

Double and triple bonds are, of course, stronger than single bonds. They are not two or three times stronger, respectively, because the strength per bond is somewhat less than that of a comparable single bond.

Bonding Capacity. In all the cases above, each carbon atom shares a total of four electron pairs and thus has exactly four covalent bonds. The number of bonds formed by an atom is sometimes called the *valence*, although this term is no longer in universal use. The number of bonds, or valence, represents the *combining capacity* of an element for a simple one-bond element like hydrogen. Carbon has a valence of four, which means it can bond with a maximum of four hydrogen atoms, as in methane, CH_4. Chlorine has a valence of one, as shown by hydrogen chloride, HCl.

The number of covalent bonds that an atom will normally form is related to the element's position in the periodic table. Group IVA elements, such as carbon (C) and silicon (Si), have four outer electrons and are four short of a complete octet. Their normal valence is thus four. Group VA elements, such as nitrogen (N) and phosphorus (P), have five outer electrons, $\cdot\overset{\displaystyle\cdot}{\underset{\displaystyle\cdot\cdot}{N}}\cdot$ and $\cdot\overset{\displaystyle\cdot}{\underset{\displaystyle\cdot\cdot}{P}}\cdot$, three short of an octet.

They normally form three bonds. For instance, ammonia (NH_3), a substance often found in oxygen-deficient, polluted water, has the bonding structure

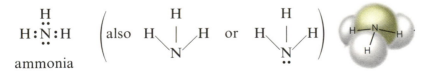

ammonia

Of the eight electrons surrounding the nitrogen atom, five were already present and three are contributed by the three hydrogen atoms.

Group VIA elements, such as oxygen and sulfur, have six outer-shell electrons, $:\ddot{\overset{\cdot}{O}}\cdot$ and $:\ddot{\overset{\cdot}{S}}\cdot$, and normally share two of them to form two bonds. Examples are water, H_2O, and hydrogen sulfide, H_2S, the latter a foul-smelling pollutant known popularly as "rotten-egg gas."

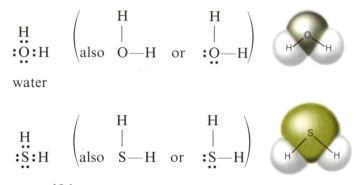

water

hydrogen sulfide

The halogens all have seven outer electrons and usually form one bond, as in the hydrogen chloride (HCl) example mentioned earlier.

Table 3-3 summarizes the common hydrogen compounds of the nonmetals. This table emphasizes the systematic variation of atom ratio with groups or families of the periodic table.

Table 3-3. COVALENT COMPOUNDS OF NONMETALS WITH HYDROGEN

	Group IVA	Group VA	Group VIA	Group VIIA
Normal bonding capacity (valence)	4	3	2	1
Common compounds with hydrogen	CH_4 SiH_4	NH_3 PH_3 AsH_3	H_2O H_2S H_2Se	HF HCl HBr HI

COORDINATE COVALENT BONDS

A *coordinate covalent bond* is a special kind of covalent bond in which one atom contributes both of the shared electrons. We use as an example sulfuric acid (H_2SO_4), which splits into ions important in the chemical balance of natural waters. This molecule can be imagined as forming from its atoms in the following way:

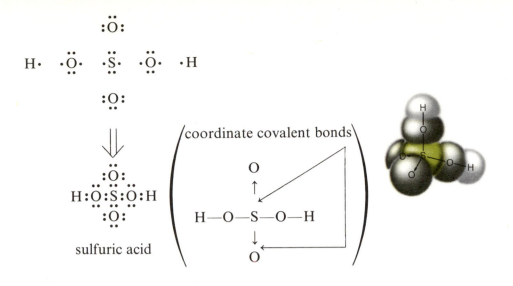

sulfuric acid

Here the sulfur atom contributes a single electron to the electron pairs left and right, but it contributes both electrons to the pairs bonding it above and below. The latter are the coordinate covalent bonds. A coordinate covalent bond is often (but not always) shown as an arrow rather than a line, as illustrated in the line formula at the lower right. The arrow points from the atom donating the electron pair to the receiving atom.

Once formed, the coordinate covalent bond acts much as would any other covalent bond involving a shared electron pair. However, its peculiar origin generally changes the normal bonding number of the participating atoms. Note that the oxygen atoms above and below sulfur have a complete octet with only one bond, less than the normal two. The sulfur atom has four bonds, an increase from the two bonds normally formed.[2]

RESONANCE BONDING

Resonance is a phenomenon in which two or more bonding structures of a molecule blend together to determine the molecule's total characteristics. Sulfur dioxide (SO_2), a notable air pollutant, is one

---

[2]With normal covalent bonding we found that each bond adds one more electron toward each atom's octet quota. However, an atom at the receiving end of a coordinate covalent bond gets two new electrons per bond, not one. Its shell is thus completed with exactly half the normal number of bonds. If this atom is oxygen, which has space in its outer shell for two more electrons, shell saturation is achieved with a single coordinate covalent bond. (See the sulfuric acid (H_2SO_4) example above.) *(Footnote continued on next page.)*

of many molecules that exhibit resonance. Imagine the two ways shown below for assembling SO_2 from its atoms.

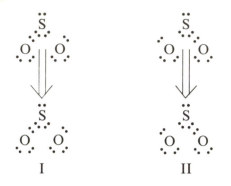

I   II

The sulfur (S) atom in structure I has double bonded with the left-hand oxygen (O) atom and single bonded (through coordinate covalence) with the right-hand O atom. In structure II, the double and single bonds are reversed in position. Each atom has an octet of electrons either way. We can interconvert two structures by a simple electron shift. Thus structure I becomes structure II if electron pairs are shifted in and out of bonds as follows:

I   II

The donor atom, on the other hand, can form a total of four bonds. First it forms the normal number of covalent bonds, a number sufficient to fill its vacancies and achieve an octet. Then any electron pairs left over may be "donated" to form coordinate covalent bonds. Thus four bonds can be formed whether the atom is in group VA, VIA, or VIIA. Below we show how this bonding is achieved for phosphorus (P), sulfur (S), and chlorine (Cl).

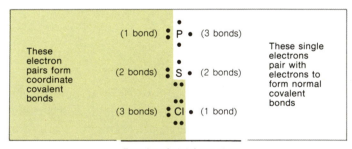

Four bonds total per atom

Sulfur (S) and phosphorus (P) frequently form coordinate covalent bonds with oxygen (O). Invariably, oxygen is the acceptor of the extra electron pair, as above. Many of these substances, including the controversial phosphates and certain insecticides, are environmentally significant and will be treated in detail later. Also discussed will be some important coordinate covalent compounds involving metal atoms.

To change back to bonding structure I, we reverse this electron shift.

When two or more different bonding structures of a molecule can be achieved by shifting electrons, as above, the molecule is said to be a *resonance hybrid* of the contributing structures. A resonance hybrid is indicated by a double-headed arrow between separate bonding structures. Sulfur dioxide is thus a resonance hybrid, represented by

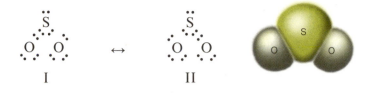

Shown in terms of line formulas, the resonance hybrid is

The bonding structure of sulfur dioxide must always be represented by this dual resonance structure: a single bonding structure does not adequately describe the molecule.

Our discussion implies that electrons merely shift back and forth in resonance hybrids like sulfur dioxide. Such is not the case. Sulfur dioxide has a single and unique bonding structure with characteristics generally intermediate between those of the contributing structures. Both structures contribute, but neither one to the exclusion of the other.

Resonance is a quantum mechanical phenomenon, as is all chemical bonding. Unlike most chemically bonded molecules, resonance hybrids cannot be adequately pictured by single bonding structures. Several are needed to characterize them properly.

Resonance tends to stabilize molecules—to lower their energy and to make them less subject to chemical change. Sulfur dioxide is more stable than we would predict from either of the contributing structures alone. However, in time, exposed to the atmosphere, SO_2 does change. It converts to sulfur trioxide (SO_3), another resonance hybrid and also a pollutant, more severe in its effects than SO_2. We shall describe such chemical conversions in the next chapter.

THE THREE-ELECTRON BOND

The *three-electron bond* is an unusual form of chemical bonding which occurs in only a few substances among nature's millions. We present it here only because these substances include some of the most important on earth, including molecular oxygen and the oxides of nitrogen.

In the three-electron bond, we have an exception to the bonding principles which govern nearly all other chemical bonds. First of all, electrons do not pair up in the normal fashion to give paired-electron covalent bonds; and second, the octet rule is violated. (These exceptions can be explained by some unique quantum mechanical phenomena involving resonance.)

In the three-electron bond, three electrons instead of the normal two participate. This bond, having an odd number (3) of electrons, is found mainly in those few molecules that have an odd number of electrons. The major exception is diatomic oxygen, O_2, a prominent molecule having an even number (16) of electrons. We shall now use O_2 to illustrate three-electron bonding.

Logically we would expect the oxygen atoms of O_2 to attach together with a double bond, as follows:

$$:\ddot{\text{O}}\cdot \quad \cdot\ddot{\text{O}}: \rightarrow :\ddot{\text{O}}::\ddot{\text{O}}:$$

$$(\text{O}=\text{O})$$

This bonding structure provides each atom with an octet of electrons, and it is consistent with the normal formation of two bonds by oxygen atoms. When electrons are paired in this way, the substance is nonmagnetic, but oxygen is peculiar among nonmetal elements because it is in fact magnetic—it is weakly attracted to magnets. Oxygen bonding has therefore been explained in terms of the unusual three-electron bond, unpaired and thus magnetic. We may draw this bond as

$$:\text{O}\genfrac{}{}{0pt}{}{\cdots}{\cdots}\text{O}:$$

where each three-electron bond is shown as three successive electron dots, $\cdots$, and the covalent bond is a short line, ——, as usual. Oxygen is bonded with two of these unusual bonds and one normal electron-pair bond. Each three-electron bond is about one half as strong as a normal electron-pair bond. Thus total bond strength (equal to $\frac{1}{2} + \frac{1}{2} + 1 = 2$ normal bonds) and the consequent stability of the O_2 molecule are about equal to the value expected for double-bonded oxygen, $\text{O}=\text{O}$. The three-electron bond imparts no unusual characteristics to O_2 except to make it magnetic, a property of no particular environmental significance.

The three-electron bond represents a more important bonding concept with the oxides of nitrogen (to be described in a later section) because a normal covalent structure cannot even be imagined for these unusual molecules, which have an odd number of electrons.

BOND POLARITY

Although atoms with covalent bonds share electrons between them, the electrons are generally pulled slightly closer to the atom that has the greater electron affinity. This shift in electrons creates a small negative charge on the high-affinity atom, leaving behind an equal positive charge on the other atom. A bond with charge separation of this form is called a *polar bond*. A whole molecule that has charge separation is called a *polar molecule*. A molecule without polar bonds is *nonpolar*.

Consider hydrogen chloride (HCl), a polar molecule. Chlorine (Cl) has the high affinity for electrons that is typical of elements toward the right side of the periodic table. (This tendency was explained in connection with ion formation in the last chapter.) The chlorine atom is therefore able to draw electrons somewhat toward itself when it combines with hydrogen:

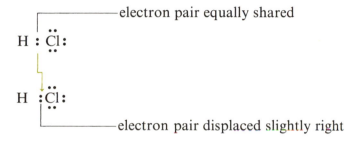

electron pair equally shared

electron pair displaced slightly right

When the electrons are equally shared, as in the top electron dot structure, both ends of the molecule are electrically neutral. The atoms started out neutral, and if they share equally they gain as much electron charge as they give up. However, if the electrons are nudged to the right by the excessive electron appetite of chlorine, the chlorine atom becomes slightly negative. The hydrogen atom, with a single positive proton, but now with slightly less than one offsetting negative charge, acquires a residue of positive charge. This imbalance makes the HCl molecule polar, with charges separated as follows:

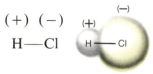

$$(+) \quad (-)$$
$$\text{H}\!-\!\text{Cl}$$

Here and later the charges shown in parentheses, $(+)$ and $(-)$, are partial charges, not whole charge units like those existing on ions. Indeed the atoms have not become ions, because the electron transfer is partial, not complete. The bond is still covalent, but it is said to have *partial ionic character*.

Elements like chlorine which have a high electron affinity are said to be *electronegative*. Elements with low electron affinity are *electropositive.* The electronegativity of an element can be judged in part by how close it is to the right in the periodic table. However, a precise scale of electronegativity values has been developed by Linus Pauling, winner of two Nobel prizes (Chemistry, 1954; Peace, 1962). Electronegativity values on this scale are shown in Table 3-4 for the first four periods of the periodic table, excluding the transition metals (and, of course, the inert gases). Clearly, electronegativity is greatest in the upper right-hand corner of the periodic table. Fluorine (F) is most electronegative, followed by oxygen (O), nitrogen (N), chlorine (Cl), bromine (Br), sulfur (S), and carbon (C).

Table 3-4. ELECTRONEGATIVITY VALUES OF SOME MAJOR ELEMENTS

IA	IIA	IIIA	IVA	VA	VIA	VIIA
1 H 2.1						
3 Li 1.0	4 Be 1.5	5 B 2.0	6 C 2.5	7 N 3.0	8 O 3.5	9 F 4.0
11 Na 1.0	12 Mg 1.2	13 Al 1.5	14 Si 1.8	15 P 2.1	16 S 2.5	17 Cl 3.0
19 K 0.8	20 Ca 1.0	31 Ga 1.6	32 Ge 1.7	33 As 2.0	34 Se 2.4	35 Br 2.8

With the help of the electronegativity chart, we can predict the direction in which electrons will shift in a covalent bond. Furthermore, we can find how far they will shift from the magnitude of the difference in electronegativity values. If the bonding atoms have an electronegativity difference greater than about 0.8, the bond is highly polar. (When the difference in electronegativity exceeds 1.7, the bond is more ionic than covalent in character.)

Consider the water molecule, H_2O. Oxygen is $3.5 - 2.1 = 1.4$ electronegativity units greater than hydrogen. This fact suggests, in accord with what we have been saying, that (1) oxygen acquires a negative charge residue and therefore hydrogen a positive charge, and (2) the bond is highly polar. Both deductions are correct. The charge separation on the water molecule is represented here:

It is an intensely polar molecule, a fact that has decisive influence on the significant properties of water on earth.

By contrast, the hydrogen molecule, H—H, is nonpolar. The two hydrogen atoms attract electrons with exactly equal vigor, neither gaining an advantage over the other. In a molecule like methane, CH_4, slight charge separation is expected because the electronegativity values of carbon (2.5) and hydrogen (2.1) differ by $2.5 - 2.1 = 0.4$. Because this difference is small, and because the four hydrogen atoms are distributed symmetrically around the carbon atom so that no outer part of the molecule can gain a charge advantage at the expense of another outer part, methane, too, is nonpolar. Other organic substances containing only carbon and hydrogen (the hydrocarbons) are also nonpolar, for similar reasons. Petroleum and various oils are mixtures of such substances, and they are thus nonpolar materials.

INTERMOLECULAR FORCES

In Chapter 1 we emphasized the importance of intermolecular forces, sometimes called van der Waals forces or *physical forces*, which attract all molecules and atoms to one another. Because of these forces, all molecules are "sticky," with a tendency to cling to one another. However, the strength of intermolecular forces and thus the "stickiness" of molecules varies widely.

Several independent mechanisms are responsible for intermolecular forces. We are now in a position to understand the principal mechanisms giving rise to these weak but important forces and to understand how they are related to the structure of molecules.

How Forces Arise. Since a polar molecule has permanently charged ends, the positive end of one molecule will attract the negative end of another nearby. The molecules will be drawn by electrostatic attraction into orientations where many contacts of positive and negative ends will be made. For instance, two randomly placed HCl molecules may be drawn together and oriented as shown in Figure 3-4. The attractive forces generated in this fashion are called *orientation forces*. They are a major component of intermolecular forces for

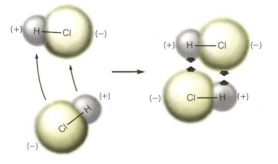

Figure 3-4. Orientation forces drawing polar HCl molecules together.

polar molecules. Such forces clearly do not exist for nonpolar molecules, which have no charged ends to attract one another.

Water molecules attract one another very strongly, as mentioned in Chapter 1. This attraction is a result of the intense polarity of the water molecule. Strong orientation forces occur, but in addition there is an intermolecular force of similar origin, called *hydrogen bonding*. Hydrogen bonding occurs only when the positive residue is on a hydrogen atom and the negative residue is on one of three small, highly electronegative atoms: fluorine, oxygen, or nitrogen. It is the strongest and most important of the intermolecular forces.

Even nonpolar molecules, such as those in oil, attract one another weakly. They do so because of another type of intermolecular attraction called *dispersion forces* or *London forces*. These forces arise with the pulsing of electrons back and forth within their electron clouds. A momentary electron displacement in one direction in one atom combined with an opposite displacement in another atom results in a positive-negative polar attraction, much as was described for polar molecules. However, since these charge displacements are very brief and rather uncoordinated between atoms, the forces are weak, only about one tenth as strong as a hydrogen bond. A small segment of the nonpolar molecule is therefore attracted only weakly to neighboring molecules. If the nonpolar molecule is large, with many atoms, like the molecules of tar or asphalt, the sum of these many weak forces may together constitute a strong force between molecules, discouraging the splitting off of molecules and hence hindering the vaporization of the material.

Finally, there are *induction forces* which attract polar molecules to all other molecules, polar or nonpolar. The positive charge at one end of a polar molecule will tug at the electrons of adjacent molecules, pulling them slightly to one side of the nucleus. This side will therefore become slightly negative, leaving the other side positive. By this means the adjacent molecule becomes temporarily polar, a polarity induced (hence the name "induction" force) strictly by the electrical forces of the permanently polar molecule. The two molecules side by side are now both polar, one on a temporary and one on a permanent basis. They attract one another by positive-negative attractive forces, just as permanently polar molecules do.

The relative forces acting between different kinds of molecular pairs are summarized in Table 3-5. This table shows that the strongest forces, as well as the largest number of force types, attract polar molecules together.

A Rule About Solubility. We noted in Chapter 1 that some liquids would not dissolve appreciably in others because of unequal intermolecular forces. Oil, we observed, would not dissolve in water because the relatively strong forces between water molecules would not allow the penetration of the molecules of oil. Table 3-5 implies

Table 3-5. THE TYPES OF INTERMOLECULAR FORCES ACTING
BETWEEN ADJACENT MOLECULES

Nature of Adjacent Molecules	Hydrogen Bonding	Orientation Forces	Induction Forces	Dispersion Forces
polar-polar	a	√	√	√
polar-nonpolar			√	√
nonpolar-nonpolar				√

ᵃHydrogen bonding is a force acting only between polar molecules of a limited number of types. Included are the molecules in water (H_2O), hydrogen fluoride (HF), ammonia (NH_3), and a few other oxygen, fluorine, and nitrogen-containing molecules.

that this phenomenon must be general for polar and nonpolar molecules. Highly polar molecules simply attract one another with such an arsenal of forces that nonpolar molecules cannot burrow between. Therefore polar liquids and nonpolar liquids often fail to dissolve in one another. However, one polar liquid will ordinarily dissolve another, and one nonpolar liquid will mix with other non-polar liquids because unequal and thus discriminating forces do not exist in liquids of like polarity. These tendencies are summarized in the timeworn chemical principle "like dissolves like." It admits some exceptions, but nonetheless it is an exceedingly useful rule. It explains a multitude of environmental phenomena, including the observation that nonpolar insecticides like DDT seek out the non-polar fatty tissues of living organisms.

Exercises

1. To achieve an "octet" electronic structure, atoms must in effect gain or lose electrons.

    a. Use the periodic table to predict how many electrons the following atoms must gain to achieve octet status: oxygen (O), chlorine (Cl), carbon (C), phosphorus (P), and silicon (Si).
    b. Predict how many electrons the following atoms must lose to achieve octet status: sodium (Na), calcium (Ca), beryllium (Be), aluminum (Al), and potassium (K).

2. Write the chemical formula for and give the proper name to the ionic compounds formed when these atoms combine:

   a. potassium and iodine
   b. magnesium and bromine
   c. strontium and fluorine
   d. sodium and sulfur
   e. magnesium and oxygen

3. Give the chemical formula and name of the ionic compounds formed when these atoms combine:

   a. aluminum and fluorine
   b. aluminum and oxygen
   c. iron(II) and chlorine
   d. iron(III) and oxygen
   e. copper(II) and fluorine

4. Write down the chemical symbols (including the charge) for the ions which form the following ionic compounds: NaF, Na_2O, BeO, $AlCl_2$, and Al_2S_3.

5. Show the line formulas for the following molecules:

   a. hydrogen bromide (HBr)
   b. diatomic fluorine (F_2)
   c. carbon tetrachloride (CCl_4)
   d. silane (SiH_4)
   e. propane (C_3H_8, the three carbon atoms linked in a row)

6. Show the detailed electron dot covalent bonding structure of the following molecules.

   a. hydrogen fluoride (HF)
   b. arsine (AsH_3)
   c. diatomic chlorine (Cl_2)
   d. phosphorus trichloride (PCl_3)
   e. phosphine (PH_3)

7. Show the detailed electron dot covalent bonding structure of butene,

   an organic compound with the line structure $\begin{array}{ccccccccccc} & & H & & & & & & H & & \\ & & | & & & & & & | & & \\ H & - & C & - & C & = & C & - & C & - & H. \\ & & | & & | & & | & & | & & \\ & & H & & H & & H & & H & & \end{array}$

8. A coordinate covalent bond exists in the molecule $\begin{array}{ccccccc} & & H & & F & & \\ & & \cdot\cdot & & \cdot\cdot & & \\ H & : & N & : & B & : & F \\ & & \cdot\cdot & & \cdot\cdot & & \\ & & H & & F & & \end{array}$. Which bond is it? (Hint: This compound can be formed by the combination of ammonia, $\begin{array}{cccc} & & H & \\ & & \cdot\cdot & \\ H & : & N & : \\ & & \cdot\cdot & \\ & & H & \end{array}$, and boron trifluoride, $\begin{array}{ccc} F & & \\ \cdot\cdot & & \\ B & : & F \\ \cdot\cdot & & \\ F & & \end{array}$. Imagine how electrons pair as these two join together.)

9. Would you expect the following molecules to be highly polar or relatively nonpolar: $SiCl_4$, NCl_3, PCl_3, $BrCl$, BrF, NF_3, HBr, AsH_3, SiH_4, H_2Se?

10. For bonds in the following molecules, find the electronegativity difference of the participating atoms. Predict (1) whether the bond is highly polar or relatively nonpolar, and (2) which atom is positive and which negative.

   a. HF   d. PH_3
   b. NH_3 e. H_2S
   c. CCl_4

11. Use electronegativity values to rank the following molecules by polarity, the most polar first: HCl, H_2S, HBr, H_2O, CCl_4, CH_4, HF, NH_3, SiH_4, PH_3.

12. Are the following processes caused by chemical bonds or by intermolecular forces?

   a. Two hydrogen atoms are pulled together to form a hydrogen molecule (H_2).
   b. Two hydrogen molecules are attracted together.
   c. A water molecule is attracted to an ammonia molecule.
   d. A sodium ion and a chlorine ion are pulled together by electrostatic forces.
   e. Water vapor condenses to liquid water.

13. Name all the different types of intermolecular forces (Table 3-5) that would attract the following molecules to each other:

   a. H_2O, H_2O d. HCl, HCl
   b. H_2, H_2 e. HCl, H_2
   c. H_2O, Cl_2

14. All molecules are attracted together to some degree, including the pairs (a) H_2O to CH_4, (b) H_2O to NH_3, (c) CH_4 to H_2. Using Table 3-5, deduce which of these molecular pairs is most strongly attracted together and which is most weakly attracted. Give your reasons.

15. The degree to which oxygen (O_2) will dissolve in water (H_2O) is one of the most critical factors in the water environment of earth (Chapter 8). Do you expect O_2 to have a high solubility or a low solubility in H_2O? (Think the question through carefully, then turn to Chapter 8 to confirm your answer.)

16. When a molecule has two or more different kinds of chemical bonds, each with its own degree of polarity, the molecule has an overall polarity level that is between the individual bond polarities. With this fact in mind, would you predict that pentane (below), with two different kinds of bonds to be considered, would be polar or nonpolar?

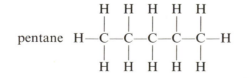

17. Fatty tissues are nonpolar. They are composed of large molecules (Chapter 5). A typical fatty molecule from animal tissues contains 53 bonds joining carbon atoms together, 110 bonds joining hydrogen atoms to carbon atoms, and 12 bonds joining carbon and oxygen atoms. By assuming that the overall polarity of the molecule is approximately equal to the average polarity of its bonds, can you explain why fatty materials are regarded as nonpolar?

18. Chlorinated hydrocarbon insecticides like DDT (see Chapter 10) commonly have covalent bonds linking together (a) carbon (C) atoms and chlorine (Cl) atoms, (b) carbon atoms and hydrogen (H) atoms, and (c) two different carbon atoms. Use the table of electronegativity values to predict whether any of these bonds are polar. Would you expect the insecticides to be highly polar or not? Does this result explain why they tend to accumulate in nonpolar fatty tissues? Give your reasons.

Glossary

Charge balance the rule that total positive and negative charges must balance, or be exactly equal.

Coordinate covalent bond a covalent bond in which one atom contributes both of the shared electrons.

Covalent bond a bond in which two atoms share two electrons. Each atom contributes one electron to the bond. Also called a *shared electron-pair bond*.

Dispersion forces an attraction between molecules that arises from momentary polarity created when electrons pulse back and forth within electron clouds. Also called *London forces*.

Double bond the sharing of two electron pairs between two atoms to form covalent bonds.

Electronegative describes an element that has a high electron affinity.

Electropositive describes an element that has a low electron affinity.

Hydrogen bonding strong attraction between a hydrogen atom on one molecule and a fluorine, oxygen, or nitrogen atom on another; strongest of the intermolecular forces.

Induction forces attraction of a polar molecule to any other molecule because of *induced* polarity in the other molecule.

Ionic bond bond between positive and negative ions resulting from strong electrostatic attraction.

Line formula a representation of the bonding structure of a molecule in which a line is used to show each covalent bond.

London forces dispersion forces.

Multiple bonds more than one covalent bond between atoms, resulting from sharing of more than one pair of electrons.

Nonpolar describes a bond or molecule without charge separation.

Octet rule the inclination of atoms to form closed outer shells, usually of eight electrons.

Orientation forces forces that cause molecules to arrange themselves so that many contacts of positive and negative ends will be made.

Polar describes a bond or molecule that has charge separation.

Resonance a phenomenon in which two or more bonding structures are blended to describe a molecule's total characteristics.

Resonance hybrid a molecule whose structure must be described by use of resonance.

Shared electron-pair bond a covalent bond.

Three-electron bond a bond having three electrons that occurs usually in molecules having an odd number of electrons.

Triple bond the sharing of three electron pairs between two atoms to form three covalent bonds.

Valence the number of bonds formed by an atom. Also called the *combining capacity*.

CALIFORNIA REDWOODS Muir Woods, Marin
County, in Northern California. (*Photo by Richard F.
Conrat/Photofind.*)

4 THE STRUCTURE OF ENVIRONMENTAL MATTER

MATTER, MOLECULES, AND BONDS

At the beginning of the last chapter, we saw that chemical bonding has a crucial role in the architecture of matter. We then explained the nature and diversity of chemical bonds. Our goal in such a study of chemical bonding is to understand matter—the final architectural product of bonding. The properties of matter hinge on the bonding structure of the molecules and ions from which it is composed. Here we shall look specifically at the structure of the most important molecules and ions in the matter that has environmental impact. We shall deal with major components of our air, water, and biological environment.

In our illustrations of chemical bonding in the last chapter, we stumbled across a number of important environmental substances, such as water (H_2O), oxygen (O_2), and sodium chloride (NaCl). In this chapter, we shall emphasize the molecules because of their own singular importance, not because they happen to be a by-product of a discussion on bonding. Clearly, however, we shall need the general concepts of bonding and the type of networks formed by bonds among atoms to understand molecular architecture, even though the detailed nature of the bonding itself will no longer be our focus.

SIZE, SHAPE, AND STABILITY OF MOLECULES

A molecule is a cluster of atoms having a rather definite size and shape and a certain level of stability in its arrangement of atoms. The size is determined largely by the size and number of the constitutent atoms and by how snugly they are pulled together by the chemical bonding between. The shape is determined by the pattern of the underlying quantum mechanical electron clouds. Stability is related to the strength of bonds between atoms. All three factors can vary considerably.

Molecular Size. The size of underlying atoms was discussed in the last chapter. The biggest atoms are those with many electron shells, found toward the bottom of the periodic table. There is also a trend to increasing size as one moves left in the periodic table. However, even very big atoms such as cesium (Cs) are a tiny 5 A (5 × 10^{-8} cm) across; the littlest is hydrogen (H), and it is just under 1 A.

The compactness of atoms bonded together is specified by the *bond length*, the distance between the centers of the bonded atoms. Since the center of an atom is the nucleus, this atomic spacing can also be called the *internuclear distance*.

If atoms were hard spheres, like miniature billiard balls, the bond length would be the sum of the atomic radii, as shown in Figure 4-1. In fact, however, the space within atoms is filled by electron clouds, somewhat wispy and soft. The powerful attraction of a chemical bond compresses the atoms where they meet, drawing them closer to one another than if they were hard spheres. (See Figure 4-1.)

Most bond lengths are in the vicinity of 1 A to 3 A. The H—Cl bond in hydrogen chloride is 1.27 A. The O—H bond in water is 0.96 A.

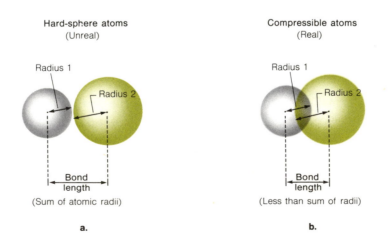

Figure 4-1. Real atoms are compressible and are "squashed" together by the force of a chemical bond.

Bonds of a given type have a rather constant bond length that is independent of the molecule they are in. Carbon-carbon single bonds are common to nearly all organic molecules and are rather uniformly 1.54 A. Carbon-hydrogen bonds in these substances are 1.10 A.

Double and triple bonds draw the atoms together more tightly than do single bonds. The $C=C$ bond in ethylene is 1.33 A, and the $C\equiv C$ bond in acetylene is 1.20 A.

The overall size of molecules depends on bond lengths and upon the number of atoms extending from end to end. Small molecules with just a few atoms are usually in the range from 3 A to 10 A across. "Giant" molecules with thousands of atoms may be 100 A to 1000 A long, depending on how the atoms are grouped together. However, we should keep in mind that among molecules even the "giants" are extremely small, still well out of the reach of a powerful light microscope.

Molecular Shape. Atoms attach to one another in various geometrical arrangements, which leads to diverse molecular shapes. Some molecules are globular clusters, some are long and thin, a number are coiled, and still others are flat. Many are long and have flexibility, like chains. Some of these many configurations are shown in Figure 4-2.

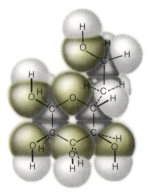

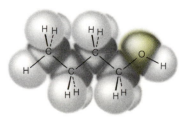

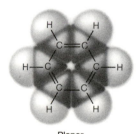

Linear
(Acetylene, C_2H_2)

Planar
(Benzene, C_6H_6)

Zigzag
(Butyl alcohol, C_4H_9OH)

Complex
($C_7H_{14}O_6$)

Figure 4-2. Some typical molecular configurations.

The overall molecular shape and many other important properties of molecules and the matter they constitute, are determined by *bond angles*, the angle between lines joining a central atom with two satellite atoms. For instance, water has a bond angle of 105°.

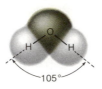

It is a bent or *nonlinear* molecule.

In compounds of carbon, the four bonds of carbon usually extend at an angle of 109° to one another, the angle of the symmetrical regular tetrahedron. This arrangement is illustrated for methane in Figure 4-3.

Some molecules are *linear*—all atomic centers lie on a straight line. Diatomic (two-atom) molecules are automatically linear, since a straight line can be passed through any two centers. A few triatomic (three-atom) molecules, such as carbon dioxide (CO_2) and mercury (II) chloride ($HgCl_2$), are linear. Molecules with more atoms are rarely linear. Acetylene (C_2H_2), a linear molecule of four atoms (Figure 4-2), is unusual in this regard.

Whether triatomic molecules are bent or linear has great impact on their properties. If the water molecule were linear, it would cease to have polar characteristics, because charge could not be drawn off toward one side. Orientation forces and hydrogen bonding would be weakened, and because of their significance, the entire chemistry of life would be drastically altered or altogether extinguished.

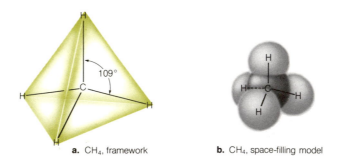

a. CH_4, framework   **b.** CH_4, space-filling model

Figure 4-3. The four bonds of carbon invariably point to the corners of a regular tetrahedron. Here this arrangement is shown for methane (CH_4), for which we can imagine a tetrahedron with a carbon (C) atom at the center and a hydrogen (H) atom at each corner. C—H directions are shown by solid lines. All bonds intersect at 109° angles. A space-filling model, showing scale size and position of atoms, is at right.

THE STRUCTURE OF ENVIRONMENTAL MATTER

Molecular Stability. One other item we must discuss is the stability of molecules and the relationship of stability to the strength of chemical bonds. A strong bond means that atoms in a molecule are held together with strong forces, and the molecule will not be easily pulled apart. Such molecules are relatively stable against change.

What is more to the point thermodynamically, strong bonds, in pulling atoms together, lower the potential energy of molecules. This reduction happens in the same way that the force of gravity, in pulling an object to earth, lowers the potential energy.

In Chapter 1 we found that natural change tends to occur in the direction of lower potential energy. A molecule already low in potential energy has, therefore, little inclination to undergo chemical change. While we shall discuss chemical change much more thoroughly in the next chapter, we wish to point out here that a relationship does exist between strong bonding and chemical stability, for it helps explain the environmental persistence of many molecules.

ARRANGEMENT OF ATOMS AND BONDS IN MOLECULES

Some important molecules are bonded together in an unavoidably complex way. To the beginning student, atoms and bonds in these molecules seem to be placed haphazardly, defying any simple order. How would anyone ever guess that one of the resonance structures of sulfur dioxide (SO_2) was O=S—O instead of equally plausible S(—O)—O, or perhaps O—S—O, O—S—O, or even O—S=O? Fortunately, there are rules that can help us decide among various bonding structures. These admit many exceptions, so they must be used with caution. However, they are a good deal better than the confusion which often accompanies a student's first exposure to molecular structure.

Bonding Rule 1: Total Number of Bonds. The sum total of bonds within a molecule can be calculated rather easily if all its bonded atoms obey the octet rule, as is usual. First we count the total number of electrons needed to give each individual atom eight electrons in its outer shell (two for hydrogen). We call this number the "electron demand." For SO_2, this number is 24 because each of the three atoms needs eight electrons.

Next we count the total number of outer-shell electrons in the molecule. Call this number the "electron supply." If we use SO_2 as an example again, there are eighteen outer-shell electrons; six brought into the molecule by each atom of sulfur and oxygen.

In terms of these two quantities, the number of bonds is given simply as[1]

$$\text{number of bonds} = \frac{\text{electron demand} - \text{electron supply}}{2}$$

This equation predicts, for instance, that SO_2 has $(24 - 18)/2 = 6/2 = 3$ bonds, which is in exact accord with the accepted resonance structure $O \diagup^S \diagdown O$.

Rule 2: Isoelectronic Structures.   Molecules and ions that have an identical number of atoms and outer-shell electrons are called *isoelectronic*. Isoelectronic chemical species nearly always have the same bonding structures. If the structure of one is known, the others can be deduced with reasonable confidence.

A simple example of isoelectronic structures is provided by two molecules we have already discussed: water (H_2O) and hydrogen sulfide (H_2S). Both oxygen and sulfur, group VIA elements, have six outer-shell electrons. Combined with two electrons for the pair of hydrogen atoms, each molecule has a total of eight outer-shell electrons, an identical number. In accord with the rule, the bonding structures are the same.

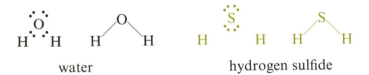

water   hydrogen sulfide

The bonding similarity of isoelectronic structures will be exploited more fully later, particularly in the section on polyatomic ions.

Bonding Rule 3: Atomic Arrangements.   Predicting the arrangement of atoms in molecules is perhaps the most difficult task of all for the student new to chemistry. It is also the hardest to codify into rules. A few simple principles, stated below, will be of some help until the student gains experience with molecules of various kinds.

As mentioned before, carbon atoms tend to link together into chainlike molecular structures. These molecules will be discussed in detail in the section on organic molecules. What is of more significance here, molecules without carbon atoms (or with at most one carbon atom) tend to have a central-atom structure: one particular atom

---

[1]The logic of this equation goes back to our premise than an electron-pair bond is formed when two electrons, shared between atoms, both count toward the octet quota of both atoms. That is, each bonding electron fills two electron slots at once. For reference, suppose that no electrons were shared between atoms in a molecule. The quantity "electron demand − electron supply" would then represent the number of vacancies short of octet completion. Each time an unshared electron becomes a shared (bonding) electron, it fills one vacancy. In this way each vacancy is filled by a bonding electron. The number of bonding electrons therefore equals exactly the number of vacancies, or the number of bonding electrons = electron demand − electron supply. It takes two electrons to make a bond: therefore there are half as many bonds as there are bonding electrons; or the number of bonds = (electron demand − electron supply)/2.

clearly is the core of the molecule. It is surrounded by and bonded to other atoms and groups of atoms. Examples of central-atom structures abound in the molecules we have already discussed.

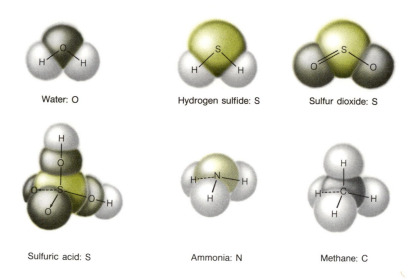

Water: O   Hydrogen sulfide: S   Sulfur dioxide: S

Sulfuric acid: S   Ammonia: N   Methane: C

Figure 4-4. Examples of central-atom structures. (Central atom identified below each formula.)

Several are shown in Figure 4-4, with the atom that has the central role indicated immediately after the molecule's name. Inspection of these examples (and many more to come) shows the following tendencies. First, the central atom tends to be the only one of its kind—the unique atom—in the molecule. In CH_4, for example, the one atom of carbon is central; the four atoms of hydrogen are satellites. Second, chemical evidence shows that identical satellite atoms tend to have identical bonding. For instance, the hydrogen atoms within any one of the molecules H_2O, H_2S, NH_3, or CH_4 are bonded with exact equality (equal bond length, the same bond strength, etc.). The oxygen atoms in SO_2 do not appear to be equal in our picture (Figure 4-4) because one is double bonded and one single bonded. The equality rule is not violated, however, if we recall that SO_2 is a resonance hybrid of two structures, which gives the two oxygen atoms equal participation in double bonding

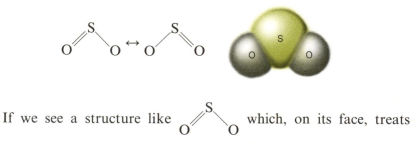

If we see a structure like which, on its face, treats

equal atoms unequally, it is a good indication that this structure is but one of several resonance structures. Resonance, in cases like this, is a master equalizer.

Note that we do not expect all oxygen (O) atoms in sulfuric acid (H_2SO_4) to be bonded identically. The two that are linked to hydrogen (H) atoms are distinctly different from the two bonded only to sulfur (S).

While exceptions to the above guidelines occur fairly often, they may still be a great help to the student in facing what appears at first to be a jungle of bonds and molecules. The best approach to understanding molecular structure is to apply these rules to all the molecular and ionic structures shown in the next few sections, noting which obey the rules and which do not, and looking for any special circumstances that may explain the latter.

MOLECULES PROMINENT IN THE ATMOSPHERE

Some molecules in the atmosphere, like oxygen (O_2) and carbon dioxide (CO_2), play a vital role in sustaining life. Others, like sulfur dioxide (SO_2) and carbon monoxide (CO), constitute a threat to living systems. Whether the molecules are vital or threatening, their common characteristic is that they are small. They are of minute size because they contain usually only two to four atoms per molecule.

Indeed these substances would not be inclined to enter the atmosphere at all unless their molecules were small. Other things being equal, the intermolecular forces binding molecules together and to earth are weaker if molecules are smaller (recall pp. 101–103 in Chapter 3). Molecules with two to four atoms are ordinarily held by such weak forces that they readily escape to the atmosphere. It is not surprising, then, that the atmosphere and its principal pollutants are composed almost entirely of these small molecules.

Another characteristic of prominent atmospheric molecules is that they are composed of atoms of nonmetals. Hence atmospheric chemistry is largely localized to the upper right-hand corner of the periodic table, as was mentioned in the last chapter. The reason for this limitation is that only nonmetals are inclined to bond chemically as small molecules. Both pure metals and the compounds of metals and nonmetals tend to exist in solid crystalline form, as we pointed out earlier. Their occasional entry into the atmosphere is in the form of small particles blown by the winds. Their airy residence, while sometimes significant, is usually temporary, because gravity and the rains drag them back to earth.

The bonding structure of a number of small but important atmospheric molecules was discussed in earlier sections. Included were water (H_2O), sulfur dioxide (SO_2), hydrogen sulfide (H_2S), and ammonia (NH_3). Next, we shall describe the structural features of other major species.

Molecules with Simple Bonding. The atmosphere is 78 weight percent nitrogen (N). Atmospheric nitrogen exists mainly in the form of the diatomic molecule N_2. Nitrogen atoms, with other members of group VA, normally form three bonds. In diatomic nitrogen, these bonds attach to one another as shown,

$$:\!\overset{\displaystyle .}{\underset{\displaystyle .}{N}}\!\cdot \quad \cdot\overset{\displaystyle .}{N}\!: \longrightarrow :\!N\!\vdots\!\vdots\!N\!: \ (N\equiv N)$$

diatomic nitrogen

giving a triple bond and an octet electron structure. The triple bond of N_2 is very strong. For this reason, diatomic nitrogen is a stable molecule in our environment, seldom undergoing chemical change.

Carbon monoxide (CO) is a major air pollutant. It has the same number of outer-shell electrons (ten) and atoms (two) as N_2. It is therefore isoelectronic with N_2 and has an identical bonding structure. However, oxygen, with two more electrons than carbon, is the sole contributor of one electron pair

$$:\!\overset{\displaystyle .}{\underset{\displaystyle .}{C}} \quad :\!\overset{\displaystyle .}{\underset{\displaystyle .}{O}}\!: \longrightarrow :\!C\!\vdots\!\vdots\!O\!: \ (C\equiv O)$$

carbon monoxide

Essentially, then, CO has one coordinate covalent bond and two regular covalent bonds, but it is not necessary to distinguish between them here. Note only that the normal bonding capacities for carbon and oxygen (four and two, respectively) have been altered in CO because of coordinate covalence. The triple bond in CO is very strong, and CO, like N_2, is a relatively stable molecule. It is the most common of the carbon-containing species recently found to be occupying the vastness of space.

Carbon dioxide (CO_2) is much less abundant than nitrogen in the earth's atmosphere, but it plays a major role in the earth's heat balance and climate through the greenhouse effect (see Chapter 6). It is also necessary for plant photosynthesis. Its bonding structure is

$$:\!\overset{\displaystyle ..}{\underset{\displaystyle ..}{O}}\!::\!C\!::\!\overset{\displaystyle ..}{\underset{\displaystyle ..}{O}}\!: \ (O\!=\!C\!=\!O)$$

carbon dioxide

This structure obeys the octet rule and is also consistent with the normal four-bond and two-bond capacities of carbon and oxygen.

Molecules with Complex Bonding. Some of the most important molecules in our atmosphere have relatively complex bonding arrangements. Resonance structures and the so called three-electron bond are significant features. The oxygen-containing molecules are particularly important and also particularly nonsimple. Recall that even ordinary diatomic oxygen, O_2—perhaps the second most important molecule after water (H_2O) in the environment of earth— has the uncommon octet-violating three-electron structure represented by

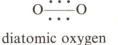

diatomic oxygen

Here three dots in a row ($\cdot \cdot \cdot$) represent the three-electron bond. Other oxygen-containing molecules with three-electron bonds are the oxides of nitrogen. These unusual molecules have an odd number of electrons.

Very few molecules with an odd number of electrons are at all stable in nature. This fact may be verified by counting electrons on molecules discussed earlier in this chapter. The two oxides of nitrogen prominent in air pollution, NO and NO_2, are exceptions. Consider first nitrogen oxide, or nitric oxide (NO). It is an *odd molecule* because nitrogen contributes 7 electrons and oxygen 8, a total of 15 (11 of them in the outer shells). The NO molecule is made stable with the help of a three-electron bond.

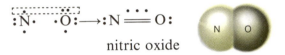

nitric oxide

This bond, at half normal bond strength, supplements the two covalent bonds, giving a total strength equivalent to two and a half bonds, with strength enough to give NO reasonable chemical stability. However, it is slowly converted to nitrogen dioxide by interaction with atmospheric oxygen (O_2) (Chapter 7).

Nitrogen dioxide (NO_2) is likewise an odd molecule, having a total of 23 electrons (17 in outer shells). It too utilizes a three-electron bonding structure. There are two equivalent resonance structures, each having a three-electron bond. Together they give the resonance hybrid for NO_2.

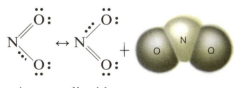

nitrogen dioxide

This molecule is bent, the bond angle being 134°.

Nitrogen dioxide is an odd-electron molecule which runs somewhat more true to form. It is less stable and is thus inclined to undergo chemical change. Normally two NO_2 molecules will combine to form dinitrogen tetroxide (N_2O_4), a molecule with $2 \times 23 = 46$ electrons, an even number. However, at atmospheric concentrations the NO_2 molecules are too dilute to find and combine with one another readily. Instead they resort to other reactions, producing photochemical smog and causing a deterioration of air quality. These reactions will be discussed in Chapter 7.

Other major oxygen-containing compounds avoid the three-electron bond, but some of them are resonance hybrids. Sulfur dioxide (SO_2), sulfur trioxide (SO_3), and ozone (O_3) are notable examples. Ozone and sulfur dioxide have exactly the same number of outer-shell electrons because oxygen and sulfur are in the same chemical family, group VIA. They are therefore isoelectronic. As a consequence, O_3 bonds in exactly the same way as SO_2 (shown earlier to be 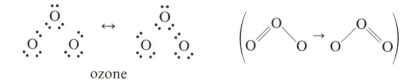)—namely, by means of resonance between the structures

ozone

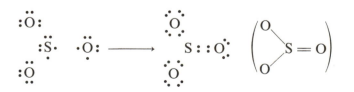

Resonance adds stability to ozone, but with an average of only one and a half bonds per bonding position, it does not rank high in stability. Partly for this reason it is highly reactive, tending to combine with and disrupt many organic molecules, including those in living systems. Because of this tendency, it is a toxic and dangerous air pollutant.

Sulfur trioxide (SO_3) is a resonance hybrid like sulfur dioxide and ozone. With four atoms rather than three, it is slightly more complex, having three contributing resonance structures rather than two. These structures are assembled from atoms as follows:

The resonance hybrid is[2]

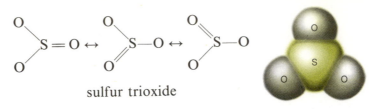

sulfur trioxide

The bond lengths and bond angles of prominent atmospheric molecules are shown in Table 4-1.

The Naming of Nonmetallic Compounds. While the student with an environmental interest may already know the names of some of the preceding molecules, there is a systematic naming method which helps with those that are not known. In compounds of two non-metals, a category which includes all the compounds shown in this section, a prefix designates the number of atoms of each kind in the compound. The prefixes used are:

mono	= 1	hexa	= 6
di	= 2	hepta	= 9
tri	= 3	octa	= 8
tetra	= 4	nona	= 9
penta	= 5	deca	= 10

Only the prefix mono may be dropped; the absence of a prefix before an element therefore indicates *one* atom of that element in the molecule. The more electropositive of the two nonmetals is named first and presented first in the chemical formula. The nonmetal named second is given the ending *-ide*. Examples are:

carbon *mon*oxide	CO	nitrogen oxide	NO
carbon *di*oxide	CO_2	hydrogen chloride	HCl
sulfur *di*oxide	SO_2	nitrogen *di*oxide	NO_2
sulfur *tri*oxide	SO_3	*di*nitrogen *tetr*oxide	N_2O_4

---

[2]The structure of SO_3 can be found from our bonding rules. The central position of the S atom (rule 3) and the presence of $(32 - 24)/2 = 4$ bonds (rule 1) lead to the structure

$$\begin{matrix} O \\\ \ \ \ S{=}O \\ O \end{matrix}$$

. The O atoms are equivalent (rule 3) only if the double bond rotates among the three oxygens. A resonance hybrid involving exactly three contributing structures, $\begin{matrix} O \\\ \ \ \ S{=}O \\ O \end{matrix}$

$\leftrightarrow \begin{matrix} O \\\ \ \ \ S{-}O \\ O \end{matrix} \leftrightarrow \begin{matrix} O \\\ \ \ \ S{-}O \\ O \end{matrix}$, is therefore the logical result.

THE STRUCTURE OF ENVIRONMENTAL MATTER

Table 4-1.   BOND LENGTHS AND BOND ANGLES OF MAJOR ATMOSPHERIC MOLECULES

Molecules	Bond Length, A	Bond Angle	Structure
Water (H_2O)	0.96	105°	
Diatomic oxygen (O_2)	1.21	(linear)	
Ozone (O_3)	1.28	117°	
Sulfur dioxide (SO_2)	1.43	120°	
Sulfur trioxide (SO_3)	1.43	120°	
Hydrogen sulfide (H_2S)	1.35	92°	
Carbon monoxide (CO)	1.13	(linear)	
Carbon dioxide (CO_2)	1.16	(linear)	
Diatomic nitrogen (N_2)	1.10	(linear)	
Nitrogen oxide (NO)	1.15	(linear)	
Nitrogen dioxide (NO_2)	1.19	134°	
Ammonia (NH_3)	1.01	107°	

Occasionally, trivial (nonsystematic) names have become ingrained in common usage, such as nitric oxide instead of the more correct nitrogen oxide. Ammonia (NH_3) and water (H_2O) are trivial names that everyone uses. In ammonia the chemical formula is also backwards, with the most electronegative element first, NH_3 instead of H_3N. This formula, too, is universally accepted.

POLYATOMIC IONS IN WATER

Polyatomic ions are formed when entire clusters of atoms rather than single atoms become charged. They are among the most important factors affecting water quality.

Many negative polyatomic ions are related to (and may be considered derivatives of) prominent acids. We shall discuss ions of this nature first, starting with the derivatives of sulfuric acid.

Acid-Related Polyatomic Ions. Earlier we showed that the electron dot structure of sulfuric acid (H_2SO_4) was

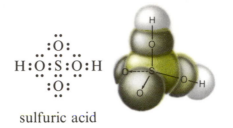

sulfuric acid

The substance H_2SO_4 is classified as an *acid* because it can lose hydrogen ions (H^+), the principal hallmark of acids. For H_2SO_4, this loss is represented by

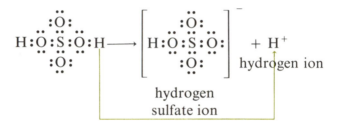

sulfuric acid

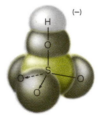

That is, the hydrogen atom splits away without its electron. It is thus a positive ion, H^+. The group left behind is now in possession of the excess electron, and it is thus negatively charged. This charged group is a polyatomic ion, the so-called hydrogen sulfate ion, chemical formula HSO_4^-.

In some circumstances a second hydrogen ion can split off:

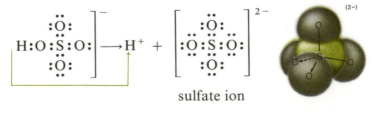

sulfate ion

This H^+, too, departs without an electron, leaving now a total of two excess electrons on the polyatomic ion.[3] This ion is the sulfate ion, $SO_4{}^{2-}$.

Another important group of ions are derivatives of phosphoric acid, H_3PO_4. By successive losses of H^+, the polyatomic ions $H_2PO_4{}^-$, $HPO_4{}^{2-}$, and $PO_4{}^{3-}$ are formed. The latter is the phosphate ion, now a symbol of water pollution and lake eutrophication.

The structure of $PO_4{}^{3-}$ is easy to figure out if we note that it is isoelectronic with the sulfate ion, $SO_4{}^{2-}$. Phosphorus, element 15, has one less electron than sulfur, element 16, but the phosphate ion $(PO_4{}^{3-})$ has an extra negative charge (three compared to two) to compensate. Therefore $PO_4{}^{3-}$ and $SO_4{}^{2-}$ have an identical number of electrons (and, of course, atoms), and they bond in identical fashion. By the same token, $HPO_4{}^{2-}$ and $HSO_4{}^-$ are isoelectronic and bond the same way, as do $H_2PO_4{}^-$ and H_2SO_4. The isoelectronic structures, side by side, are

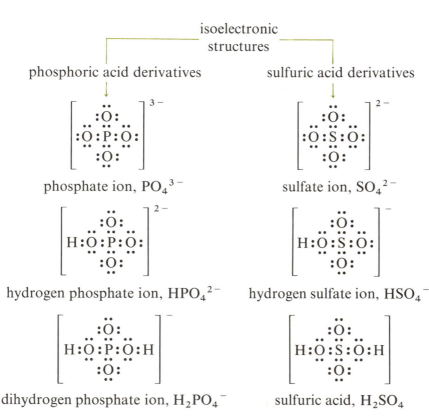

isoelectronic
structures

phosphoric acid derivatives

phosphate ion, $PO_4{}^{3-}$

hydrogen phosphate ion, $HPO_4{}^{2-}$

dihydrogen phosphate ion, $H_2PO_4{}^-$

sulfuric acid derivatives

sulfate ion, $SO_4{}^{2-}$

hydrogen sulfate ion, $HSO_4{}^-$

sulfuric acid, H_2SO_4

---

[3]Polyatomic ions, as well as molecules, generally bond according to the common bonding rules. The sulfate ion $(SO_4{}^{2-})$ has a central atom, the S atom. All oxygen atoms are equivalent. The number of bonds is calculated according to the formula that the number of bonds = (electron demand − electron supply)/2. The electron demand—to fill the octet of all five atoms— is $5 \times 8 = 40$. The electron supply includes the six electrons brought in by each of the five atoms, $5 \times 6 = 30$, plus (for polyatomic ions only) the ionic charge of 2, which represents two surplus electrons. Total electron supply is 32. These numbers substituted in the above equation give $(40 − 32)/2 = 8/2 = 4$ chemical bonds. This number is correct, as seen from the $SO_4{}^{2-}$ electron dot structure.

The presence and role of the phosphorus-based ions in natural waters will be discussed in a subsequent chapter.

Another important polyatomic ion series, which has much to do with the carbon dioxide balance on earth, is the derivatives of carbonic acid, H_2CO_3.

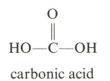

carbonic acid

These derivatives include the hydrogen carbonate (sometimes called bicarbonate) ion, HCO_3^-, and the carbonate ion, CO_3^{2-}. The latter is isoelectronic with sulfur trioxide (SO_3), and as expected, it has the same bonding structure.

Other important polyatomic ions include those derived from the following acids.

Nitric acid (HNO_3). The loss of H^+ yields the nitrate ion, NO_3^-, isoelectronic with both sulfur trioxide (SO_3) and the carbonate ion (CO_3^{2-}).

Nitrous acid (HNO_2). This acid yields the nitrite ion, NO_2^-, isoelectronic with sulfur dioxide (SO_2).

Sulfurous acid (H_2SO_3). This acid yields the hydrogen sulfite ion, HSO_3^-, and sulfite ion, SO_3^{2-}, neither one isoelectronic with other molecules discussed.

Hydrogen sulfide (H_2S). This weak acid reluctantly loses H^+ to form the hydrogen sulfide ion, HS^-, and the sulfide ion, S^{2-}, the latter no longer polyatomic. The ion HS^- is isoelectronic with hydrogen chloride (HCl).

Water itself can lose H^+ to a very slight degree (such a slight degree that water is not acidic) to form one of the most important polyatomic ions, the hydroxide ion, OH^-.

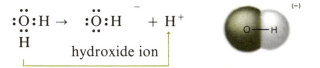

hydroxide ion

The OH^- ion is characteristic of *bases*, or *alkalis*, just as H^+ is characteristic of acids. Bases are the chemical opposites of acids. Both cannot exist simultaneously because large numbers of H^+ and OH^- ions mixed together combine to form water (H_2O). This process is called *neutralization*. Water itself is neither acidic nor basic, because on those rare and temporary occasions when it splits into H^+ and OH^-, the two are formed in equal amounts. These amounts are very small for, as we noted, substantial concentrations of H^+ and OH^- cannot exist together.

A summary of the important polyatomic ions with negative charge is given in Table 4-2. As we have seen, their atoms bond

Table 4-2.   MAJOR POLYATOMIC ANIONS IN NATURAL WATER

Ion	Name	Parent Acid
OH^-	Hydroxide	
PO_4^{3-}	Phosphate	
HPO_4^{2-}	Hydrogen phosphate	Phosphoric acid (H_3PO_4)
$H_2PO_4^-$	Dihydrogen phosphate	
SO_4^{2-}	Sulfate	
HSO_4^-	Hydrogen sulfate	Sulfuric acid (H_2SO_4)
SO_3^{2-}	Sulfite	
HSO_3^-	Hydrogen sulfite	Sulfurous acid (H_2SO_3)
HS^-	Hydrogen sulfide	
S^{2-}	Sulfide	Hydrogen sulfide (H_2S)
CO_3^{2-}	Carbonate	
HCO_3^-	Hydrogen carbonate (bicarbonate)	Carbonic acid (H_2CO_3)
NO_3^-	Nitrate	Nitric acid (HNO_3)
NO_2^-	Nitrite	Nitrous acid (HNO_2)

together by covalence, coordinate covalence, and resonance structures, just as do those of uncharged molecules. We can find out the structures of most polyatomic ions by comparing them to uncharged molecules using the isoelectronic principle.

Compounds in which cations are ionically bonded to negative polyatomic ions are named simply, incorporating the name of the anion last. The prefixes, di, tri, and so on are used *only* when necessary. Examples:

Potassium sulfate	K_2SO_4
Potassium hydrogen sulfate	$KHSO_4$
Sodium phosphate	Na_3PO_4
Disodium hydrogen phosphate	Na_2HPO_4
Sodium dihydrogen phosphate	NaH_2PO_4

Positive Polyatomic Ions.   Polyatomic ions with positive charge are less common than those with negative charge. A few important ones exist, most notably the hydronium ion, H_3O^+. Chemical investigations have shown that the hydrogen ion (H^+) does not exist alone in solution, but instead is chemically bonded to a water molecule:

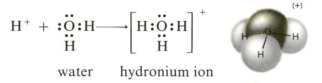

water   hydronium ion

We may think of the water molecule as little more than a carrier of H^+, able to release the H^+ ion whenever required by the chemistry of the system. Therefore we often symbolize acids by the simple H^+ rather than by the true chemical form, the polyatomic ion H_3O^+.

Another important polyatomic ion with positive charge is the ammonium ion, NH_4^+. This ion is isoelectronic with methane (CH_4), and therefore it has the simple methanelike structure

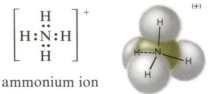

ammonium ion

This structure can be visualized as resulting from the combination of ammonia (NH_3) and a hydrogen ion (H^+).

ROLE OF ORGANIC MATTER

Organic matter is commonly taken to be any carbon-containing material whose ultimate source is living organisms, both plant and animal. It is paper, wood, wool, cotton, petroleum, and coal. It is lettuce, apples, broccoli, beans, milk, and meat. Clearly it is the food, the fuel, and the fabric of human existence.

Organic matter is also dead algae, rotting leaves, sewage, garbage, and slaughterhouse effluent. It is a culture for the growth of disease organisms, and it robs natural water of its oxygen, rendering it putrid and unfit for consumption or recreation. Without question, organic matter plays a major role in the environmental problems of this age.

Organic matter falls under the wide umbrella of *organic chemistry*. Organic chemistry is the chemistry of carbon-containing compounds. These compounds may or may not be derived from living systems. They include those in presently living organisms, those derived from organisms once alive, and those carbon-containing molecules synthesized or altered by man.

Nearly all organic molecules contain considerable hydrogen, usually linked by covalent bonds to various carbon atoms in the molecule. There is a small class of organic substances with no hydrogen-carbon bonds, a class which includes such environmental heavies as carbon dioxide (CO_2), carbon monoxide (CO), and the carbonate ion (CO_3^{2-}). Despite their organic character, these few molecules and ions were discussed with inorganic compounds earlier because their chemical bonding most resembles that of certain inorganic substances.

An important branch of organic chemistry is *biochemistry*. This branch deals specifically with chemical processes that go on in living organisms. Biochemistry is of particular concern to the student of environment, for the ultimate effect of pollution and other ecological intrusions is a disturbance of the chemical processes in living systems. A brief description of biochemical substances will appear in the last section of this chapter.

Carbon is but one element in a periodic system that contains over a hundred. Its abundance in the crust of the earth is an undistinguished 0.08 percent, under one part per thousand by weight. Yet of the more than one million different kinds of molecules known to man, involving bonding combinations of the different atoms of the periodic system, over 90 percent contain carbon. Some of these organic substances are of unique importance to environment and life, as has been pointed out. It is little wonder, then, that the study of carbon-containing molecules—organic chemistry—is one of the four major fields of chemical science.

ORGANIC MOLECULES

In the previous chapter we mentioned that carbon atoms bond readily to one another. The stable carbon-carbon link is responsible for the great number of organic molecules in existence. By linking together in various ways, carbon atoms form a rich profusion of chain and ring molecules. Because there are four bonds per carbon atom, branches of the molecule can spread out in various directions, forming a diversity of intricate three-dimensional molecular structures. And since carbon can bond to other atoms as well, organic molecules are often spiced with a variety of other bonded atoms, especially hydrogen, oxygen, nitrogen, sulfur, and chlorine. The different molecular arrangements are almost endless, a fact that explains why a million or so kinds of organic molecules are known to man.

The bonding versatility of carbon is utilized most dramatically in the essential and often complex molecules of living systems. Linked carbon atoms serve as the structural backbone of most of these vital molecules. Examples will be shown in the next section.

The unique ability of carbon to form large molecules through carbon-carbon bonds is due to several coinciding factors. Foremost is carbon's position halfway across the periodic table: its outer shell (with four electrons) is exactly half filled. In this central position it has little inclination to form ionic bonds, a type of bond that ordinarily exists only between atoms at opposite extremes of the periodic table. Instead it forms covalent bonds. These bonds usually are as strong between adjacent carbon atoms as they would be between carbon and other kinds of atoms. Furthermore, with four outer-shell vacancies, carbon can form four of these covalent bonds.

This number is unusually high among atoms and encourages carbon to establish linkages in several directions and therefore to build an extended molecular structure.

Organic molecules of various kinds may be grouped into a number of major classes and subclasses, which we now describe.

Hydrocarbons. Compounds containing only hydrogen and carbon are called *hydrocarbons*. Petroleum, gasoline, and natural gas are composed mainly of hydrocarbons. Thus hydrocarbons have a central role in providing the energy of civilization. In addition, they may become pollutants, particularly in oil spills and as ingredients in the formation of photochemical smog.

If no double or triple bonds exist between its carbon atoms, the hydrocarbon is called an *alkane* or *paraffin hydrocarbon*. The first 10 in a series of chain alkane molecules are shown in Table 4-3. The first two members, methane (CH_4) and ethane (C_2H_6), were shown in Chapter 3. Members of this series are called *straight-chain* alkanes because carbon atoms link in serial fashion. The actual

Table 4-3. **THE FIRST TEN ALKANES**

Number of Carbon Atoms	Formula	Name	Structural Formula
1	CH_4	Methane	methane structure
2	C_2H_6	Ethane	ethane structure
3	C_3H_8	Propane	propane structure
4	C_4H_{10}	Butane	butane structure
5	C_5H_{12}	Pentane	pentane structure

geometry of the carbon backbone is zig-zag, since the tetrahedral bond angle (109°) forces carbon atoms out of line.

By contrast to the straight-chain alkanes, the *branched-chain* alkanes have chain parts that intersect at certain carbon atoms. An example is isooctane, the reference substance for octane ratings, which ranks by definition as a "100 octane" fuel.

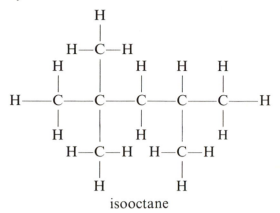

isooctane

Table 4-3. **THE FIRST TEN ALKANES** *(Continued)*

Number of Carbon Atoms	Formula	Name	Structural Formula
6	C_6H_{14}	Hexane	H—C—C—C—C—C—C—H
7	C_7H_{16}	Heptane	H—C—C—C—C—C—C—C—H
8	C_8H_{18}	Octane	H—C—C—C—C—C—C—C—C—H
9	C_9H_{20}	Nonane	H—C—C—C—C—C—C—C—C—C—H
10	$C_{10}H_{22}$	Decane	H—C—C—C—C—C—C—C—C—C—C—H

The chemical formula of isooctane (C_8H_{18}) is the same as that of the straight-chain octane (C_8H_{18}). Molecules like these, with identical atomic composition but different bonding structure, are called *isomers*.

Tens of thousands of branched-chain molecules exist because branches of different numbers can have different lengths and can connect at different places. We need not study these variations in detail.

When a hydrocarbon molecule has one or more carbon-carbon double bonds, it is called an *olefin* (oil former) or an *alkene*. Ethylene (C_2H_4), mentioned earlier, is an example, as is butene (C_4H_9):

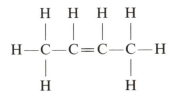

Many other olefin molecules exist, some prominent in forming photochemical smog.

If there is a carbon-carbon triple bond in the hydrocarbon, it is an *alkyne*. Acetylene (C_2H_2) is the simplest alkyne.

Molecules with carbon-carbon double and triple bonds are collectively called *unsaturated*. They contain fewer hydrogen atoms than do corresponding alkane chains because extra bonds are used tying carbon atoms together rather than bonding hydrogens. They could hold more hydrogen (or other) atoms than they do and are thus "unsaturated" with respect to hydrogen. By contrast, singly bonded molecules, able to hold no more hydrogen, are *saturated*.

Aromatic Hydrocarbons. The *aromatic hydrocarbons* are a special, environmentally important class of ring compounds with alternating double bonds in which resonance plays a major bonding role. The simplest and most important aromatic hydrocarbon is benzene (C_6H_6), a resonance hybrid in which double bonds are shifted one position around the ring in the two resonance structures:

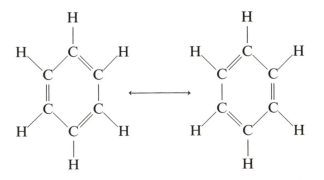

For simplicity, benzene and other aromatics are often represented by

THE STRUCTURE OF ENVIRONMENTAL MATTER

condensed structural formulas, which in this case are simplified line formulas of the type:

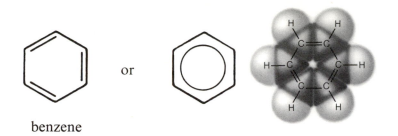

or

benzene

The carbon and hydrogen atoms at the corners of the hexagon are not shown, and it is understood that there are other resonance structures.

Aromatic rings may be joined together to yield *polynuclear aromatics*. Napthalene ($C_{10}H_8$), a moth repellant sometimes found in mothballs, is the simplest of these. Like benzene, it is a resonance hybrid, but there are more contributing structures, three in this case, because there are more rings in which electron shifts can occur.

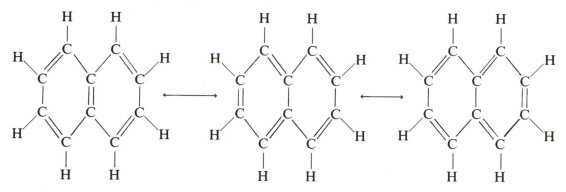

Condensed structural formulas for napthalene, along with the three-ring aromatic anthracene and the four-ring pyrene, are as follows:

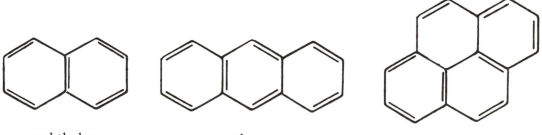

naphthalene   anthracene   pyrene

Certain polynuclear aromatics are *carcinogenic* (cancer producing), a point of great concern when they are environmental contaminants. The major carcinogens have as a framework a four-ring aromatic hydrocarbon, benzanthracene.

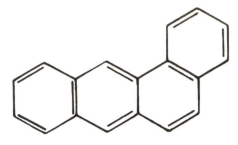

benzanthracene

The best-known environmental carcinogen, benzopyrene (more technically known as benzo(a) pyrene or 3,4-benzpyrene), consists of benzanthracene with one added benzene-type ring, which is shown below filled in with dashed lines:

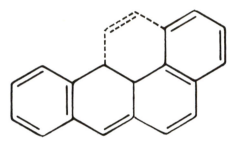

benzopyrene

The name "benzopyrene" is given because the molecule can be visualized as made up of pyrene with an added benzene-type ring—the ring to the far left. However, it is most useful here to consider benzopyrene as a relative of benzanthracene rather than of pyrene, for in this way it falls into the same structural category as do many other carcinogens having a benzanthracene core, including those below and on the facing page.

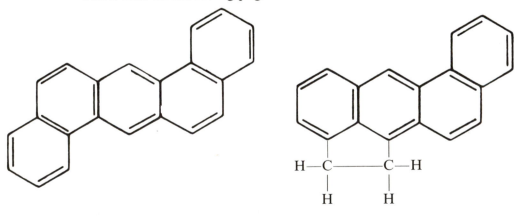

dibenzanthracene   cholanthrene

THE STRUCTURE OF ENVIRONMENTAL MATTER

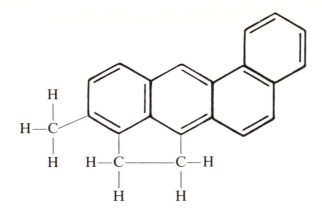

methylcholanthrene

Hydrocarbon + Functional Groups. Hydrocarbon molecules are nonpolar and for the most part rather unreactive and unobtrusive. If some foreign atom or group of atoms replaces a hydrogen atom bonded to carbon, this chemical quietness can change. Many kinds of foreign atoms can replace any number of hydrogen atoms on each and every hydrocarbon. The situation is typical of organic molecules, with a nearly infinite number of molecular bonding arrangements possible. Here we shall look at some simple but rather important cases in which there is but one substitution for hydrogen.

If a chlorine atom replaces a hydrogen atom on ethane, we get the structure

$$H-\overset{\overset{\displaystyle H}{|}}{\underset{\underset{\displaystyle H}{|}}{C}}-\overset{\overset{\displaystyle H}{|}}{\underset{\underset{\displaystyle H}{|}}{C}}-Cl \qquad (CH_3-CH_2-Cl)$$

ethyl chloride

This consists of a hydrocarbon *radical*, represented by the condensed structural formula CH_3-CH_2-, and a *functional group*, in this case $-Cl$, bonded together. The dashes extending to one side of radicals and functional groups may be viewed as single electrons, any of which can combine with any other single electron (shown by another dash) to form a covalent (electron-pair) bond.

Any hydrocarbon molecule that is short one hydrogen atom can serve as an organic radical. These radicals are given the general symbol $R-$. If the radical is an alkane, we form its specific name by deleting *-ane* and adding *-yl* to the hydrocarbon's name. Some examples are given in Table 4-4.

Various functional groups are attached to radicals by covalent bonds. They give the hydrocarbon specific and interesting properties,

Table 4-4. SOME HYDROCARBON RADICALS
(General symbol, R—)

Formula and Structure	Name
CH_3- (CH_3-)	Methyl
C_2H_5- ($CH_3—CH_2-$)	Ethyl
C_3H_7- ($CH_3—CH_2—CH_2-$)	Propyl

iso C_3H_7- ($CH_3—CH-$
     $\mid$
    CH_3)
	Isopropyl

C_4H_7- ($CH_3—CH_2—CH_2—CH_2-$)	Butyl
C_6H_5-	Phenyl

and they usually make it more reactive chemically and more active in its effects on living organisms. The major functional groups are listed in Table 4-5, along with examples of their presence in the environment. Many of these examples will be discussed later in an environmental context.

The functional groups, when attached to various radicals, yield compounds that have the common names ethyl alcohol ($CH_3—CH_2—OH$), methyl amine ($CH_3—NH_2$), and so on. The simple organic acids, containing carboxyl (—COOH) groups which readily release their hydrogen as the acid hydrogen ion H^+, are not so simply named. The important acid $CH_3—COOH$ is called acetic acid. There are, of course, systematic names for all organic compounds, but they will not be detailed here. Nor shall we elaborate on the vast number of *polyfunctional* molecules, containing two or more functional groups each. Most polyfunctional compounds take on some of the properties characteristic of each of the attached functional groups.

We likewise omit for the present those many organic molecules having atoms like oxygen, nitrogen, sulfur, or phosphorus inserted in the carbon chain. Some specific examples, however, will be noted later because of their environmental impact.

THE MOLECULES OF LIFE

The profound complexity of life reflects a profound complexity in the structure and activities of molecules in living systems. Almost in a literal sense, life is chemistry. Behind each action, each breath, and each thought, millions of molecules and ions are at work, unleashing energy and providing regulation of a most precise kind.

The simplest structural unit of most living systems is the cell. Although microscopic, each cell is like a complex chemical factory,

Table 4-5.   PRINCIPAL FUNCTIONAL GROUPS[a]

Formula and Structure	Name of Functional Group	Class Name of Organic Molecules Formed	Environmental Examples
—Cl	Chloride	Organochlorines	DDT and related insecticides
—OH (—O—H)	Hydroxyl	Alcohols	Air and water pollutants
—NH$_2$ $\left(-N\begin{smallmatrix}H\\\\H\end{smallmatrix}\right)$	Amino	Amines	Air and water pollutants; gives water bad taste
—SH (—S—H)	Sulfhydryl	Mercaptans	Odorous air pollutants; a part of enzyme molecules attacked by heavy metals
$>CO \left(>C=O\right)$	Carbonyl	Aldehydes, ketones	Air pollutants from auto exhaust
—COOH $\left(-\overset{\overset{O}{\|\|}}{C}-OH\right)$	Carboxyl	Acids	Influences acidity of natural waters; in NTA molecule
—SO$_3$H $\left(-\overset{\overset{O}{\uparrow}}{\underset{\downarrow}{S}}-OH \atop O\right)$	Sulfo	Sulfonic acids	Sodium salt (sodium in place of H) is major detergent constituent
—COOR $\left(-\overset{\overset{O}{\|\|}}{C}-O-R\right)$	Ester	Esters	In animal fats used in making soap
—NO$_2$ $\left(-N\overset{\nearrow O}{\searrow_O} \leftrightarrow -N\overset{\nwarrow O}{\diagdown_O}\right)$	Nitro	Nitro compounds	Some insecticides
—NHCOO— $\left(-\overset{H}{\underset{}{N}}-\overset{\overset{O}{\|\|}}{C}-O-\right)$	Carbamate	Carbamates	Core of carbamate insecticides

[a]Always bonded covalently to one or more organic radicals, R —, R′ —, and so on.

converting an assortment of raw materials into end products along well-regulated chemical assembly lines. At the same time the cell derives useful energy from these raw materials. In addition, the cell contains the genetic code, information (in chemical form) that is sufficiently detailed to construct an entire organism. How can this vast activity and information storage occur in a single cell 10^{-3}

to 10^{-4} cm across, a speck not even visible to the eye? These seemingly unbelievable feats become plausible only when we realize that the cell factory is working with atoms and molecules. Tiny as the cell is, atoms are ten thousand times smaller, mere dwarfs that are readily accommodated in vast numbers in this tiny chemical factory.

The delicate balance of the cell microfactory and its supply routes can be upset by foreign chemicals which in any way alter the chemical machinery. Untoward chemical bonds formed between foreign molecules and the molecular regulators can disturb chemical functioning and lead to sickness or death. Moreover, the regulating molecules, important as they are, may not be abundant in the cell. As with an inadequately guarded city, these few key molecules can be inactivated by a small invading force, and chaos may result. Some substances are toxic in minute amounts precisely because relatively few of their molecules are enough to chemically attack and overwhelm certain key molecules of which there are few in the cell.

Through eons of evolution, cells have adjusted to their molecular environment, developing machinery to utilize, reject, or coexist with most environmental chemicals at their natural level. As man's activities increasingly alter the surroundings, introducing myriad new environmental chemicals and multiplying the concentration of some of those long present, the carefully tuned machinery of life is showing proportional malfunctions. Some of these problems are highly visible, as when intense air pollution episodes increase the immediate death rate. Others are of a smoldering sort, barely observable, at whose exact nature and full extent we can only guess.

Life's chemical machinery is exceedingly complex, much of it still beyond any man's intellectual grasp. Our inability to understand the chemical basis of cancer is a glaring example of everything we do not know. Nonetheless, by studying a few of the major cogs in this marvelous machinery, we can gain a better appreciation of its functioning and of its susceptibility to environmental disruption.

Several major classes of molecules are involved in the machinery of life. Foremost among them are carbohydrates, fats, proteins (including enzymes), and nucleic acids. Every class includes a great number of chemically distinct molecules, each with a unique function. For instance, there are probably 10,000 or more different kinds of protein molecules, each playing a unique role in the human organism.

Here we shall highlight some of the major features of molecules that are integral to living systems.

Carbohydrates. *Carbohydrates* are sugarlike substances. They have a general significance to all living organisms, for the sugars, particularly glucose, are the basic fuel of life, produced by green-plant photosynthesis. Through photosynthesis and the sugarlike fuels, the energy of sunlight is converted into the energy of fossil fuels. The latter will be discussed at some length in Chapter 6, where it will be shown that photosynthesis is the ultimate source of atmospheric oxygen, as well as of the fossil fuels that run our industrial civilization.

Glucose, $C_6H_{12}O_6$, is the chief product of photosynthesis. It exists in two forms, one being an open chain and the other a ring formed by the chemical joining of the ends of the open chain.

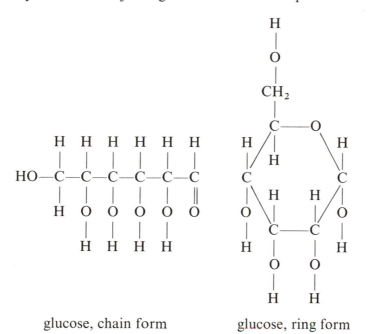

glucose, chain form   glucose, ring form

Both, of course, have the same number of atoms of each kind— 6 carbons (C), 12 hydrogens (H), and 6 oxygens (O)—giving identical formulas, $C_6H_{12}O_6$. These and other carbohydrates are characterized by an abundance of hydroxyl (—OH) groups.

Frequently, glucose rings are joined to one another by chemical bonds to form long molecular strands. Starch and cellulose, two of the most common carbohydrate substances, are composed of glucose strands.

Carbohydrates, the most abundant molecular class (besides water) in plants, constitute about 70 percent by weight of the average diet of man. However, they do not accumulate in man to beyond a few percent. After they are ingested, most carbohydrates either are "burned" to provide energy or are chemically converted to fats.

Fats. *Fats* are nonpolar, oil-like materials with many vital functions. They are a food and, deposited in various parts of the body, they serve as a means for storing food energy. They give structure to the oil-like membranes that surround each cell. These membranes not only hold the cell intact, but also control the ingress and egress of thousands of substances involved in the cell's chemical factory. The membrane is like a guard at the gate, blocking the entry of saboteurs (toxic compounds), allowing raw materials (foods) to enter, and preventing the loss of vital machinery (proteins and so on).

The most common fatty materials are the so called triglycerides, whose chemical structure can be represented by

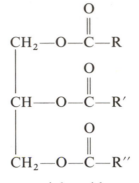

a triglyceride

The oxygen atoms are participating in an ester linkage (ester =

$$-\overset{\overset{\displaystyle O}{\|}}{C}-O-,$$ last section), which serves to hold parts of the molecule together. The symbols R, R′, and R″ stand for various hydrocarbon radicals. These radicals contain carbon backbones that usually have 13 to 17 linked carbon atoms. Two common examples are

$$-CH_2-CH_2-CH_2-CH_2-CH_2-CH_2-CH_2-CH_2-CH_2\diagdown$$
$$CH_3-CH_2-CH_2-CH_2-CH_2-CH_2-CH_2\diagup CH_2$$

stearic radical, $-C_{17}H_{35}$

$$-CH_2-CH_2-CH_2-CH_2-CH_2-CH_2-CH_2-CH=CH\diagdown$$
$$CH_3-CH_2-CH_2-CH_2-CH_2-CH_2-CH_2\diagup CH_2$$

oleic radical, $-C_{17}H_{33}$

Some of the radicals, such as oleic, are unsaturated (containing double bonds) and some, such as stearic, are saturated. They give, respectively, the unsaturated and saturated fats so discussed in nutritional circles. It is thought that less cholesterol, a fatty substance manufactured by the cell machinery which deposits in veins, is made with a diet of unsaturated fats.

The hydrocarbon chains of fatty materials are obviously quite long and make up the great bulk of the fatty molecules. For this reason, fats are much like hydrocarbons—they are slippery and oil like. More important, they are nonpolar, as hydrocarbons are, a fact that explains why fatty materials accumulate DDT and other nonpolar toxins that tend to dissolve only in nonpolar materials. (Recall from pp. 101–103 in Chapter 3 the rule that "like dissolves like.")

Proteins. *Proteins* are giant molecules that contain thousands of atoms. Many-atom molecules such as these are known as *macro-*

     THE STRUCTURE OF ENVIRONMENTAL MATTER

molecules or *polymers*. Macromolecules of many kinds have vital functions in living systems, as will become increasingly clear.

Protein molecules are formed by the joining together or *polymerization* of many amino acids, small molecules containing both amine ($-NH_2$) and organic acid ($-COOH$) functional groups. Most of the significant amino acids have the general structure

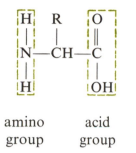

amino   acid
group   group

amino acid

The radical $-R$ is usually a small group containing one to nine carbon atoms. Some of these radicals contains hydroxyl ($-OH$) groups, sulfhydryl ($-SH$) groups, additional organic acid and amino groups, and so on.

Twenty such radicals are commonly found in proteins: hence there are 20 distinct amino acids that serve as building blocks for protein molecules. Several of them are shown in Figure 4-5.

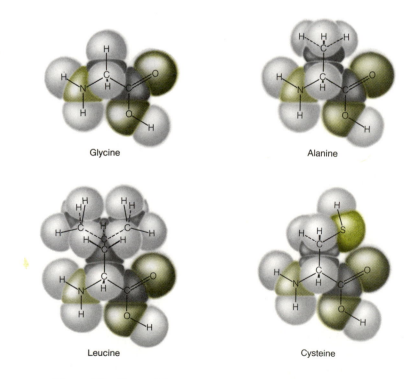

Glycine

Alanine

Leucine

Cysteine

Figure 4-5. Four of the twenty amino acids common in proteins.

Typically, hundreds of amino acid units are joined together (polymerized) to form a complete protein molecule. A small segment of the protein chain is shown below.

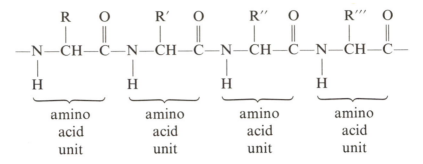

Note that the amino acid units are not complete amino acids—a hydrogen (—H) has been lost from the amino (—NH_2) group, and a hydroxyl (—OH) from the carboxyl group (—COOH). Their removal has reduced by one the bonds to the nitrogen atom in —NH_2 and to the carbon atom in —COOH. To make up for this deficit in bonds, new covalent bonds are formed between adjacent amino acid groups, forging them together. The —H and —OH groups combine to form byproduct water, HOH or H_2O.

The thousands of distinct proteins differ from one another in their *primary structure*: that is, in the number and sequence of the 20 different kinds of amino acid units linked together.

Protein "chains" do not flop around loosely as we would expect of a chain: instead, most of them are precisely wound, like a spiral staircase, into an *alpha helix*. A few lie in parallel rows called the *sheet structure*. This organized *secondary structure* is brought about by intermolecular forces (mostly hydrogen bonds) which hold the close-lying loops of the helix together.

Similar forces cause the entire spiral to fold or loop into a definite *tertiary structure*. They may also cause separate protein chains to associate with each other to form a *quaternary structure*.

The higher structural levels—secondary, tertiary, and quaternary—are, like the primary structure itself, crucial to protein functioning. While the higher structural levels are ordinarily controlled by intermolecular forces, a few covalent bonds aid significantly in organizing them. These bonds are between sulfur atoms on different cysteine units (Figure 4-5). Hydrogen atoms are split out of the sulfhydryl (—SH) group of cysteine units, and the bond that that sulfur had with hydrogen is replaced by a bond to another sulfur atom on a different cysteine unit. The sulfur-sulfur bond is called the *disulfide bridge*. It is shown in Figure 4-6. Disulfide bridges are not abundant in protein molecules. Some proteins, such as hemoglobin, have none at all; others, such as insulin, have three or so; a few, such as gamma globulin, have up to 25 per molecule. Because covalent bonds are much stronger than intermolecular forces, these disulfide bridges influence a molecule's structure wherever they exist.

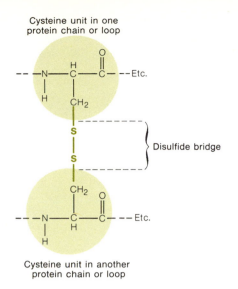

Figure 4-6. The disulfide bridge holding segments of protein chains together.

Proteins have many roles integral to living systems. In animals they provide much of the gross structure, such as that of hair, skin, fingernails, and tendons. Proteins in muscle cells contract to provide muscle power. Other proteins transport oxygen through the body (hemoglobin) and protect organisms from viruses (antibodies), and so on.

Of all the proteins, none is quite so central to the act of living as are the *enzymes*. These proteins promote and control the hundreds of chemical reactions of the cell's chemical factory. We shall refer to enzymes many times in this text, for they are susceptible to disruptions of a most serious kind by environmental contaminants such as mercury and insecticides.

Nucleic acids. The *nucleic acids* are key molecules responsible for hereditary characteristics and their transmission. One kind of nucleic acid contains in its makeup the genetic code—the elaborate blueprints defining the detailed structural features and chemical functioning of organisms. Other nucleic acids transmit the genetic blueprints into chemical action, controlling the synthesis of proteins (including enzymes), which in turn control all chemical processes in the cell.

The specific type of nucleic acid which carries the genetic code is known as *DNA* (deoxyribonucleic acid). DNA consists of enormous macromolecules, some containing between a million and 100 million atoms. The genetic code is locked into the bonding arrangement of these multitudinous atoms. The precise arrangement is different for each individual among the higher organisms, thus ensuring that every creature—every human being—is unique on earth.

The nucleic acids are polymers in which molecular units of three basic types are joined in regular fashion. These three are the phosphate group or linkage, a sugar group, and a so called "base" group. In DNA, there are four distinct but closely related bases: adenine, guanine, cytosine, and thymine. They have structures as shown in Figure 4-7.

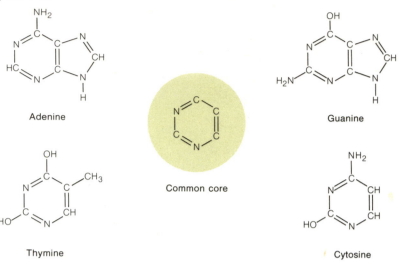

Figure 4-7. The four "bases" found in DNA.

The bases are resonating structures that somewhat resemble benzene or naphthalene (the various resonance forms are not shown here). Notice, however, that nitrogen appears in the rings along with carbon.

When one of these bases chemically bonds to the sugar molecule deoxyribose, a *nucleoside* is formed. For instance, the combination of the base adenine and the sugar deoxyribose yields the nucleoside deoxyadenosine, shown in Figure 4-8. There is one such nucleoside for each of the four bases.

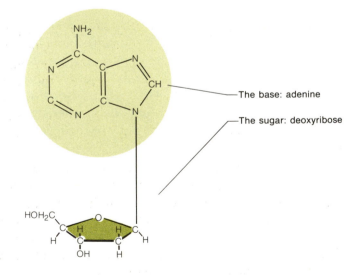

The nucleoside: deoxyadenosine

Figure 4-8. Nucleosides are the chemical combination of a base and the sugar deoxyribose.

The chemical backbone of DNA is forged by phosphate linkages between nucleosides. A small segment of a DNA strand appears in Figure 4-9.

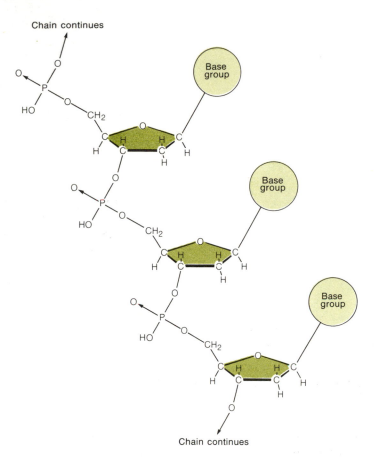

Figure 4-9. A short segment of a DNA strand.

The thousands of base groups attached to a DNA strand are a mixture of the four bases mentioned above. *The unique sequence of bases in each organism's DNA constitutes its genetic code.*

Finally, of critical importance, two long DNA strands spiral around each other in the famed *double helix* (Figure 4-10). The two are held together by hydrogen bonds between their respective bases. The two DNA strands exactly complement each other in base sequence, and when a new strand is formed, it also must complement the old. In this way the base sequence (genetic code) is preserved and passed from generation to generation.

The structure and genetic function of the double helix was unraveled by James Watson and Francis Crick in 1953. For this profound work these men received the Nobel Prize in 1962.

Another group of nucleic acids of great importance are the RNA's (ribonucleic acids). These substances are involved in translating the genetic code of DNA into specific chemical activity. Structurally, they are much like DNA, the chief difference being that the sugar in the

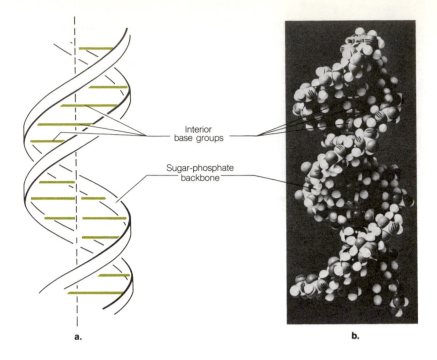

Interior
base groups

Sugar-phosphate
backbone

a.   b.

Figure 4-10. The "double helix" structure of the DNA molecule. In (a), the double ribbons show schematically the configuration of the chemical backbone of DNA strands. In (b), a short segment of double helix is shown with all its atoms, reflecting the chemical intricacy of DNA.

RNA backbone is ribose rather than the closely related deoxyribose (hence the name *ribo*nucleic acid, as opposed to *deoxyribo*nucleic acid).

It is clear that the chemical structure of nucleic acids is of utmost importance to an organism and to all of its descendents. Any environmental agent that threatens to change that structure is a hazard of the first magnitude. Several such agents will be mentioned in subsequent chapters. Foremost among these are radioactive substances, which emit particles of such high energy that parts of these key molecules can be literally ripped asunder.

Exercises

1. Recall that carbon-carbon bond lengths are 1.54 A for single bonds, 1.33 A for double bonds, and 1.20 A for triple bonds. Basing your decision on this trend and on knowledge of molecular bonding structures, assign one of the three carbon-oxygen bond lengths 1.43 A, 1.22 A, and 1.13 A to the carbon-oxygen bonds in each of the following three molecules: the aldehyde known as acetaldehyde (CH_3CHO), the common alcohol called ethyl alcohol (C_2H_5OH), and carbon monoxide (CO).

2. Pick the one "odd" molecule out of the following assortment: PCl_3, CH_3OH, NH_3, CCl_4, ClO_2, HCl, N_2O, and N_2O_5.

3. For each of the ions and molecules on the left, select an ion or molecule that is isoelectronic with it from the list on the right.

   a. NH_3 A. OH^-
   b. HI B. CN^-
   c. $SO_3{}^{2-}$ C. SO_3
   d. $NO_3{}^-$ D. H_3O^+
   e. CO E. $ClO_3{}^-$

4. Write the chemical formula for substances having the following names:

   a. dinitrogen oxide
   b. dinitrogen pentoxide
   c. phosphorus trichloride
   d. sodium hydrogen carbonate (sodium bicarbonate)
   e. potassium dihydrogen phosphate.

5. Use the bonding rules (especially 1 and 3) to deduce the bonding structure of the sulfite ion, $SO_3{}^{2-}$.

6. Use bonding rules 1 and 3 to deduce the bonding structure of the perchlorate ion, ClO_4^-. Confirm it by locating one (or more) isoelectronic ions whose structure is shown on pp. 121–126.

7. Deduce the bonding structure of the carbonate ion, $CO_3{}^{2-}$, in two independent ways: (a) use bonding rules 1 and 3, and (b) use bonding rule 2, noting the electronic similarity of $CO_3{}^{2-}$ and SO_3.

8. Use the isoelectronic principle to write the structures of the following polyatomic ions: nitrate ion, $NO_3{}^-$; nitrite ion, $NO_2{}^-$; and hydrogen sulfide ion, HS^-. (Hint: On pp. 121–126 it was noted that these ions are isoelectronic with sulfur trioxide, sulfur dioxide, and hydrogen chloride, respectively.)

9. Octane and isooctane are two different isomers shown in the text that have the same chemical formula, C_8H_{18}. Draw three other isomers of C_8H_{18}. (Altogether, there are 18 isomers of C_8H_{18}. The numerous bonding arrangements possible with the single combination of 8 carbon atoms and 18 hydrogen atoms illustrates why, when all combinations of carbon atoms, hydrogen atoms, and other atoms are considered, there are as many as one million organic substances.)

10. Propyl alcohol (C_3H_7OH) has two isomers. Show the bonding structures by drawing line formulas for both isomers.

11. Predict which of the following structures is more likely to be carcinogenic. Why?

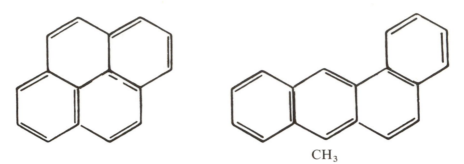

CH$_3$

12. Some typical organic molecules are

$CH \equiv CH$ $CH_3—CH_2—CH_2—CH_3$ $CH_3—CH = CH_2$
acetylene   butane   propene

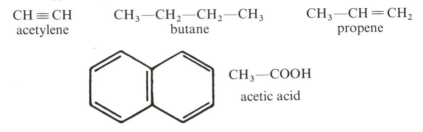

$CH_3—COOH$
acetic acid

napthalene

Indicate all of the categories, below, into which each molecule falls.

a. hydrocarbons   f. unsaturated compounds
b. alkanes   g. aromatics
c. olefins (alkenes)   h. polynuclear aromatics
d. alkynes   i. carcinogens
e. saturated compounds   j. organic acids

13. Show the complete bonding structures (that is, show every chemical
bond) for the following organic molecules by drawing line formulas:
methyl chloride, ethyl mercaptan, butyl alcohol, propyl amine, and
benzene sulfonic acid (read as "benzyl sulfonic acid").

14. The organic molecules $C_4H_9—Cl$, $C_4H_9—OH$ and $C_4H_9—I$ differ in
polarity. On this basis (consult Chapter 3), can you explain their
respective solubilities in water of 0.06 percent, 7.9 percent, and 0.02
percent? (All percentages are by weight.)

15. Propose a bonding structure consistent with the formula C_3H_6.

16. Classify each of the molecular structures shown below into one of
the following classes: amine, acid, amino acid, olefin, alkane, and
alcohol.

a. $CH_3—CH_2—CH_3$ d. $CH_3—NH_2$

b. $CH_3—CH_2—OH$ e. $R—C—NH_2$
                                          $COOH$

c. $R—C—OH$ f. $CH_3—CH = CH—CH_3$
     ‖
     O

17. Match each of the chemical characteristics on the right to the most appropriate class of substances on the left.

a. carbohydrates
b. fats
c. proteins
d. nucleic acids

A. contain many phosphorus atoms
B. contain many hydroxyl groups
C. consist of linked amino acids
D. contain long hydrocarbon chains

Glossary

Acid a compound that can lose a hydrogen ion (H^+).

Alkali a base.

Alkane a hydrocarbon with no double or triple bonds between atoms. Also called a *paraffin*.

Alkene a hydrocarbon with at least one double bond but no triple bond between atoms. Also called an *olefin*.

Alkyne a hydrocarbon with at least one triple bond between carbon atoms.

Alpha helix the spiral structure into which most proteins are wound; a secondary structure.

Aromatic hydrocarbon a hydrocarbon with one or more rings of six carbon atoms joined by alternating double and single bonds in which resonance plays a major role.

Base a compound that can lose a hydroxide ion (OH^-). Also called an *alkali*.

Biochemistry the branch of organic chemistry that deals with the chemical processes that occur in living organisms.

Branched-chain alkane an alkane in which some carbon atoms are bonded to more than two other carbon atoms.

Bond angle the angle between the lines joining a central atom with two satellite atoms.

Bond length the distance between the nuclei of two atoms that are bonded to each other. Also called *internuclear distance*.

Carbohydrates sugarlike substances with great significance to all living organisms.

Carcinogenic cancer producing.

DNA deoxyribonucleic acid; the type of nucleic acid that carries the genetic code.

Disulfide bridge a sulfur-sulfur bond bridging parts of a protein molecule together.

Double helix spiral structure of a pair of DNA or RNA molecules.

Enzymes proteins that promote and control the hundreds of chemical reactions in the cell's chemical factory.

Fats nonpolar, oil-like materials that have many vital functions in a cell.

Functional group atom or group of atoms that makes the molecule a member of a certain family of organic compounds and usually defines its properties.

Hydrocarbon a compound containing only carbon and hydrogen.

Isoelectronic describes molecules or ions that have an identical number of atoms and of outer-shell electrons.

Isomers molecules with identical chemical formulas but different bonding structures.

Linear describes a molecule in which the nuclear centers form a straight line.

Macromolecule a molecule made up of many atoms; also called a *polymer*.

Neutralization the combination of hydrogen ion (H^+) from an acid with hydroxide ion (OH^-) from a base to form neutral water (H_2O).

Nonlinear describes a molecule in which nuclear centers do not form a straight line; a bent molecule.

Nucleic acids molecules responsible for hereditary characteristics and their transmission.

Nucleoside chemical combination of a base and the sugar deoxyribose.

Odd molecule a molecule with an odd number of electrons.

Olefin an alkene.

Organic chemistry chemistry of carbon-containing compounds.

Organic matter carbon-containing material whose ultimate source is living organisms.

Polyatomic ion an ion composed of more than one atom bonded together.

Polymer a macromolecule having repeating molecular unity.

Polymerization process in which many small similar molecules join together to form a polymer.

Polynuclear aromatic a compound made up of fused aromatic rings.

Primary structure the number and sequence of amino acids in proteins.

Proteins giant molecules containing thousands of atoms and having important roles in living systems.

Quaternary structure protein structure caused by association of separate protein chains.

Radical a hydrocarbon molecule lacking one hydrogen atom.

RNA ribonucleic acids, which translate genetic information into chemical activity.

Saturated describes a hydrocarbon that can bond to no more hydrogen atoms; a molecule with only single carbon-carbon bonds.

Secondary structure protein structure resulting from intermolecular forces that hold nearby parts of protein molecules together.

Sheet structure a form of secondary structure in which the protein segments lie in parallel rows.

Straight-chain alkane an alkane with no carbon bonded to more than two other carbons.

Tertiary structure the overall shape of a protein molecule, caused by intermolecular forces which make the protein spiral, fold, or loop in a definite manner.

Unsaturated describes a hydrocarbon that can react with hydrogen to form bonds; a hydrocarbon having double or triple bonds.

BRAZIL'S FALLS OF IGNACA
(*Photo by Dr. George B. Water/Rapho Guillumette.*)

5 MOLECULAR MOTION AND CHEMICAL CHANGE

THE MOTION OF MOLECULES AND ATOMS

Molecules and atoms are in constant motion. In a gas, molecules dart about erratically. They collide with other molecules, rebound, then collide again. Between collisions they are spinning and tumbling through space. The molecules in a liquid or a solid are more crowded, so they cannot fly around freely. Nonetheless they are moving just as rapidly, colliding and rebounding back and forth in the small space they find between their nearest neighbors.

Atoms that are chemically bonded in molecules have their own motion—back and forth within the molecule. Chemical bonds lock atoms into their proper location and do not permit great freedom of motion. However, within the narrow confines of movement permitted by chemical bonds, atoms vibrate vigorously.

The three basic kinds of motion, all alluded to above, are straight-line motion through space, or *translational motion*; tumbling, or *rotational motion*; and *vibrational motion*. These are illustrated in Figure 5-1 on the next page.

Molecules and atoms are not slow in their various forms of motion. As an average, they move slightly faster than a jet airplane. This velocity is roughly that of sound in air, which is indeed no surprise, since molecules by their motion cause the transmission of sound.

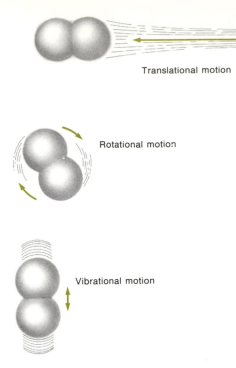

Translational motion

Rotational motion

Vibrational motion

Figure 5-1. The three basic kinds of atomic and molecular motions.

The average translational velocity of nitrogen molecules in our atmosphere is just under 500 m/sec. At any instant, some are slower and some are faster. A very few are momentarily much faster: about one in a trillion is traveling over five times the average velocity, in excess of 2500 m/sec. These super-fast molecules strike other molecules with shattering violence. These collisions can rupture chemical bonds and lead to chemical change, as will be described in the section on chemical reactions.

Molecular velocity increases with temperature. In fact, *heat energy* is nothing more than the energy of motion (kinetic energy) of atoms and molecules. For this reason, molecular motion is sometimes called the *thermal motion* of molecules. When heat is added to a system, its molecules increase in velocity, they acquire more energy, and the temperature goes up as a reflection of increased molecular activity.[1]

---

[1]Temperature, which reflects the vigor of molecular motion, is generally measured on one of two scales: Fahrenheit and centigrade (or celsius). The centigrade scale is used in most of the world; the Fahrenheit scale is employed throughout the United States. The centigrade temperature, t_C, can be obtained from the Fahrenheit temperature, t_F, by the formula $t_C = (5/9)(t_F - 32)$. Water boils at 212 degrees Fahrenheit (°F) and freezes at 32°F. From the formula we determine that boiling and freezing on the centigrade scale occur at 100 degrees centigrade (°C) and 0°C. The so-called absolute temperature scales are arranged so that a zero temperature occurs at the hypothetical point where all molecular motion ceases. There are two absolute scales of temperature: Temperature on the Rankine scale is obtained by adding 460 to the Fahrenheit temperature, and temperature on the Kelvin scale is obtained by adding 273 to the centigrade temperature.

The ceaseless motion of atoms and molecules is responsible for the major physical and chemical changes that occur on this earth.

VAPORIZATION AND PHYSICAL CHANGE

We learned in Chapter 1 that by vaporization, molecules leave a liquid to become airborne. They are able to tear loose from the liquid's surface despite the fact that they are tied to neighboring liquid molecules by universal intermolecular forces. They can do so because of incessant molecular motion. Molecules in a liquid are in agitated motion back and forth, and when those near the surface get, momentarily, enough energy to break the intermolecular links, they become free to leave the liquid. Thus vapor is formed.

Vapor Pressure and Volatility. Vapor molecules, through incessant collisions, exert an outward force on their surroundings. The force exerted by vapor per unit area is called the *vapor pressure*.[2] Vapor pressure increases with the concentration of vapor because of the increased number of bombarding vapor molecules.

The ease with which a substance vaporizes is called the *volatility* of the substance. Volatility is measured by vapor pressure. If the vapor pressure is high, the substance is called *volatile*. If the pressure is very small, it is *nonvolatile*. Most air-pollutant molecules are highly volatile, or they would not have entered the air in the first place.

Volatility is determined by two opposing influences. Molecular motion is acting to free the bound molecules; intermolecular forces try to retain them. If molecular motion is increased by heating, more molecules will break free and vapor pressure will increase. Boiling occurs when molecules in a liquid become agitated enough to create a vapor with the same pressure as that of the surrounding atmosphere. Vapor bubbles can then expand because of their high pressure, pushing back the surrounding water and atmosphere. We associate the rapid bubble expansion with boiling. Whether brought to boiling or not, all substances have a vapor pressure that increases with temperature, because with increasing molecular motion, more molecules break free.

If the intermolecular restraining forces are relatively large, making escape difficult, volatility tends to be low. Intermolecular

---

[2]All molecules exert a force on their surroundings because with each collision they tend to push molecules in the vicinity away. In a container, molecules bombard the wall, pushing it outward. Technically, this push occurs because the molecules' outward momentum is transferred to the wall upon collision. The force exerted per unit area is known as the *pressure*. Several standard units of pressure exist. One *atmosphere* (atm) is the mean pressure of air at sea level. One *bar* is a metric unit close to the atmosphere in value, equivalent to 0.987 atm. One *mm of mercury* (also called *torr*) is the pressure necessary to push a column of liquid mercury (Hg) 1 mm upward against gravity; 760 mm of Hg make 1 atm of pressure. One often sees atmospheric pressure expressed in inches of mercury. The conversion from mm to in. shows that 1 atm is equivalent to 760 mm Hg $\times$ 0.0394 in./mm, which equals roughly 30 in. of mercury.

forces are relatively large for polar molecules, and they are particularly large for polar molecules united by hydrogen bonds, the strongest of the intermolecular forces (Chapter 3). Water (H_2O) is far less volatile than other similarly small molecules such as hydrogen sulfide (H_2S), sulfur dioxide (SO_2), nitrogen dioxide (NO_2), or carbon dioxide (CO_2) because of its strong hydrogen bonding. This force explains why water molecules stick together as a liquid at ordinary temperatures whereas the other substances are gases. However, the hydrogen bonds are not strong enough to retain water molecules entirely: water is fortunately a moderately volatile substance at ordinary temperatures. Its vapor pressure is high enough so that much vapor penetrates the atmosphere, enough to bring water from the oceans to the continents to make them habitable.

The Effect of Molecular Size. Intermolecular forces increase with the size of molecules. Big molecules "stick" to other molecules over their entire surface, so that the total restraining force is considerable even though the intermolecular forces are weak for a given segment of the molecule. Even oil-like hydrocarbon molecules with the weakest kind of intermolecular forces, dispersion forces (Chapter 3), do not vaporize very much if they are large enough. Various lube oils and asphaltic materials from petroleum are made up mostly of long-chain hydrocarbon molecules that may have up to 70 linked carbon atoms. They are nonvolatile because of their large size. Gasoline, consisting mainly of shorter chains with five to twelve carbon atoms, is moderately volatile. It can vaporize directly to the atmosphere in appreciable amounts from fuel tanks and carburetors, thus contributing significantly to air pollution by hydrocarbons.

In general, air-pollutant molecules are among the smallest molecules known, commonly having only two or three atoms. Examples are CO, NO, NO_2, and SO_2. This small size explains why most air pollutants are so volatile, tending to form gases and thus seeking out the atmosphere.

Melting. The melting process is in some ways similar to vaporization. The perfect order of a solid crystal, established by intermolecular forces which draw molecules (or ions) into closely packed unions, can be disrupted by increasing temperature. At the melting point the thermal agitation of molecules becomes so violent that the precise regularity of the molecular array is broken up. At this point the more random molecular arrangement of a liquid is formed. With cooling, the molecules again achieve a close and regular packing: this process is called freezing.

Vaporization and melting are *physical changes*. Molecules remain intact when they go through physical changes such as vaporization. The bonding arrangement of atoms within the molecule is identical before and after the change occurs. If, however, a process occurs in which chemical bonds are broken or rearranged, it is termed *chemical change* or *chemical reaction*.

CHEMICAL REACTIONS

A chemical reaction is the process by which new molecules are formed from old. It occurs by the rearrangement of atoms in and among molecules. Chemical reactions produce protein and other molecules necessary to life and also most of the pollutant molecules that harm life. Photosynthesis, which is the ultimate source of all food, is a series of chemical reactions, as is the process of organic decay. In general, any system, natural or man made, from which new chemical compounds emerge has utilized some form of chemical reaction.

The chemical reactions controlled by man generally have one of two functions: they either produce useful substances, such as drugs and plastics, or, in the form of combustion, they provide abundant energy, a product in great demand throughout the world. Most air pollutants are formed by the chemical reactions that occur in combustion.

A third incentive for controlling chemical reactions is now emerging: the control of pollution. Many pollutants, if caught at the source, can be made to chemically react in such a way as to render them harmless, as we shall see later on.

How Chemical Reactions Occur.   The rearrangement of atoms by a chemical reaction generally occurs by the breaking of old chemical bonds and the formation of new ones. This change occurs by different mechanisms for different reactions. The most direct mechanism is exemplified by the reaction in which nitric oxide (NO) is formed from diatomic nitrogen (N_2) and oxygen (O_2). Present evidence suggests that this reaction occurs as N_2 and O_2 molecules slam into one another, as shown in Figure 5-2 (p. 156). The bonds holding the atoms of N_2 and O_2 together are ruptured, and simultaneously new bonds are formed between N and O atoms.

Other chemical reactions proceed by stages. When hydrogen (H_2) and chlorine (Cl_2) react to form hydrogen chloride (HCl), for example, the Cl_2 molecule first must split into two Cl atoms; a Cl atom will then collide with an H_2 molecule to form HCl and leave behind an H atom; and the H atom will collide with Cl_2 to form another HCl molecule and a Cl atom. The Cl atom starts the cycle again, and we have a *chain reaction* of repetitive cycles involving free, reactive atoms (H and Cl). Other complex mechanisms occur in which bonds are broken and formed in still different sequences, but the details need not concern us here. What does concern us is that for all these mechanisms, reaction can begin only with the rupture

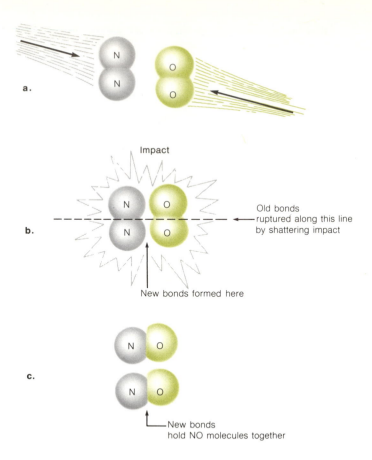

Impact

Old bonds
ruptured along this line
by shattering impact

New bonds formed here

New bonds
hold NO molecules together

Figure 5-2. Nitric oxide (NO) is formed by exceptionally hard collisions between nitrogen (N_2) and oxygen (O_2) molecules. Bonds between N and O start to form upon collision, as shown in (*a*), and within about a trillionth of a second (*b*) the old bonds are broken and the new bonds are completed. Two nitric oxide molecules result.

of some chemical bond or bonds. This rupture requires an initial input of energy into the reacting molecules,[3] an energy whose magnitude determines the all-important speed of chemical reaction.

The "starter" energy needed to get the reaction underway is called *activation energy*. Considerable energy is needed to rupture most chemical bonds, and therefore the molecules must momentarily accumulate a large activation energy before reaction can occur. The principal energy source is the thermal motion of molecules. Collision with the occasional molecule that has energy far above average provides the energy necessary for bond rupture. If the bonds are very strong and the activation energy high, a long time may go by before a molecule chances along with enough energy to cause

[3] Although energy may need to be supplied to stimulate chemical reactions, in the end most reactions release more energy than they consume. The chemical bonds formed in a reaction are commonly stronger than those broken, and they provide a net release of energy, as heat and sometimes as light. We shall discuss energy-releasing reactions (called exothermic reactions) shortly.

chemical reaction. If the bonds are weak, reaction can occur very rapidly. Every reaction has its own pace.[4] Some reactions, like those in explosions, are very fast, while others require millions of years. Later we shall discuss the speed of reaction and its consequences for certain environmental contaminants.

Nearly all chemical reactions are speeded up greatly by an increase in temperature. The increased molecular motion causes molecules to collide with more impact, providing an abundance of activation energy, which is needed if fast reaction is to occur. The effect of temperature is so profound that normally the speed of reaction will approximately double with a mere $10°$ increase in the centigrade temperature (an $18°$ rise on the Fahrenheit scale).

Catalysts. Special substances called *catalysts* speed up chemical reactions without themselves being altered. They do so by aiding the shift of electrons from old bonds to new, thus reducing the activation energy required for reaction. Catalysts are "electron brokers," so called because of their help in moving electrons within molecules to form new bonds.

Many catalysts are transition metals. Platinum is the outstanding example. It is used widely in industry, particularly to reform hydrocarbon molecules from petroleum to increase the gasoline yield and octane number.

Most of the environmental reactions described in later chapters proceed with the aid of some catalyst. The catalyst may or may not be mentioned; occasionally it is unknown, being perhaps some atmospheric or water contaminant, or the enzyme from a living organism.

Enzymes are biological catalysts (Chapter 4). They promote and regulate the rates of vast numbers of chemical reactions in living systems. They are responsible for the synthesis of key molecules such as proteins and hormones, for the energy-producing reactions of food metabolism, and for reactions that rid the body of toxic substances and waste products. By driving and governing chemical reactions, they are basic to the processes of life.

Enzymes control many important environmental reactions,

---

[4]The pace of a reaction is defined by the *reaction rate*, r, the number of molecular reactions per sec. Usually the greater the concentration, C, of reacting molecules, the greater the reaction rate. Sometimes reaction rates are proportional to concentration: $r = kC$. Such reactions are called *first-order reactions,* because their rate (r) increases with the first power of concentration (C). The constant of proportionality, k, is called the *rate constant.* With a *second-order reaction*, $r = kC^2$, the rate increases with the second power of concentration. There are reactions of other orders also.

As reaction occurs, reacting molecules disappear and their concentration (C) diminishes. The equations show that most reactions slow down with decreasing C. Therefore the last molecules may hang on a very long time before they react, and a residue may persist long after most of the reacting substance has been used up. This pattern holds for environmental contaminants such as insecticides and air pollutants. It also holds for radioactive substances, which, though they originate in nuclear rather than chemical reactions, disappear by the same first-order rate law. For such a rate law, a constant *half-life* (the time needed for half the remaining substance to disappear) describes the depletion process. This subject will be explored in Chapter 11.

particularly those in water and soil. Decay, photosynthesis, and insecticide breakdown are but a few examples.

Enzymes are giant protein molecules, as we noted in Chapter 4. Most of them contain a thousand or more chemically bonded atoms. They are very specific catalysts, each having a particular reaction to promote. Their specificity comes in part from an intricate shape that will exactly fit the reactant molecule, as a lock fits a particular key. By this *lock and key* effect, they entrap specific molecules and cause them to react. Since there are hundreds of chemical reactions in living systems, there are hundreds of enzymes to make them go.

The life process is an incredibly complex series of chemical reactions. Enzymes promote and regulate those reactions. Enzyme functioning and enzyme concentrations must obviously be properly maintained in living systems. A number of pollutants upset this delicate enzyme balance. Among them are many heavy metals like mercury (Hg) and most of the insecticides like DDT and parathion. The chemical nature of these disturbances will be discussed in Chapters 9 and 10.

REACTION TYPES

The diverse reactions on this earth fall into various categories. Often a reaction will have several prominent features that qualify it for membership in several distinct categories. Here we define the most important types of reactions. We shall use them later in describing certain reactions in the environment.

1. Oxidation Reactions. In a specific sense, an *oxidation reaction* is one in which elements or compounds combine with oxygen (O)[5]. Atoms involved in an oxidation reaction form new chemical bonds with oxygen, replacing the bonds that link them to other kinds of atoms. For instance, when methane (CH_4) is oxidized (as when natural gas is burned), the carbon (C) atom gives up its bond to hydrogen (H) atoms and forms bonds with oxygen (O), yielding carbon dioxide (CO_2). Similarly, the hydrogen in methane ceases bonding with carbon and bonds with oxygen as water (H_2O). These oxidations are illustrated in Figure 5-3.

Oxidation reactions affecting the environment are abundant and critical. All burning (combustion) processes are based on chemical oxidation. Nearly all important air pollutants (SO_2, SO_3, CO, CO_2, NO, NO_2, particulates, and so on) are direct or indirect products of

[5]In a more general sense, oxidation is a process in which electrons are pulled partially or completely away from atoms. Oxygen pulls electrons away from every element to which it chemically bonds except fluorine, because it is more electronegative than all elements but fluorine. Thus when any element but fluorine combines with oxygen, it is oxidized in the process. By this general definition, oxidation may entail a loss of electrons by any element to another element that is significantly more electronegative. However, oxygen is by far the most common substance causing serious electron losses among elements, and in this book we shall treat "oxidation" and "combination with oxygen" as synonymous.

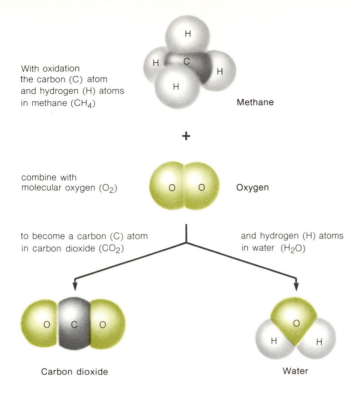

With oxidation
the carbon (C) atom
and hydrogen (H) atoms
in methane (CH_4)

Methane

+

combine with
molecular oxygen (O_2)

Oxygen

to become a carbon (C) atom
in carbon dioxide (CO_2)

and hydrogen (H) atoms
in water (H_2O)

Carbon dioxide

Water

Figure 5-3. Oxidation reactions.

combustive oxidation—either in automobiles or industry. Since oxidation is now the main energy source for civilization, it is currently the principal cause of thermal pollution. Also, oxidation is the main tool used by nature (and man) for destroying toxic organic water pollutants and cleaning up water. Finally, the oxidation of food-stuffs, called *respiration*, provides the bulk of energy needed for the functioning of all living organisms, including man.

2. Reduction Reactions. *Reduction* is the opposite of oxidation. In reduction, oxygen is taken away from an element and most often replaced by hydrogen. The single most important environmental process is photosynthesis, a series of reduction reactions in which oxygen (O) is stripped from the carbon (C) of carbon dioxide (CO_2), leaving the *reduced carbon* or *organic carbon* that is typically found in organic matter, and that is so necessary to living systems.

3. Hydrolysis Reactions. Hydrolysis is a chemical reaction with water. Hydrolysis reactions are responsible for the breakdown of organophosphorus insecticides. This and other environmental ex-amples will be shown later on.

4. Photochemical Reactions. In photochemical reactions, the necessary "starter" energy or activation energy comes from electro-magnetic radiation. Recall from Chapter 2 that this radiation can transfer its energy to atomic systems. The intense electromagnetic radiation of sunlight causes many photochemical reactions. Some of these reactions, if the proper chemical soup is present, lead to photochemical smog. Other photochemical reactions help degrade pesticides in the environment. Photosynthesis is a photochemical reaction promoted by sunlight, as well as being a reduction reaction.

5. Decomposition Reactions. The reaction that is going on when molecules split into smaller molecular units is called a decomposition reaction. Many pesticides and environmental contaminants are eventually rendered harmless by decomposition reactions.

6. Exothermic Reactions. Most chemical reactions are accompanied by the release of heat energy. Such reactions are termed exothermic reactions. In this kind of reaction, the bonds formed are stronger than the bonds broken. Heat is released as atoms are pulled into new bonding arrangements by the strong chemical bonding forces (just as frictional heat is released in the end when matter is pulled to earth by gravitational forces).

Since oxygen forms bonds of extraordinary strength with most elements, most oxidation reactions are strongly exothermic. The combustive oxidation of gasoline, for example, is an exothermic reaction in which some of the energy that is released propels the automobile. The burning of coal in a power plant is similar: the chemical energy is converted to high-temperature heat, which then changes in part to electrical energy.

7. Endothermic Reactions. Endothermic reactions absorb heat energy: they are exactly the opposite of exothermic reactions. Endothermic chemical change is less important than exothermic change, but it is listed here for completeness. Endothermic reactions are uncommon, because heat tends to be produced rather than consumed by natural change on earth. We shall explain this point next.

CHEMICAL REACTIONS AND EQUILIBRIUM

Chemical reactions are one form of natural change. The broad thermodynamic criteria for all forms of natural change were discussed in Chapter 1. Recall that natural processes are pushed along by heat release and entropy (randomness) gain. We find accordingly that most reactions are exothermic, in accord with the normal expectation of heat release. A few reactions are mildly endothermic, and they occur only because they involve an entropy gain that overwhelms the heat criterion. Usually, however, the heat criterion is the dominant influence in chemical reactions.

Reactions that are strongly exothermic tend to proceed vigorously until nearly all the reacting molecules are consumed. Oxidation reactions are generally of this type. When the reactant is converted virtually 100 percent to product, the reaction is said to have *gone to completion.* However, entropy discourages the total and absolute conversion of reactant molecules to product molecules. There is more randomness (chaos) when a few reactant molecules remain to mix haphazardly with product molecules than when all molecules are uniformly of the product variety. Therefore the degree of chemical changeover approaches 100 percent only when the reaction is encouraged by a large heat release, enough to override the inclination of entropy to maintain a well balanced product-reactant mixture. Reactions that do not release considerable heat usually stop well short of 100 percent conversion.

In short, all reactions stop at some point—the point where the thermodynamic driving forces of heat and entropy balance out. This balance point is the point of equilibrium. *Equilibrium*, then, is a condition in which no further change occurs because of perfect thermodynamic balance.

Even very slow changes are ruled out at equilibrium: we must be careful to distinguish between very slow reactions and those whose progress has ceased entirely because of equilibrium. Slow reactions grind inexorably toward equilibrium. They can be speeded by a catalyst. But once the reaction is at equilibrium, change ceases entirely, with or without a catalyst.

Two Examples. Two noteworthy reactions are shown in Figure 5-4: they illustrate how equilibrium points can vary drastically. The

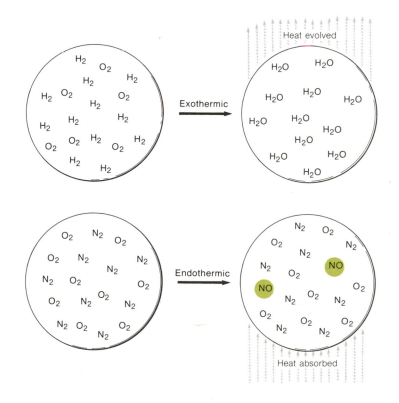

Figure 5-4. Two chemical reactions, the first exothermic and proceeding virtually to completion (100 percent H_2O), and the second endothermic, producing only a minor fraction of product molecules, NO.

oxidation of hydrogen (H$_2$), to which we alluded earlier, is strongly exothermic, producing a hot flame as a consequence of the large heat release. This reaction proceeds virtually to completion, as do most strongly exothermic reactions, producing nearly 100 percent product, water (H$_2$O). This reaction is to be contrasted with the oxidation of nitrogen (N$_2$) to nitric oxide (NO), unusual as an oxidation reaction because it is strongly endothermic. (The triple bond between nitrogen atoms in N$_2$ is so extraordinarily strong that more energy is expended ripping N$_2$ and O$_2$ apart than is released as the new N-to-O bonds are formed.) Endothermic reactions do not usually get far, and this reaction is no exception, a fortunate circumstance, for otherwise the abundant N$_2$ and O$_2$ of the atmosphere would form intolerable quantities of toxic NO. In an automobile engine the reaction fares slightly better, because at high temperatures the thrust of entropy competes better with the inhibiting effect of endothermicity. Quite generally, with increasing temperatures, more chemical species that are products of endothermic reactions appear at measurable concentrations, providing entropy by their mixing. The equilibrium level of NO follows this rule, increasing from a negligible concentration at ordinary temperatures to a concentration level that is over 10 percent of the O$_2$ level in air (3 percent of the air itself) at 4500 °F, the combustion temperature reached in an internal-combustion engine. These increased levels explain why NO is a prominent pollutant from combustion sources, either automotive or power plants. However, there is much more to the NO story, as will be explained in Chapter 7.

In summary, for some reactions (particularly those that are strongly exothermic), equilibrium is reached very close to completion, and for others it is reached at some specific point far short of completion. Furthermore, the degree to which a reaction is exothermic depends on the strength of chemical bonds, which depends in turn on bonding structure, resonance, and other features discussed in Chapter 3. Although we cannot do a complete quantitative study of the relationship of environmental equilibrium to chemical bonding and thermodynamics, we have at least explored here the flavor of the scientific arguments underlying the behavior of our chemical environment.

Physical Processes at Equilibrium. Physical as well as chemical processes have equilibrium end points. Vaporization, for instance, occurs until the vapor reaches its equilibrium pressure, the so called vapor pressure. For volatile substances the equilibrium vapor pressure is high, and for nonvolatile substances it is low.

Dynamic Equilibrium. At the molecular level, equilibrium is dynamic, not static. When liquid water comes to equilibrium with its vapor, the number of molecules vaporizing every second exactly equals the number returning to the liquid per second. No net change occurs because the opposing processes match perfectly. The chemical reaction at equilibrium has just as many reactant-to-product reactions per second as it does product-to-reactant changes. The dynamic balance of opposing processes is characteristic of all equilibrium in molecular systems.

The Steady State. Related to the equilibrium concept is the notion of steady state. Like equilibrium, the term *steady state* designates a condition in which there is no net change with time. However, in contrast to the equilibrium situation, molecules or other entities at a steady-state level disappear along a pathway that is different from the one by which they appeared. The amount of water in an overflowing bucket is in steady state, with water pouring in from above at the same rate that it is spilling over to below.

The classic example of the steady-state condition is the one describing future populations and resources. Proponents of zero population growth point out that on a finite earth, birthrate and deathrate must eventually match, giving a steady-state population. By the same reasoning, it can be argued that the limited resources of the earth will all have to reach a steady state of utilization, with the production of goods exactly matching the rate they are removed for recycle. This system is the *steady-state earth*, a major goal of most environmentalists.

CHEMICAL EQUATIONS

Chemical reactions are represented by *chemical equations*, which show the atoms and molecules which react and the products which emerge. An arrow is used to point to the products of the reaction, showing which way the process is going. For example, the burning of charcoal, which is principally carbon (C), in an excess of air can be represented by

$$C + O_2 \rightarrow CO_2 \tag{1}$$

This statement means that one atom of carbon can combine with one molecule of oxygen (O_2) to form one molecule of carbon dioxide (CO_2). As charcoal is burned, this elementary reaction happens millions of times over. The ratio of C atoms to O_2 molecules stays constant, fixed always by this elementary atomic and molecular reaction. For now, we shall focus on the underlying reaction, but because the ratio of atoms and molecules in it stays constant as the reaction is repeated over and over, we can, as we shall show later, extend our reasoning to vast quantities of environmental materials.

Note that with chemical equations we are using chemical formulas, such as O_2, in a slightly different sense from before. Previously, "O_2" stood for molecular oxygen present in any amount. In a chemical equation, the symbol O_2 is a quantitative expression: it stands for exactly one molecule of diatomic oxygen. Two molecules are denoted by $2O_2$, three by $3O_2$, and so on. Do not confuse coefficients (like the 3 in $3O_2$) with subscripts: the distinction is shown below.

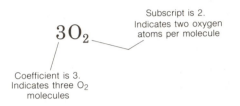

It is an essential principle of all chemical reactions that atoms are neither created nor destroyed by the reaction process. They merely change positions in molecules. Therefore each atom that enters a chemical reaction must be accounted for at the end. In the above reaction, equation (1), each reacting carbon (C) atom appears subsequently as a product C atom in the carbon dioxide (CO_2) molecule. Likewise, the two oxygen (O) atoms of the single oxygen (O_2) molecule emerge as two O atoms in CO_2. A properly written chemical equation, in which all atoms appear in equal numbers on both sides, is called a *balanced equation*. The reaction in our example is balanced.

Oxidation of N_2 to NO. Consider now the formation of the common pollutant nitric (or nitrogen) oxide (NO) through a chemical reaction. When atmospheric nitrogen (N_2) and oxygen (O_2) are subjected to high temperature—as we have seen occurs in an internal-combustion engine or the combustion chamber of a coal-fired power plant—some N_2 is oxidized by O_2 to NO. On first inclination we would write this reaction in the simplest way, without coefficients:

$$N_2 + O_2 \longrightarrow NO \quad \text{(unbalanced)} \qquad (2)$$

| 2 N atoms | 1 N atom |
| 2 O atoms | 1 O atom |

We write *unbalanced* by this equation to indicate that, in this simple form, some atoms are not accounted for. This fact is confirmed by the inventory of the numbers of atoms, written directly below the equation. The inventory shows, for example, that two nitrogen (N) atoms from N_2 enter the reaction, but only one emerges in the NO molecule. Similarly, two oxygen (O) atoms are consumed, but only one appears in the product. Nature does not lose or misplace atoms in this way, so the equation, as written, is incorrect. Chemical equations like this, made incorrect by a faulty inventory of atoms, are always called "unbalanced." In order to keep track of substances entering and leaving the environment via chemical reactions, it is necessary that we balance such equations. We can do so as follows.

Inspection of the atom inventory of reaction (2) shows that only half enough atoms of N and O emerge. One more of each is required.

This requirement will be met if we write 2NO in place of NO on the right of reaction (2).

$$N_2 + O_2 \longrightarrow 2NO \quad \text{(balanced)} \qquad (3)$$

2 N atoms   2 N atoms
2 O atoms   2 O atoms

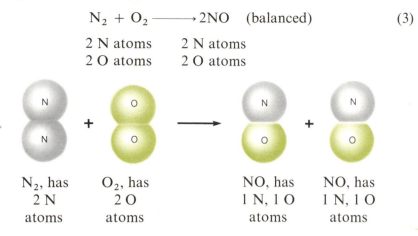

N_2, has O_2, has   NO, has   NO, has
   2 N   2 O   1 N, 1 O   1 N, 1 O
  atoms   atoms   atoms   atoms

The atom inventory—always the ultimate criterion of a balanced reaction—shows that N and O atoms exist in equal numbers (namely, two) on left and right. This fact shows us that the equation is now balanced—the only type of equation that can ever occur in nature. The correctness of the equation is illustrated further by the diagram below the balanced equation. This diagram shows in a pictorial way the entry of two N atoms and two O atoms into a reaction producing two NO molecules that contain the same two N and two O atoms. (See also Figure 5-2.)

Oxidation of H_2 to H_2O. Another example is provided by the burning (rapid oxidation) of diatomic hydrogen (H_2). Hydrogen (H_2) has been suggested as a fuel which, if produced by nuclear power, could largely replace fossil fuels when the latter run out. Energy would be produced as H_2 combined with atmospheric oxygen (O_2) to form water (H_2O), a reaction written offhand as

$$H_2 + O_2 \longrightarrow H_2O \quad \text{(unbalanced)} \qquad (4)$$

2 H atoms   2 H atoms
2 O atoms   1 O atom

Again, the simplest way of writing the chemical equation, using only one molecule of each type, leads to an unbalanced equation. Hydrogen (H) atoms are in proper balance with two H atoms on each side, but the oxygen (O) atoms are unbalanced because two O atoms enter while only one emerges in the single H_2O molecule.

We can correct that O imbalance by assuming that two H_2O molecules are formed. Again, we indicate this assumption by the coefficient 2 in front of H_2O:

$$H_2 + O_2 \longrightarrow 2H_2O \quad \text{(unbalanced)} \qquad (5)$$

2 H atoms   4 H atoms
2 O atoms   2 O atoms

Now, however, the H atoms do not balance, with two on the left and four (two from each of the two H_2O molecules) on the right. We can correct this imbalance if we enter two hydrogen molecules, $2H_2$, on the left instead of one, which gives a total now of four hydrogen atoms on both left and right:

$$2H_2 + O_2 \longrightarrow 2H_2O \quad \text{(balanced)} \qquad (6)$$

4 H atoms   4 H atoms
2 O atoms   2 O atoms

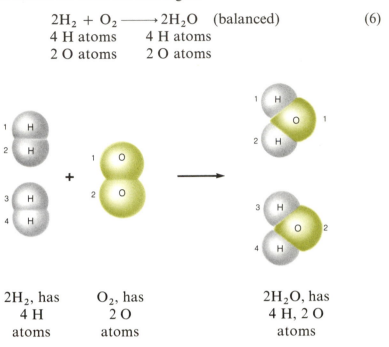

$2H_2$, has	O_2, has	$2H_2O$, has
4 H	2 O	4 H, 2 O
atoms	atoms	atoms

The equation in this form is properly balanced because H and O atoms both balance. The equations tell us correctly that two H_2 molecules and one O_2 molecule unite to form two H_2O molecules. This scheme is illustrated beneath the chemical equation, where each atom is numbered and followed through the reaction.

Combustion of Octane. As a final example, we shall balance the equation for the combustion (like burning, a rapid oxidation) of octane (C_8H_{18}), a common constituent of automobile fuel. The complete combustion of a hydrocarbon always yields carbon dioxide (CO_2) and water (H_2O).[6] In simplest terms,

$$C_8H_{18} + O_2 \longrightarrow CO_2 + H_2O \quad \text{(unbalanced)} \qquad (7)$$

8 C atoms   1 C atom
18 H atoms   2 H atoms
2 O atoms   3 O atoms

With this equation, we must follow three kinds of atoms, C, H, and O, balancing each. We focus first on the large molecule of octane (C_8H_{18}). The eight carbon atoms of a single C_8H_{18} molecule all end

---

[6]Unfortunately, complete combustion does not always occur in a car engine, and many other chemicals, some environmentally offensive, such as carbon monoxide (CO) and various aldehydes (R—CHO), are produced.

up in CO_2 (since carbon appears nowhere else on the right): hence there must be at least eight CO_2's, for which we shall write $8CO_2$. Similarly, the 18 H atoms all become a part of H_2O, requiring 9 water molecules, $9H_2O$. When the equation is written to reflect these findings, we have

$$C_8H_{18} + O_2 \longrightarrow 8CO_2 + 9H_2O \text{ (unbalanced)} \qquad (8)$$

8 C atoms	8 C atoms
18 H atoms	18 H atoms
2 O atoms	25 O atoms

Indeed the C and H atoms, both originating in the octane molecule, balance, but we have neglected to account for O. We need at least 25 O atoms to produce the 9 H_2O's (a step that takes 9 O) and the 8 CO_2's (taking the remaining 16 O's). It takes $12\frac{1}{2}$ O_2's to produce 25 O atoms. Hence we can balance oxygen by writing the equation as

$$C_8H_{18} + 12\frac{1}{2}O_2 \longrightarrow 8CO_2 + 9H_2O \text{ (balanced)} \qquad (9)$$

8 C atoms	8 C atoms
18 H atoms	18 H atoms
25 O atoms	25 O atoms

This equation is entirely balanced, but one flaw remains. In writing $12\frac{1}{2}O_2$, we imply that one half of an O_2 molecule can exist. Molecules, of course, exist only in whole units. Therefore it is preferable to clear all fractions like $\frac{1}{2}$ and $\frac{1}{3}$ in balanced equations to avoid this implication. (Chemists occasionally retain fractions with the clear understanding that they represent relative amounts of substances, not the actual molecule-by-molecule reaction.) Wherever the fraction $\frac{1}{2}$ appears, it can be cleared by multiplication by 2. We must be careful to multiply *all* coefficients by two so that all atoms throughout the equation increase in equal proportion, none getting out of balance. With this operation the chemical equation of (9) becomes

$$2C_8H_{18} + 25O_2 \rightarrow 16CO_2 + 18H_2O \text{ (balanced)} \qquad (10)$$

16 C atoms	16 C atoms
36 H atoms	36 H atoms
50 O atoms	50 O atoms

The equation is now in its final, preferred form, balanced with respect to all three kinds of atoms, and having no fractions for coefficients.

ATOMIC AND MOLECULAR WEIGHTS

The mass of pollutants produced or consumed by chemical reaction depends on the masses of the pollutant atoms and molecules. We shall discuss these masses now.

Different kinds of atoms have different masses. For instance, an atom of sulfur (S) weighs roughly twice as much as an atom of oxygen (O). The weight of atoms is given by the *atomic weight*. The atomic weight is a relative weight in which the common carbon (C) atom is assigned the standard mass of exactly 12.0. Oxygen atoms are one third again as heavy as C atoms, so the atomic weight of O is $1.33 \times 12.0 = 16.0$ (rounded off to three figures). Hydrogen (H) atoms are about one twelfth as heavy, which leads to the atomic weight $\frac{1}{12} \times 12 = 1.0$ (approximate, so rounded off to two figures). Atoms of all the other elements have atomic weights listed on the same C scale. The atomic weights of some important elements, many of which are involved in environmental problems, are listed in Table 5-1. A more complete tabulation appears inside the back cover.

Table 5-1. ATOMIC WEIGHTS OF SOME PROMINENT ELEMENTS

Element and Atomic Number	Atomic Weight	Element and Atomic Number	Atomic Weight
Hydrogen (H), 1	1.01	Argon (Ar), 18	39.9
Helium (He), 2	4.00	Potassium (K), 19	39.1
Carbon (C), 6	12.0	Calcium (Ca), 20	40.1
Nitrogen (N), 7	14.0	Iron (Fe), 26	55.8
Oxygen (O), 8	16.0	Copper (Cu), 29	63.5
Fluorine (F), 9	19.0	Arsenic (As), 33	74.9
Sodium (Na), 11	23.0	Strontium (Sr), 38	87.6
Magnesium (Mg), 12	24.3	Cadmium (Cd), 48	112.4
Aluminum (Al), 13	27.0	Iodine (I), 53	126.9
Silicon (Si), 14	28.1	Mercury (Hg), 80	200.6
Phosphorous (P), 15	31.0	Lead (Pb), 82	207.2
Sulfur (S), 16	32.1	Uranium (U), 92	238.0
Chlorine (Cl), 17	35.5		

For mass calculations (next section), the mass of atoms must be expressed complete with the proper units. The units which go with atomic weights are called *atomic mass units*, abbreviated *amu*. Thus a single O atom has a mass of 16.0 amu, and so on.

Basis of the Atomic Weight Scale. It is natural to ask why atomic weights are listed on an arbitrary scale with C the standard at exactly 12.0. Why not report the real weights of the atoms in pounds or grams? Remember that atoms are extremely small and that their weight is correspondingly tiny. For example, a single H atom has the mass 0.00000000000000000000000169 g. This number is so small that it is extremely awkward to work with. Even in the shorthand exponential notation, this weight is in the altogether too cumbersome form 1.69×10^{-24} g, which means that the decimal place in 1.69 is moved an impressive 24 digits to the left. It is simpler to use the atomic weight scale outlined above.

Atomic weight is closely related to another mass scale, the mass number. The *mass number*, denoted by symbol A, is the sum of protons and neutrons in the nucleus. Recall that these two particles account for nearly all the atom's mass, since electrons are light by comparison. It would seem simplest to use the mass number for our universal scale of atomic mass, since to get it one must merely count and add protons and neutrons. Unfortunately, this approach will not work. The principal reason is that atoms of an element may have a variable number of neutrons: thus several mass numbers become associated with the element.

Consider sulfur (S), an element whose atomic number of 16 means that all S atoms have 16 protons. Most (95 percent) of S atoms have 16 neutrons (mass number A = 16 + 16 = 32). About 4 percent have 18 neutrons (A = 16 + 18 = 34). Somewhat under 1 percent have 17 neutrons (A = 16 + 17 = 33), and a tinier fraction still have 20 neutrons (A = 16 + 20 = 36). These atoms are the different isotopes of sulfur, *isotopes* being atoms with the same atomic number (Z) but different mass numbers (A). Sulfur has four natural isotopes and thus four different mass numbers. Certainly mass number is not suitable as a mass scale because there are too many values of A to choose from; only one value is wanted. At the very least, one would have to compromise this scale by taking a weighted average of the mass numbers for an element. This approach would work reasonably well, but it still would not provide a totally accurate scale for atomic masses.[7] By contrast the atomic weight scale is accurate, because it is based on the mass of a single isotope of carbon (C) that is absolutely constant throughout the natural world. This isotope, the most common one of carbon, is one with six protons and six neutrons (A = 12). This standard is chosen because carbon is present in so many important molecules (all organic molecules) and the isotope can be measured with great accuracy by use of a *mass spectrometer*. When the standard atomic weight of 12.0 is assigned to this isotope, the two scales, atomic weight and average mass number, are made to coincide approximately.

Molecular Weights. The weights of molecules are also expressed on the atomic weight scale. To get the *molecular weight* (MW), we simply add up the weights of the constituent atoms in the molecule. For instance, carbon monoxide (CO) has a molecular weight of 28.0, calculated as follows:

$$
\begin{array}{lll}
\text{weight of 1 C atom in amu } (1 \times 12.0) & = & 12.0 \\
\text{weight of 1 O atom in amu } (1 \times 16.0) & = & \underline{16.0} \\
\text{MW} & = & 28.0
\end{array}
$$

---

[7]The failure of the mass number, even when averaged, to provide an accurate atomic mass scale is due to several factors. First, the weight of a nucleus does not equal exactly the mass of its protons and neutrons. The nucleus has a tremendous binding energy—the energy lost as the nucleus forms. This lost energy has a small amount of lost mass associated with it, according to the Einstein equation relating mass and energy (more about this relation when we discuss radioactivity). This *mass defect* amounts to almost 1 percent, although it varies from atom to atom. In addition, protons and neutrons do not weigh exactly the same. The neutron is heavier by about 0.1 percent. Thus for accurate work, we cannot simply add up the number of protons and neutrons to get atomic mass. We must use the atomic weight scale, which is based on the actual and not hypothetical weights of atoms.

Sulfur dioxide (SO_2) has a MW of 64.1:

weight of 1 S atom in amu (1×32.1)　　= 32.1
weight of 2 O atoms in amu (2×16.0)　= 32.0
　　　　　　　　　　　　　MW　= 64.1

More complicated molecules, such as DDT ($C_{14}H_9Cl_5$), have a higher molecular weight, but it is calculated in exactly the same fashion:

weight of 14 C atoms in amu (14×12.0) = 168
weight of 9 H atoms in amu (9×1.01)　=　9.09
weight of 5 Cl atoms in amu (5×35.5) = 178
　　　　　　　　　　　　　MW　= 355.09

Actually, we must round this number off so that no figures remain beyond the decimal, as no more significant figures are justified. Thus MW = 355.

CHEMICAL EQUATIONS—A QUANTITATIVE LINK BETWEEN POLLUTION AND RESOURCES

Almost every contaminant in the environment can be traced ultimately to a chemical reaction. The amount of the contaminant is linked inextricably by a chemical equation to some resource consumed or some other substance chemically altered. The chemical equation (with associated atomic weights) serves as a quantitative bridge connecting pollution, resources, and the level of man's activities. Here we use environmental examples to learn how to work with the unifying concepts provided by chemical equations.

Oxidation of Carbon. The oxidation of carbon (charcoal) was represented earlier by the balanced equation

$$C + O_2 \rightarrow CO_2$$

Since charcoal is nearly pure carbon (C), this equation serves admirably for the burning of charcoal. But it does much more. Coal is mostly carbon (commonly 50 to 95 percent), so this equation is a simple approximation to the combustion of coal. Many organic materials, including organic wastes, are largely carbon, so the same equation expresses the most central feature of their oxidation.[8]

---

[8]Coal and many organic wastes are complex mixtures in which few molecules are precisely known. Carbon dioxide CO_2 is usually the main product of oxidation, but much H_2O and lesser amounts of SO_2 and other gases are produced. Our simple oxidation equation, $C + O_2 \rightarrow CO_2$, has the following limitations: (1) It ignores emitted H_2O. However, the product H_2O has minimal environment impact. (2) It ignores emitted SO_2. This pollutant is usually accounted for separately. (3) It only approximates oxygen (O_2) consumption. If the organic material contains substantial oxygen, less than the estimated O_2 will be consumed; if it contains much hydrogen, as with the hydrocarbons, more than the estimated O_2 will be consumed. However, the error rarely exceeds 50 percent.

Altogether it provides a simple description of the oxidation of many organic materials throughout the environment. As such, it is one of the most important chemical equations underlying environmental changes.

In the case of the combustion of fossil carbon (for instance, the carbon of coal), $C + O_2 \rightarrow CO_2$ represents the loss of one C atom (along with one diatomic oxygen (O_2) molecule), and the gain of one carbon dioxide (CO_2) molecule. The ratio of C atoms lost to CO_2 molecules gained is always one to one. However, the mass of CO_2 produced is greater than the mass of C consumed, since the mass of the CO_2 molecule (molecular weight = 44.0) is greater than that of the consumed C atom (atomic weight = 12.0). The mass ratio of CO_2 to C is (44.0/12.0) = 11/3. Thus if 3 g of fossil carbon are consumed by this reaction, the CO_2 produced will equal

$$3 \text{ g C} \times \frac{11 \text{ amu } CO_2}{3 \text{ amu C}} = 3 \text{ g C} \times \frac{11 \text{ parts } CO_2}{3 \text{ parts C}} = 11 \text{ g } CO_2$$

With the burning of approximately six billion tons of fossil carbon in this world each year, the weight of associated CO_2 is found to be

$$6 \times 10^9 \text{ tons C} \times \frac{11 \text{ parts } CO_2}{3 \text{ parts C}} = 22 \times 10^9 \text{ tons } CO_2$$

This result shows that roughly 22 billion tons of new CO_2 enter the atmosphere annually. This amount is vast enough so that continued (and expanded) release threatens world climate through the greenhouse effect. The details of the greenhouse effect will be reserved for Chapter 6. The point here is simply that the production of a major atmospheric contaminant (CO_2) is directly related to the consumption of a resource (fossil carbon) by a chemical equation. The resource consumption is related to human population and living habits.

The simple equation $C + O_2 \rightarrow CO_2$ provides other critical environmental links. By similar reasoning, the ratio of the weight of O_2 consumed to C consumed is found to be 8/3. We shall use this critical ratio later to help us decide when organic wastes endanger the oxygen supply of water and to assess the broader question of whether our atmosphere will someday run out of oxygen because of relentless fuel combustion.

We can most easily calculate the ratios of weights in a chemical equation by writing out the weight of each substance on the atomic weight scale (here to three significant figures) below the balanced reaction:

Substances:	carbon	diatomic oxygen	carbon dioxide
Atoms reacting:	C $+$	O_2 $\rightarrow$	CO_2
Amu:	12.0	32.0	44.0

The numbers 12.0, 32.0, and 44.0 are the respective weights in atomic mass units (amu) of C, O_2, and CO_2. This accounting is an inventory of mass, similar to our earlier inventory of atoms. Just as atoms must appear in equal numbers on each side of the equation, so must the mass be equal from left to right. The logic is identical. Mass, like the atom, is neither created nor destroyed by a chemical reaction, only changed in form. Actually, for any equation that is balanced, the mass will automatically balance since the mass on each side is the sum of the atomic weights of equal numbers of atoms. Thus the above example shows a mass of $12.0 + 32.0 = 44.0$ on the left, exactly the same as 44.0 on the right. This sum is the mass of the one C atom plus that of the two O atoms appearing on each side.

From this mass inventory, all possible weight ratios can be calculated directly. For example, the weight ratio of product CO_2 to reactant C is $(44.0/12.0) = 11/3$, as deduced above. The weight ratio of consumed O_2 to consumed C is $(32.0/12.0) = 8/3$, and so on.

Sulfur Dioxide from Copper Smelters. Considerable sulfur dioxide (SO_2) pollution is often found in the vicinity of copper (Cu) smelters [lead (Pb) and zinc (Zn) smelters are also offenders]. The sulfur-rich plumes of a Western smelter are shown below. The sulfur (S) in SO_2 has its origin in certain copper ores, commonly chalcocite (Cu_2S) and chalcopyrite ($CuFeS_2$). In the simplest case, chalcocite (Cu_2S), the roasting and smelting process, in which the ore reacts with oxygen (O_2) (here we have another important oxidation reaction), can be represented by the following simplified, balanced chemical equation:

Substance:	chalcocite		diatomic oxygen		copper metal		sulfur dioxide
Atoms reacting:	Cu_2S	$+$	O_2	$\rightarrow$	$2Cu$	$+$	SO_2
Amu:	159		32.0		127		64.1
	total, 191 amu				total, 191 amu		

Air pollution of a Western copper smelter. (*Photo by Alexis Kelner.*)

The mass inventory is calculated by the addition of atomic weights, as we described before.

From the foregoing equation we can determine the amount of SO_2 produced in the smelting of Cu_2S ore. For example, the weight ratio of SO_2 and Cu is $(64.1/127) = 0.505$, barely over $1/2$. Thus for each ton of copper metal produced, roughly half a ton of SO_2 is released. In a like manner we can calculate that each ton of Cu_2S ore smelted yields about $2/5$ ton of SO_2. Again we see how a chemical equation provides the connections among resource (Cu_2S) consumption, industrial (Cu) production, and pollution (SO_2).

Chalcopyrite ($CuFeS_2$) rather than chalcocite (Cu_2S) is the principal mineral processed at most smelters. The roasting and smelting of chalcopyrite ($CuFeS_2$) can be described by the following balanced equation and its mass inventory:

Substances:	chalco-pyrite	diatomic oxygen	copper metal	iron (II) oxide	sulfur dioxide
Atoms reacting:	$2CuFeS_2$ +	$5O_2$ →	$2Cu$	$2FeO$	$4SO_2$
Amu:	367	160	127	144	256

In this case we see that the weight ratio of SO_2 to Cu metal is $(256/127)$, just slightly over 2. Clearly the processing of chalcopyrite ($CuFeS_2$) ore has far more serious implications for SO_2 pollution than does the processing of chalcocite (Cu_2S) ore, for which, as we just saw, the SO_2-to-Cu ratio is only $1/2$. Per ton of Cu produced, four times as much SO_2 (two tons instead of one half ton) is produced in the chalcopyrite case.[9]

Much of the SO_2 from copper smelters is captured before release to the environment. However, the production levels of copper smelters are so high that release of even a small fraction of the produced SO_2 is significant and has created considerable controversy in the vicinity of copper smelters in the western United States.

The Kennecott Utah smelter, located in the Salt Lake valley of Utah, is in many respects typical of copper smelters and will be used as an example to show the nature and magnitude of the problem. Two unfortunate accidents of geography have brought unusual public pressure to bear on this smelter: first, the Salt Lake valley suffers frequent inversions which temporarily trap any emitted pollutant near ground level, increasing its concentration; and, second, the valley is so situated and endowed that it has attracted a population of nearly one half million, in sharp contrast to the isolated location of many western smelters.

The Utah smelter produces around 900 tons of copper metal daily, about 750 of those tons coming from the smelting of ores in

---

[9]We might have guessed this answer in the first place if we had noted that $CuFeS_2$ has a four times greater ratio of sulfur atoms to copper atoms than does Cu_2S; two to one compared with one to two.

which chalcopyrite ($CuFeS_2$) is the dominant mineral. Therefore the attendant daily SO_2 production is roughly

$$750 \text{ tons Cu} \times \frac{2 \text{ parts } SO_2}{1 \text{ part Cu}} = 1500 \text{ tons } SO_2$$

an impressive quantity of SO_2. However, most of it is captured and converted to sulfuric acid (H_2SO_4).

Sulfur Dioxide Cleanup. Some copper smelters capture a considerable fraction of the sulfur dioxide (SO_2) they produce before it reaches the environment. Most commonly, the SO_2 is chemically converted to sulfuric acid (H_2SO_4), but chemical transformation to elemental sulfur (S) is also possible.

The chemical change from SO_2 to H_2SO_4 can be represented by the balanced equation and mass inventory below.

Substances:	sulfur dioxide		oxygen		water		sulfuric acid
Atoms reacting:	$2SO_2$	$+$	O_2	$+$	$2H_2O$	$\rightarrow$	$2H_2SO_4$
Amu:	128		32.0		36.0		196

The ratio of the mass of H_2SO_4 produced to SO_2 consumed is seen to be $(196/128) = 1.53$, or rounded to two significant figures, 1.5, which provides a convenient working number. Thus roughly 1.5 tons of pure H_2SO_4 can be manufactured and sold for each ton of SO_2 recovered. (Industrial H_2SO_4 is actually diluted with a small amount of water, usually 2 to 4 percent.)

As an example, the Kennecott Utah smelter captures nearly two thirds of the 1500 tons of SO_2 produced daily, a record for the copper industry. Hence about 1000 tons of SO_2 are retained and utilized each day out of 1500 tons total. (The remainder goes into the atmosphere.) From this figure, the daily sulfuric acid production can be found as approximately

$$1000 \text{ tons } SO_2 \times \frac{1.5 \text{ parts } H_2SO_4}{1 \text{ part } SO_2} = 1500 \text{ tons } H_2SO_4$$

Although the sulfuric acid market is currently depressed, more so in some areas than others, a price as low as a few dollars (about $17 in 1972) a ton brings substantial daily revenues and helps offset the cost of pollution control. Indeed, the earliest sulfuric acid plants in the western states were feasible on economic grounds alone, partial pollution control being a beneficial side effect.

Weight Ratios from Chemical Formulas. Occasionally, environmental quantities can be obtained without a balanced chemical equation, especially if we wish to deal only with certain elements within a compound. For instance, we know that chalcopyrite ($CuFeS_2$) is valuable because it yields the element copper (Cu). The amount of copper can be found by a mass analysis of $CuFeS_2$, element by element. This analysis comes out as follows: one Cu atom is 63.5 amu; one iron (Fe) atom is 55.8 amu; and two sulfur (S) atoms are 64.2 amu. Their total is 183.5 amu (round to 184 amu). These numbers are represented symbolically below.

Substance:	chalcopyrite
Chemical formula:	$CuFeS_2$
Mass of each element, in amu:	Cu: 63.5
	Fe: 55.8
	S: 44.2
Total mass, in amu:	$CuFeS_2$: 184

The ratio of the mass of Cu to chalcopyrite is seen immediately to be 63.5/184 = 0.345. With a certain amount of chalcopyrite mineral, say exactly 2000 tons, we can use this ratio to calculate the extractable copper:

$$2000 \text{ tons } CuFeS_2 \times \frac{0.354 \text{ parts Cu}}{1 \text{ part } CuFeS_2} = 708 \text{ tons Cu}$$

We can also calculate the ratio of the mass of extractable S to that of Cu: 64.2/63.5 = a value close to one. If the sulfur is known to enter the atmosphere as SO_2, then the tonnage of SO_2 can be obtained by a similar mass analysis of SO_2:

Chemical formula:	SO_2
Mass of each element, in amu:	S: 32.1
	O: 32.0
Total mass, in amu:	SO_2: 64.1

This breakdown shows, again without benefit of a chemical equation, that SO_2 and S exist in the mass ratio 64.1/32.1 = 2.00. Hence one ton of sulfur can provide two tons of sulfur dioxide:

$$1 \text{ ton S } \times \frac{2 \text{ parts } SO_2}{1 \text{ part S}} = 2 \text{ tons } SO_2$$

Since S and Cu are produced in roughly equal amounts from $CuFeS_2$ (above), the SO_2 produced has twice the mass of either the sulfur or the copper. The two-to-one mass ratio of SO_2 to Cu was reached earlier by the use of a balanced chemical equation.

While we have shown how to obtain important results without the aid of a chemical equation, we have needed to know the chemical fate of the atom (or atoms) involved (copper finally extracted as the pure element Cu, etc.). Chemical destinies are best expressed by chemical equations. We can, as shown, bypass the underlying chemical equation occasionally if we know by other means the chemical destination of the atoms. However, if the proper chemical equation for the process can be found, it is the most certain guideline for chemical mass calculations.

THE MOLE

So far we have calculated quantitative amounts by determining the weight ratio of atoms in the underlying reaction and then applying that same ratio to the reaction of large quantities of substances (involving, in total, trillions upon trillions of atomic reactions). However, chemists frequently find it convenient to use another basic unit of amount besides single atoms and molecules. This unit is called the mole. The *mole* is an amount of substance having a mass in grams equal to the atomic or molecular weight of that substance. To illustrate: the molecular weight of diatomic oxygen (O_2) is 32.0; therefore one mole of O_2 is 32.0 grams. The molecular weight of sulfur dioxide (SO_2) is 64.1; consequently one mole of SO_2 contains 64.1 grams. The atomic weight of phosphorus (P) is 31.0; a mole (for atoms, a mole is usually called a *gram-atom*) is 31.0 grams.

Since the mass of a mole increases in proportion to the mass of the molecule involved, a mole contains a fixed number of molecules. A specific example may help clarify this statement. An O_2 molecule has a mass of 32 amu (to two significant figures). The mass of SO_2 is 64 amu. Therefore each SO_2 molecule is twice as heavy as an O_2 molecule. Similarly, 10 molecules of SO_2 must be twice as heavy as 10 molecules of O_2. Generalizing, any fixed number of SO_2 molecules would be twice as heavy as the same number of O_2 molecules. A mole of SO_2 (64 g) is twice as heavy as a mole of O_2 (32 g), and it must therefore contain the same number of molecules. This universal number is known as *Avogadro's number*, and it is a staggering 6.02×10^{23} in magnitude. We might have expected it to be large, since atoms are so tiny that it takes a great number of them to weigh a few grams.

The proportionality between numbers of moles and numbers of atoms makes it possible to do chemical-environmental calculations on a mole basis. For instance, the chemical equation

$$C + O_2 \rightarrow CO_2$$

can represent the reaction of one C atom and one O_2 molecule to produce one CO_2 molecule, or it can be thought of as the reaction of one mole each of C and O_2 to produce one mole of CO_2. Either way, the fixed ratio of C, O_2, and CO_2 is one to one to one. If, for instance, 12 g of C are known to react, we recognize this amount as a mole of carbon, and we know then that 1 mole of O_2 (32 g) will react and that 1 mole of CO_2 (44 g) will be produced. Likewise, 100 moles of C (1200 g) would involve 100 moles of O_2 (3200 g) and 100 moles of

CO_2 (4400 g). The fixed ratio makes calculations possible with any number of moles.[10]

The greatest advantage of the mole unit appears when we deal with concentrations of substances, particularly in water solutions. Most common concentration units are based on the mole. For instance, a *molar solution* of some acid contains one mole of that acid in a liter of solution. The important universal index of acid and base strength, the pH unit, is based on the molar concentration concept. We shall explain these matters more fully when we deal with water, water chemistry, and water pollution in Chapter 8.

AND NOW, THE TOTAL ENVIRONMENT

The five chapters ending here have served to introduce the principles of chemistry and show the fundamental bearing they have on the environment of our small planet. With some basic chemical tools now in our possession, we are able to explore major aspects of the environment in considerable depth. We shall trace many modern-day problems to the atoms and molecules that surround us—to their chemical properties, abundance, role in nature, and uses by man.

The environment is an integrated whole, and it cannot be thought of as existing in distinct compartments. However, we always organize and learn intricate subjects in compartments. The next six chapters are six major compartments of environmental concern. In each chapter we study a class of critical processes that determine, in part, the nature of the total environment. Although we shall occasionally remind ourselves that all compartments are like strands in a large web, the steady circulation of our attention from earth to air to water to life should constantly remind us that our environment is laced with interconnections.

The six chapters following may be read in any order, and if time is short, some may be omitted. There are occasional cross references (to point out the connections between environmental compartments), but they are never essential to the flow of understanding. This plan gives the student who is short of time an opportunity to select for study those topics of greatest interest and concern.

---

[10]Moles and grams can be interconverted by the factor unit method. For example, we can calculate the grams contained in 10 moles of O_2(MW = 32.0) by the approach

$$10 \text{ moles } O_2 \times \frac{32.0 \text{ g}}{\text{mole}} = 320 \text{ g } O_2$$

Likewise, a fixed mass of O_2, say 160 g, can be converted to moles by use of the reciprocal of the conversion factor 32.0 g/1 mole—namely, 1 mole/32.0 g—giving

$$160 \text{ g } O_2 \times \frac{1 \text{ mole}}{32.0 \text{ g}} = 5 \text{ moles } O_2$$

Exercises

1. Hexane (CH_3—CH_2—CH_2—CH_2—CH_2—CH_3), a hydrocarbon found in gasoline, has a vapor pressure of 120 mm of mercury at 68°F. Predict whether the vapor pressure would rise or fall under the following circumstances. Explain.

   a. If the temperature rose to 90°F.
   b. If a hydrogen atom (—H) were removed and replaced by an ethyl radical (—CH_2—CH_3).
   c. If a methyl radical (—CH_3) were removed and replaced by a hydrogen atom (—H).
   d. If a hydrogen atom (—H) were removed and replaced by a hydroxyl group (—OH).

2. If the speed of an environmental reaction doubles for every 10° rise in centigrade temperature (°C) or an 18° rise in Fahrenheit temperature (°F), by what factor will the speed increase from (a) a summer night of 65°F to a summer day of 83°F? (b) From a winter day of 18°F to a summer day of 90°F? (c) From Antarctica at −54°F to the tropics at 108°F?

3. Decide whether the following entities (which can remain at a fixed level over at least a moderate time period) are stationary because of equilibrium or steady state.

   a. A river at a constant level, in which flow to the ocean just matches the average inflow of groundwater from rain.
   b. The fixed amount of water vapor above the liquid water in a partly full, closed container.
   c. The average level in the environment of the short-lived insecticide parathion, in which the rate of new spray applications for insect control is matched by the rate of chemical decomposition.
   d. The content of atmospheric sulfur dioxide (SO_2) above a city under fixed meteorological conditions, where emissions equal the rate of dispersal by wind currents.
   e. The nearly constant level in the worldwide atmosphere of carbon monoxide, in which its generation by automobiles (mainly) is matched by its removal by soil microorganisms (by the best present evidence).

4. Balance the following reactions, each having the environmental significance indicated. The role of some of these reactions will be discussed further in subsequent text material. (All the reactions of Problems 4 through 6 are balanced somewhere later in the book.)

   a. $CO + O_2 \rightarrow CO_2$
      Destroys carbon monoxide (CO) both in nature and in manmade pollution control devices.
   b. $SO_2 + O_2 \rightarrow SO_3$
      Sulfur dioxide (SO_2) becomes the more toxic sulfur trioxide (SO_3).
   c. $SO_3 + H_2O \rightarrow H_2SO_4$
      Sulfur trioxide (SO_3) with atmospheric water becomes sulfuric acid (H_2SO_4).
   d. $H_2SO_4 + NH_3 \rightarrow (NH_4)_2 SO_4$
      Sulfuric acid and atmospheric ammonia (NH_3) form ammonium

sulfate $(NH_4)_2SO_4$, which becomes an important particulate contaminant of the atmosphere.

e. $H_2S + SO_2 \rightarrow S + H_2O$

An industrial process (the Claus reaction) that rids plant effluents of sulfur dioxide.

5. Balance the following reactions, each having the environmental significance indicated:

   a. $C + O_2 \rightarrow CO$
   Formation of carbon monoxide (CO).

   b. $CO + SO_2 \rightarrow S + CO_2$
   A reaction used to get rid of sulfur dioxide (SO_2).

   c. $H_2 + SO_2 \rightarrow H_2S + H_2O$
   A reaction used to clean up SO_2 emissions.

   d. $CaCO_3 + SO_2 + O_2 \rightarrow CaSO_4 + CO_2$
   A reaction used to clean up SO_2 emissions.

   e. $NO_2 + H_2O \rightarrow HNO_3 + NO$
   A reaction occurring to atmospheric nitrogen dioxide (NO_2).

6. Balance the following reactions, each having the environmental significance indicated:

   a. $FeO + O_2 \rightarrow Fe_2O_3$
   Consumed oxygen (O_2) in the primitive atmosphere.

   b. $NO + O_3 \rightarrow NO_2 + O_2$
   Nitric oxide (NO) from SST removes ozone (O_3) shield from stratosphere.

   c. $CO + NO \rightarrow N_2 + CO_2$
   A reaction used to clean up nitric oxide (NO) emissions.

   d. $CH_4 + NO \rightarrow N_2 + CO_2 + H_2O$
   A reaction used to clean up NO emissions.

   e. $CO + SO_2 \rightarrow CO_2 + COS$
   A side reaction in stack gas cleanup, producing toxic carbonyl sulfide (COS).

7. Balance the following organic reactions having the general environmental significance indicated.

   a. $C_6H_{12} + O_2 \rightarrow CO_2 + H_2O$
   hexene

   The oxidation of an organic molecule, important in combustion, biological degradation, etc.

   b. $C_8H_{18} + O_2 \rightarrow CO_2 + H_2O$
   octane

   The oxidation of an organic molecule.

   c. $CH_4 + O_2 \rightarrow CH_3OH$
   methane   methyl alcohol
   The partial oxidation of an organic molecule.

   d. $C_2H_6 + O_2 \rightarrow CH_3COOH + H_2O$
   ethane   acetic acid
   The partial oxidation of an organic molecule.

e. $CO_2 + H_2O \rightarrow C_6H_{12}O_6 + O_2$
               glucose

The basic reaction of photosynthesis.

8. Classify the following reactions as oxidation, reduction, hydrolysis, or photochemical reactions:

  a. $Na_5P_3O_{10} + 2H_2O \rightarrow 2Na_2HPO_4 + NaH_2PO_4$
  b. $2H_2S + 3O_2 \rightarrow 2SO_2 + 2H_2O$
  c. $NO_2 \xrightarrow{\text{with aid of sunlight}} NO + O$
  d. $C_2H_4 + H_2 \rightarrow C_2H_6$
  e. $C + O_2 \rightarrow CO_2$

9. Calculate the molecular weights of the following important molecules.

    Common air pollutants:
  a. nitric oxide (NO)
  b. nitrogen dioxide (NO_2)
  c. sulfur dioxide (SO_2)
  d. sulfur trioxide (SO_3)
  e. ozone (O_3)
  f. hydrogen fluoride (HF)

    Molecules prominent in the environment:
  g. water (H_2O)
  h. carbon dioxide (CO_2)
  i. nitrogen (N_2)
  j. oxygen (O_2)

10. Calculate the molecular weights of the following important molecules and molecular units:

  a. ammonia (NH_3)
  b. ethyl alcohol (C_2H_5OH)
  c. phosphoric acid (H_3PO_4)
  d. uranium dioxide (UO_2)
  e. tetraethyl lead ($Pb(C_2H_5)_4$)

11. The pesticide parathion is one of the common organophosphorus insecticides, so called because it contains phosphorus and carbon atoms, among others. Its chemical formula can be written $C_{10}H_{14}PNSO_5$. Calculate the molecular weight of parathion.

12. How much chalcopyrite ($CuFeS_2$) must be used per day if this mineral is the sole source of 500 tons of copper (Cu) produced daily in a copper smelter?

13. Behind chalcopyrite, bornite (Cu_3FeS_3) is the copper ore next highest in abundance at the Kennecott Utah smelter. Its smelting may be represented by

$$2Cu_3FeS_3 + 7O_2 \rightarrow 6Cu + 2FeO + 6SO_2$$

How much sulfur dioxide gas is produced for each ton of copper produced from bornite?

14. We have seen that the refining of chalcopyrite ($CuFeS_2$) ore causes four times as much sulfur dioxide (SO_2) pollution as does the refining of chalcocite (Cu_2S) ore per ton of copper (Cu) produced. What can you imagine as some of the implications if a national policy were instituted prohibiting the opening of new chalcopyrite ($CuFeS_2$) deposits (those containing over 50 percent chalcopyrite) unless a 90 percent SO_2 cleanup could be guaranteed? Discuss possible short-term and long-term effects separately.

15. It is estimated that 500 million metric tons of carbon monoxide (CO) exist in the earth's atmosphere. How many metric tons of the element carbon (C) exist in the form of atmospheric CO?

16. Each year roughly 200 million metric tons of new carbon monoxide is released into the atmosphere, and a like amount somehow disappears (creating a steady state), possibly by way of the reaction

$$2CO + O_2 \rightarrow 2CO_2$$

Assuming this reaction represents the fate of atmospheric CO, calculate the mass of carbon dioxide (CO_2) produced yearly by this source. Would this amount be likely to greatly disturb the world's atmosphere, which presently contains about 2,500,000,000,000,000 kg of CO_2?

17. A recently developed technique (the "Nanaimo System") for reducing the amount of nitric oxide (NO) in automobile exhaust involves the injection of a stream of ammonia (NH_3) into the exhaust vapor, which converts NO to harmless nitrogen (N_2) and water (H_2O) via the reaction

$$4NH_3 + 6NO \rightarrow 5N_2 + 6H_2O$$

Automobiles typically emit about 5 g of NO per mile. How many lb of ammonia (NH_3) would be needed to clean up the NO emitted by a car in one year, assuming it is driven 10,000 miles per year?

18. It is believed that most atmospheric sulfur dioxide (SO_2) is eventually converted to sulfur trioxide (SO_3) by reaction with oxygen (O_2), as follows:

$$2SO_2 + O_2 \rightarrow 2SO_3$$

What daily weight of SO_3 eventually results from a copper smelter emitting an estimated 600 tons of SO_2 in a day? (Atomic weights: O = 16, S = 32).

19. Methyl mercury chloride, CH_3HgCl, is one of the most deadly compounds known. It is formed by microorganisms when the element mercury (Hg) is dumped into water. How much mercury would be needed to form 250 lb of CH_3HgCl? (Pertinent atomic weights are Hg = 200, Cl = 35, C = 12 and H = 1.0.)

20. How many grams are there in (a) 2 moles of CO, (b) 1.5 moles of CO_2, (c) 8 moles of N_2, (d) 50 moles of NO, and (e) 10 moles of DDT ($C_{14}H_9Cl_5$)?

21. How many moles are there in (a) 196 g of H_2SO_4, (b) 1840 g of NO_2, (c) 570 g of octane (C_8H_{18}), (d) 1 lb of Cu metal, (e) 641 metric tons of SO_2?

22. The average car, without emission controls, releases about 5 g of nitric oxide (NO) into the air for each mile driven. Using the fact that there are about 6×10^{23} molecules in one mole, calculate the number of NO molecules emitted per mile from a typical uncontrolled vehicle.

Glossary

Activation energy energy that must be supplied to start a reaction.

Atomic mass units (amu) units in which atomic masses are expressed.

Atomic weight weight of an atom according to a system in which an isotope of carbon has a mass of 12.

Balanced equation a chemical equation in which all atoms appear in equal numbers on both sides.

Catalyst a substance that speeds up a chemical reaction without itself being altered.

Chain reaction a reaction that has intermediate products that keep the reaction going.

Chemical equation a method of representing a chemical reaction, showing the amounts and kinds of molecules that react and the products that are formed.

Decomposition reaction a reaction in which molecules are split into smaller molecular units.

Enzyme a biological catalyst.

Endothermic reaction a reaction that absorbs energy, usually as heat.

Exothermic reaction a reaction that releases energy, usually as heat.

Gram atom a mole of atoms.

Hydrolysis reaction a reaction with water.

Isotopes atoms of the same atomic number but different mass numbers.

Mass number sum of the numbers of protons and neutrons in the nucleus of an atom.

Molar solution solution having one mole of a given compound in one liter of solution.

Mole amount of substance having a mass in grams equal to the molecular or atomic weight of that substance.

Molecular weight (mw) the sum of the atomic weights of the atoms in a molecule.

Nonvolatile not tending to vaporize readily; having a small vapor pressure.

Oxidation reaction reaction in which elements of compounds combine with oxygen.

Photochemical reaction reaction in which the activation energy is supplied by electromagnetic radiation, often from the sun.

Reduction reaction reaction in which oxygen is removed from a compound.

Rotational motion tumbling motion.

Steady state condition in which there is no change with time.

Thermal motion molecular motion stemming from heat.

Translational motion motion through space.

Vapor pressure force per unit area exerted by molecules in a vapor.

Vibrational motion motion in which the internuclear distance in a molecule varies repetitively.

Volatile having a high vapor pressure.

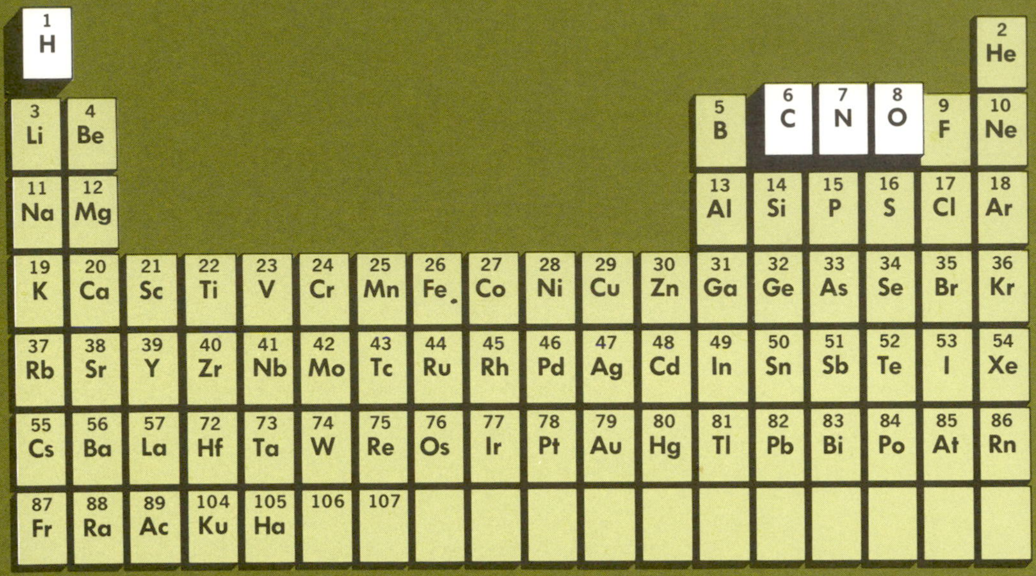

HYDROGEN, CARBON, NITROGEN, OXYGEN
The four elements shown above, combined in various molecules described in the text, are cornerstones of our priceless atmosphere. Oxygen stands out with unique significance, for reasons to be described.

6 ATMOSPHERE, OXYGEN, AND GLOBAL AIR POLLUTION

OUR PRICELESS ATMOSPHERE

Until recently the atmosphere seemed limitless and unchangeable. On the scale of man it is indeed massive, containing over 5×10^{15} metric tons of air, more than a million (10^6) tons for each man, woman, and child on earth. (All "tons" in this chapter are metric tons, 2200 lb.) It is so big that no gross changes can be imagined at the hand of man in the foreseeable future. Atmospheric oxygen, for instance, is well beyond man's immediate control and even beyond his power to destroy, as we shall show presently. However, the atmosphere is a subtle and dynamic system, interacting strongly with climate, ocean, and living systems. How long the delicate balances can be maintained under the increasing onslaught by man is unknown.

On an earth scale, the atmosphere is a thin and fragile cloak of air. Its mass is less than one millionth of the earth's 6×10^{21} metric tons. Although wisps of it extend hundreds of miles up, 75 percent of it embraces the earth in a 7-mile-thick layer, a trifling distance compared to the earth's 8000-mile diameter. Yet this airy film is far more instrumental in fixing conditions on the earth's surface and among its inhabitants than is all the great interior mass of the earth. It provides oxygen (O_2) to breathe; possesses an ozone (O_3) shield against the sun's deadly ultraviolet light; warms the earth by providing a "greenhouse" cover; releases rain to water the continents; moderates the weather; and pressurizes our living environment, as a spacecraft must be pressurized, to ensure that our body fluids do not boil away. It is a singularly hospitable environment for terrestrial life. No other planet has an atmosphere even remotely supportive of earth life. Their atmospheres are so hostile that apparently no form of life at all has been able to evolve on them. The earth's thin film of air is a unique and limited resource which should be regarded as more than an aerial cesspool for the wastes of civilization.

In the next two sections we focus on the most significant of all the gases in our hospitable atmosphere: diatomic oxygen (O_2). Gaseous O_2, having neither color nor odor, makes up 21 percent by volume of our atmosphere. It is the second most plentiful gas, falling behind nitrogen (N_2), at 78 percent, and ahead of all other gases, whose levels are under 1 percent.

An abundance of O_2 makes our atmosphere almost unique in the known universe because it creates an oxidizing environment in which our earth is bathed.[1] Nearly all other environments of the universe are dominated by reducing substances, notably hydrogen (H).

The oxidizing environment provided by O_2 makes possible the burning of fuels and the release of abundant energy. This energy drives industry and all higher living systems as well. Also, O_2 makes possible the ozone screen which protects living systems from ultraviolet radiation. Clearly, civilization and life itself depend on O_2. Furthermore, O_2 consumes by chemical reaction the organic wastes of this world, including those that pollute our water and our landscape.

We shall now examine the origin and continued availability in the atmosphere of the life-giving, waste-consuming substance, diatomic oxygen, O_2.

ORIGIN OF ATMOSPHERIC OXYGEN

The evolution of our global air blanket is one of chemistry's intriguing stories. Its fascination is not diminished because sections of the story are missing here and there, particularly toward the beginning. Much scientific sleuthing has been needed to reconstruct chemical events that occurred in the atmosphere before even the dinosaurs roamed the continents. But like dinosaurs, the primitive air left its mark in the rocks, chemical etchings that give us a glimpse of what transpired. Great bands of rock colored red by iron(III) oxide (Fe_2O_3), as an example, pinpoint eras when atmospheric oxygen (O_2) was present to oxidize all iron in accumulating sediments to this colorful form. The fiery color of the upper Grand Canyon is testament to the long

---

[1]An oxidizing environment is one with a surplus content of electronegative (electron-hungry) substances like O_2, substances which tend to combine with most elements and most organic molecules. Reducing environments have a surplus of electropositive elements such as hydrogen (H) and the metals.

geological history of atmospheric O_2, a history that began 1.8 billion years ago. Sedimentary rocks older than 1.8 billion years lack this color and reflect the oxygen-free state of the earth's original atmosphere.

It is believed that the earth was born without an atmosphere some 4.5 billion years ago. Subsequently, gases emitted from volcanoes cloaked the planet. These gases probably included water vapor (H_2O), carbon dioxide (CO_2), carbon monoxide (CO), methane (CH_4), hydrogen (H_2), sulfur dioxide (SO_2), hydrogen sulfide (H_2S), nitrogen (N_2), ammonia (NH_3), hydrogen chloride (HCl), and hydrogen fluoride (HF). Some of these gases disappeared rather soon by reacting with each other or with the bare rocks of the lifeless earth. Left behind was an atmosphere of CH_4, NH_3, H_2O, H_2, and a few lesser gases—but without oxygen. An atmosphere of this type persisted until 1.8 billion years ago. Then the most significant of all atmospheric events took place—free oxygen (O_2) began to well up from below, produced by life in the primitive seas. To understand this singular event, we must leave the atmosphere to dwell for a time with the earth-bound habitat of life.

From the early atmosphere condensed great bodies of water. These seas and the overlying atmosphere were rich in methane (CH_4), ammonia (NH_3), hydrogen (H_2), and of course, water (H_2O). The earth was at that time bombarded with ultraviolet light, high-energy radiation capable of rupturing chemical bonds. Photochemical reactions of the greatest variety took place under this constant bombardment. Chemical change was perhaps aided by lightning and the earth's natural radioactivity, much more active then than now. Complex organic molecules were formed, including amino acids and sugars, making the primitive seas a rich chemical soup. It is believed that living systems were first assembled from this supply of photochemically generated organic molecules nearly 3.5 billion years ago.

The first living organisms obtained energy by *fermentation* of the original organic soup, a reaction represented generally by

$$sugar \longrightarrow alcohol + CO_2 + energy$$

With the gradual depletion of fermentable substances, the organisms had to produce their own fermentation materials to survive. Photosynthesis, a key step in the evolution of life on earth, entered at this stage, possibly some 3 billion years ago. By photosynthesis, sugars such as glucose are produced from carbon dioxide and water by use of the energy of sunlight:

$$6CO_2 + 6H_2O \xrightarrow{\text{sunlight}} C_6H_{12}O_6 + 6O_2$$

carbon   water   glucose   oxygen
dioxide

The sugars could then be fermented to obtain the energy of life. A byproduct of the reaction was free oxygen (O_2). This substance played a curious role. We have already learned that oxygen combines avidly with many substances, consuming them by oxidation. In particular, it is inclined to react with the carbon and hydrogen of organic molecules, altering or destroying the molecules. As oxygen was first released by photosynthesis, it must have attacked the very organisms that produced it. It was most certainly an undesirable new entry into the living environment. In a broad sense, it became a major global pollutant, probably the most serious that living systems have ever encountered. In all likelihood, it was toxic enough to severely limit the population of these first simple life forms. This limit in turn limited oxygen production, because the organisms were themselves the sole source of oxygen. Life existed perhaps for millions of years in this precarious state—needing photosynthesis to supply fuel for fermentation, but gagging on the inevitable byproduct, oxygen. But the wheels of evolution grind on.

At first, soluble iron(II) compounds may have offered some protection against oxygen. They are oxidized to insoluble iron(III) oxide (Fe_2O_3) by the oxygen-consuming reaction

$$4FeO \quad + \quad O_2 \longrightarrow 2Fe_2O_3$$

iron(II)	diatomic	iron(III)
oxide	oxygen	oxide

Later, special *oxygen-mediating enzymes* arose which tied up O_2 and allowed the cell to control its disposal by reaction with surplus organic materials in the cell. With the corrosive influence of oxygen checked, life in the seas could proliferate, and its photosynthetic byproduct, oxygen, became abundant enough to permeate the atmosphere. This step brings us to the time mentioned earlier, 1.8 billion years ago, when iron(II) in seas around the world was oxidized and deposited with other sediments as red iron(III) oxide.

One major check on life still remained—the molecule-rupturing ultraviolet (UV) light still beating incessantly on the face of the earth. This light may have been integral to the creation of life by rearranging atoms to form exotic molecules, but now life was established, and it depended on the integrity of an intricate molecular structure. The random disruption of molecules was unwelcome indeed. It could not be coped with as easily as oxygen had been, because the elusive but destructive UV waves could not be so easily captured and controlled. Thus life was held deep in the primitive seas and pools, protected from ultraviolet light by overlying water. Here life evolved through the steps mentioned above, finally producing surplus oxygen, which diffused up to the atmosphere.

High in the atmosphere, some of the newly arrived molecular oxygen (O_2) was split into atoms by the ultraviolet (UV) light.

$$O_2 \xrightarrow{\text{UV light}} 2O \qquad \text{(a photochemical reaction)}$$

diatomic	atomic
oxygen	oxygen

Lone atoms, except those of the inert gases, are extremely reactive. They have lost their electron octet and recklessly seek to regain stability through chemical combination. In this hurried pilgrimage, they will react with and alter even very stable molecules. So it was in the ancient atmosphere that free oxygen atoms first reacted with neighboring oxygen molecules (O_2) to produce ozone (O_3).

$$O \quad + \quad O_2 \longrightarrow O_3$$

atomic   diatomic   ozone
oxygen   oxygen

Ozone is a molecule that readily captures energy-rich ultraviolet light, although the molecule is broken apart, *dissociated*, by the absorbed UV energy.

$$O_3 \xrightarrow{\text{UV light}} O_2 + O$$

This reaction, too, is a photochemical one. Through this process, ozone is able to remove ultraviolet light. The released oxygen atoms combine with O_2 again, and the cycle repeats.

As the parent oxygen became more plentiful in the atmosphere, ozone too increased in abundance, eventually forming an effective screen against the ultraviolet light. It is estimated that only one percent of the present atmospheric O_2 formed a screen effective enough to release life from its confinement in the watery depths. Life bloomed forth, as did byproduct O_2. This stage occurred somewhat over 0.6 billion years ago.

In the meantime, perhaps 1.3 billion years ago, although some think more recently, living systems learned (by evolutionary steps) to utilize the former nuisance, diatomic oxygen (O_2). Through oxidation (controlled burning) rather than fermentation of the organic fuel produced by photosynthesis, an abundance of energy was made available for the first time. Before, with fermentation, the energy production of organisms was marginal, something like twentyfold smaller.

The wholesale proliferation of life in the sea and finally on land was thus made possible, as we presently understand it, by five epochal chemical-biological events, listed below in chronology with approximate dates:

1. The origin of life from the complex soup of UV-formed molecules (nearly 3.5 billion years ago).
2. Fuel production by photosynthesis (3 billion years ago).
3. Enzyme protection against byproduct oxygen from photosynthesis (red-rock formation) (1.8 billion years ago)
4. Utilization of oxygen for energy needs (1.3 billion years ago).
5. Formation of the ozone screen against ultraviolet light (0.6 billion years ago).

With proliferation, an increasing amount of photosynthetic

oxygen was produced and released to the atmosphere. The oxygen-rich atmosphere of today is a product of this extended biological activity. A curve of oxygen buildup in the atmosphere over the eons is shown in Figure 6-1. The curve is approximate because oxygen levels of bygone ages cannot be obtained exactly. It shows clearly, however, that on a geological time scale, oxygen became abundant only recently, and that its emergence is closely coupled to the chemistry of life.

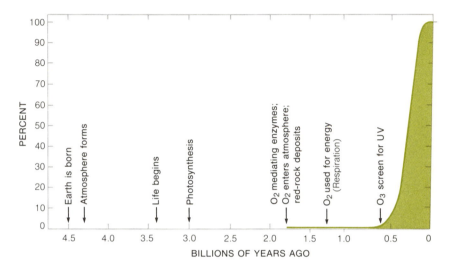

Figure 6-1. The chronology of life and the emergence of atmospheric oxygen (dates are approximate).

The history of the atmosphere reminds us that earth, air, water, and living systems affect one another profoundly. We should keep this fact in mind before making reckless moves that may in any way upset global processes. To do so without a thorough knowledge of global interactions, a knowledge which we do not at present possess, is a form of Russian roulette, with life on earth ultimately at stake.

WORLD OXYGEN SUPPLY

A recent dispute has raged over whether the atmosphere's supply of molecular oxygen is endangered. Recall that photosynthesis has produced that oxygen. It is reasoned that if photosynthesis were now sharply curtailed by some environmental blunder, oxygen depletion might be serious. Further, people have wondered if the steady burning of vast reserves of coal and oil will not unduly drain the atmosphere's oxygen. Our background in chemistry is now good enough for us to evaluate these threats rather critically. To do so, we must first examine the consequences of photosynthesis in more detail.

The Role of Photosynthesis. The glucose produced by plant photosynthesis is the major building block of living systems. It is a sugar with the structural formula shown. (See Chapter 4.)

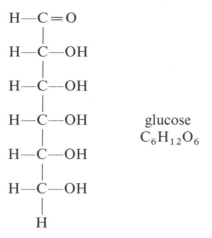

glucose
$C_6H_{12}O_6$

Once formed by photosynthesis, glucose may undergo a variety of chemical reactions by which it is transformed to the essential organic materials of life. These materials may pass from one organism to another, as when a deer crops a shrub, and the molecular form may undergo change by chemical reaction in the recipient organism, but all living material begins with photosynthesis. A few biological molecules are formed directly by photosynthesis, without the glucose stepping stone. In general, however, glucose may be considered the parent molecule of the organic materials of life, and for the sake of argument we can use glucose here to represent these complex materials.

The chemical equation for oxygen-producing photosynthesis, repeated from the last section, is

$$6CO_2 \ + \ 6H_2O \ \xrightarrow{\text{sunlight}} \ C_6H_{12}O_6 \ + \ 6O_2$$

(photosynthesis)

| carbon dioxide | water | glucose | diatomic oxygen |

The glucose produced along with O_2 has two functions. First, it fuels the activities of living systems by combining with O_2:

$$C_6H_{12}O_6 + 6O_2 \rightarrow 6CO_2 + 6H_2O + \text{(energy)} \qquad \text{(respiration)}$$

This energy-producing oxidation reaction is harnessed by living systems in a process called *respiration*. It is the reverse of photosynthesis, and it consumes exactly as much oxygen as was produced in the first place. In other words, six molecules of O_2 are produced with a glucose molecule in photosynthesis, and six are consumed as the glucose is used in respiration. There is no net gain in O_2. Photosynthesis coupled with respiration cannot therefore explain the accumulation of O_2 in the atmosphere.

The second role of glucose is to serve as a building block of living material. As glucose is produced to become plant material, O_2 is released. However, when the plant dies and decays, its bulk is consumed by oxidation, spurred on by the enzymes of microorganisms. The overall reaction is identical to that of respiration:

$$C_6H_{12}O_6 + 6O_2 \rightarrow 6CO_2 + 6H_2O \qquad \text{(decay)}$$

Again, O_2 is consumed in the same amount as was produced in the first place, and there is no net O_2 gain in the atmosphere. There is only a small and temporary O_2 release while the plant is alive, before both plant and O_2 succumb to decay.

Clearly, our reservoir of atmospheric oxygen was not produced in the photosynthesis-respiration cycle nor by the photosynthesis-decay cycle, because in each case the organic material (glucose and related substances) eventually captures and recombines with all the released O_2. This cycle is illustrated in Figure 6-2. Where, then, does atmospheric O_2 come from?

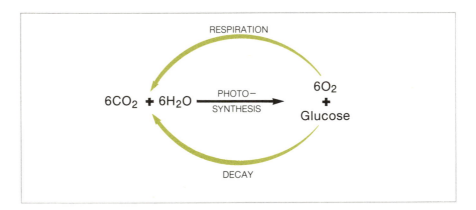

Figure 6-2. Through respiration and decay, the O_2 released in photosynthesis is returned to CO_2 and H_2O. There is no O_2 left over.

A very small fraction of plant material, estimated as one part in 10^4, settles without decay to the bottom of oceans and lakes and becomes trapped in accumulating sediment. The sediment turns to rock and the organic material is insulated from decay. It cannot recombine with the O_2 released in its formation. Therefore for each bit of organic material buried over the ages, a corresponding quantity of O_2 has been released to the atmosphere. This event, as shown in Figure 6-3, is the real source of atmospheric O_2. It would not exist if organic material had not become encapsulated in the crust of the earth, and it would now disappear if these organic materials were brought to the surface of the earth and oxidized. Since coal and petroleum are part of the buried organic debris, we want to know

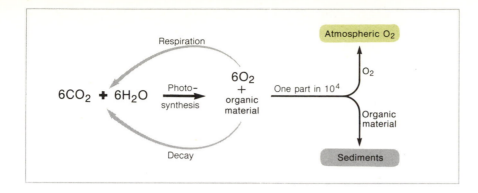

Figure 6-3. The right side of the diagram shows how the origin of atmospheric O_2 was coupled with the deposition of organic material in the crust of the earth.

whether their continued burning will consume the atmosphere's O_2. To find an answer, we must examine the mass relationships of photosynthesis.

Oxygen Mass Compared to Carbon Mass. In the basic photosynthesis reaction,

$$6CO_2 + 6H_2O \rightarrow C_6H_{12}O_6 + 6O_2 \quad \text{(photosynthesis)}$$

Partial mass
inventory, amu: $\qquad$ (C: 6 × 12) $\qquad$ (O_2: 6 × 32)

Six molecules of oxygen (O_2) are produced for each six atoms of organic carbon, a one-to-one ratio. The O_2/C mass ratio is 2 × atomic weight of O/atomic weight of C = 2 × 16/12 = 32/12 = 8/3, giving the conversion factor

$$\frac{8 \text{ parts } O_2}{3 \text{ parts } C}$$

In the reverse reaction, representing respiration or decay,

$$C_6H_{12}O_6 + 6O_2 \rightarrow 6CO_2 + 6H_2O \quad \text{(respiration, decay)}$$

O_2 and C are consumed in the same 8/3 ratio. If we somehow could retain only the basic carbon skeleton of organic materials, ridding it of H, O, and other miscellaneous atoms that later creep in, we could oxidize it (perhaps by burning) according to the basic chemical equation for organic oxidation discussed in Chapter 5.

	organic carbon	diatomic oxygen	carbon dioxide
	C	+ O_2	→ CO_2
Partial mass inventory, amu:	(C: 12)	(O_2: 32)	

Again, the O_2/C mass ratio is 8/3. We thus find that the ratio 8/3 is nearly universal for O_2/C in reactions involving organic carbon and oxygen. The two are created together in that ratio, and they disappear together in like ratio. (The importance of the ratio 8/3 was also emphasized in Chapter 5.)

By their burial, organic sediments permitted O_2 to accumulate in the atmosphere. The mass ratio of released O_2 to carbon in the sediments is presumed to be the universal ratio 8/3. As the carbon of coal and oil is recombined with O_2 by burning, oxygen in the atmosphere is used up in the 8/3 ratio; 8 tons of oxygen for each 3 tons of fossil carbon consumed. The danger of serious O_2 depletion, either by excess fossil fuel consumption or by a halt in photosynthesis, depends on the relative size of various world reserves of organic carbon and atmospheric O_2 and on their connection by the ratio 8/3. We shall discuss fossil fuel consumption first.

Fossil-Fuel Burning and the Extent of Oxygen Depletion. Estimated world reserves of recoverable fossil fuels and their carbon content are shown in Table 6-1. Such estimates are very rough, possibly off by a factor of two or three, but they are adequate for our purposes. Also shown in this table is the potential tonnage depletion of O_2 in the atmosphere, 8/3 times the carbon content.

Table 6-1 suggests that as much as 6.5 trillion (6.5×10^{12}) tons of fossil carbon may eventually be dug from the earth and burned. Potential consumption of O_2 is shown by the table to be 8/3 times this amount, or 17.3×10^{12} tons. This quantity is staggering, but the atmosphere's present reserve of O_2 amounts to 10^{15} tons, a much larger number still. The potential depletion on a fractional basis is only ($17.3 \times 10^{12}/10^{15}$) = 0.0173 (about 1.7 percent of the O_2, close to 1 part in 60 of that in the atmosphere). Therefore we may conclude that O_2 reserves in the atmosphere are so great that less than 2 percent would be depleted by the most intense and complete utilization of fossil fuels. There are other side effects of fossil-fuel combustion that cannot be so readily dismissed, as we shall see later on, but world O_2 supply is clearly not endangered by the present utilization of fossil fuels.

Table 6-1. ESTIMATED WORLDWIDE RESERVES OF RECOVERABLE FOSSIL FUELS AND FOSSIL CARBON, AND THE CORRESPONDING POTENTIAL OXYGEN DEPLETION, IN TRILLIONS OF METRIC TONS[a]

Fossil Fuel	Amount of Fuel	Amount of Carbon (C) Contained in Fuel	Potential O_2 Depletion (8/3 times amount of C)
Coal	7.6	5.7	15.2
Oil shale	0.5	0.4	1.1
Petroleum	0.2	0.2	0.5
Natural gas	0.2	0.1	0.3
Tar sand	0.1	0.1	0.2
Totals	8.6	6.5	17.3

[a]Estimates for coal reserves are from M. King Hubbert in *Resources and Man* (San Francisco: W. H. Freeman, 1969). Other reserves from Eugene K. Peterson, *Environment*, December 1970. The estimate of percentage carbon content, used to get the values in the "amount of C contained" column, is also from Peterson.

If Photosynthesis Stops? Over 50 percent (variously estimated up to 80 percent) of the world's photosynthesis occurs in the oceans, and it is mainly carried out by phytoplankton—microscopic marine plants. These plants are highly susceptible to chemical poisoning. Their photosynthesis is measurably reduced by levels of DDT as low as a few ppb (parts per billion) and by mercury (Hg) at a miniscule 0.1 ppb in water. Some people have said that if ocean phytoplankton were destroyed, over 50 percent of the world's production of new photosynthetic O_2 would be halted and we would be plunged into an oxygen crisis. This projection is incorrect. Phytoplankton use as much O_2 in respiration and decay as they produce by photosynthesis. As was shown in Figure 6-2, the photosynthesis-respiration and photosynthesis-decay cycles produce no net O_2. If the phytoplankton were killed, the only depletion of O_2 would be the small amount needed to decay the final remnants of the plant community. Living phytoplankton are estimated to contain 5 to 10 billion metric tons of organic carbon. Terminal decay would use up O_2 at 8/3 times this amount, in the vicinity of 20 billion (2×10^{10}) tons O_2. This seemingly enormous amount is only a negligible part of the 10^{15} tons of atmospheric O_2. Therefore depletion would be negligible, around 1 part in 50,000 ($= 2 \times 10^{10}/10^{15}$) of available O_2.

Land plants are somewhat more bulky than sea plants, containing in total between 10 and 100 times more organic carbon than phytoplankton. (In sea and on land, the amount of animal matter is negligible compared to plant matter.) Therefore if these plants were destroyed too, O_2 depletion still would not exceed 10^{12} tons, 1/1000 of total atmospheric O_2. This amount would put no serious dent in

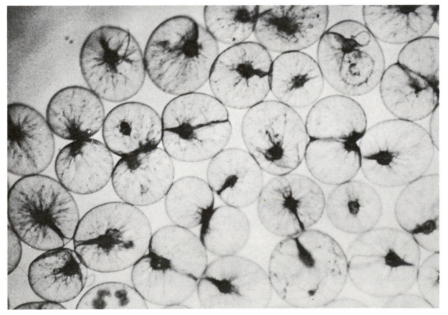

Phytoplankton. (*Photo by DeBoyd L. Smith.*)

O_2 supply, but it would, of course, terminate the ultimate source of all food and lead to the rapid extinction of life.

In summary, the 10^{15} ton reservoir of O_2 in the atmosphere is so enormous that there is no serious threat of oxygen depletion for many centuries despite the worst blunders that man can imaginably make. There are serious environmental dangers that we shall discuss in detail shortly, but an O_2 shortage is not one of them.

A Perspective on World O_2 Supply. Our present reservoir of atmospheric O_2 has been generated by the relentless action of photosynthesis and the concomitant burial of organic sediments over hundreds of millions of years. Only a small fraction of this trapped organic matter has accumulated as fossil fuels. Most organic deposits are fragments scattered in sedimentary rock throughout the crust of the earth. There they are present at about 0.4 percent by weight. There are roughly 2×10^{18} metric tons of sedimentary rock on earth, which at 0.4 percent gives 8×10^{15} tons of deposited carbon. This figure dwarfs by a factor of more than 1000 the amount of carbon in recoverable fossil fuels, 6.5×10^{12} tons (Table 6-1). The deposit of 8×10^{15} tons of organic carbon presumably led to the production of 8/3 times as much O_2,

$$8 \times 10^{15} \text{ tons C} \times \frac{8 \text{ parts } O_2}{3 \text{ parts C}} = 21 \times 10^{15} \text{ tons } O_2$$

This number, like all others in this section, is very approximate, but it provides a good working illustration of the vast quantities of O_2 evolved over geological time. This amount is 21 times more than the 10^{15} tons of O_2 now in the atmosphere. Where did it all go?

Typical seams of coal, nature's most abundant fossil fuel. If the 7.6 trillion tons in the earth's crust were burned, over 15 trillion tons of oxygen would be used up. (*Photo by Alexis Kelner.*)

It is surmised that the missing O_2, 20×10^{15} tons, more or less, has been used up in (1) oxidizing gases such as H_2S, SO_2, NH_3, and CO emitted from volcanoes over the ages and (2) oxidizing iron and sulfur in the crust of the earth. The fate of atmospheric O_2, from its photosynthetic beginning to this terminal oxidation process, is shown in Figure 6-4. The slow erosion of free oxygen by global oxidation continues today. This erosion must be matched by a steady net supply of O_2 from photosynthesis and organic sediment formation, or the atmosphere will eventually be depleted of molecular oxygen. This process would require millions of years, so in the remainder of this chapter we turn our attention to more immediate global problems.

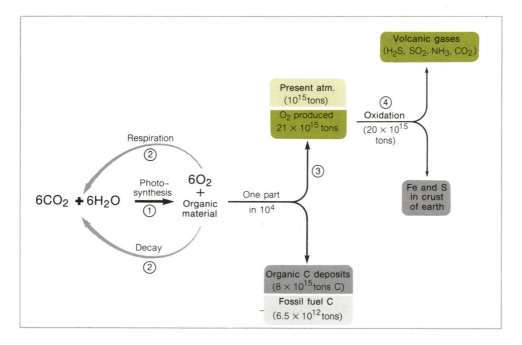

Figure 6-4. The chemical fate of O_2 produced photosynthetically (1). Most released O_2 is recycled (2) back to CO_2 and H_2O. One part in 10^4 has gone (3) to the atmosphere, and most of it has been consumed (4) in oxidizing global chemicals. The staggering amount of photosynthesis over the eons can be surmised by the fact that only one part of its products in 10^4 has alone produced vast amounts of organic sediments and atmospheric O_2. For example, we can estimate that all the coal so far used in building and energizing our industrial civilization is only one billionth of the organic material produced in geological time by photosynthesis, step (1) above.

A PERSPECTIVE ON ATMOSPHERIC GASES ENDANGERED BY MAN

Our consideration of atmospheric oxygen depletion suggest that man's activities are too trivial and limited to affect the massive global environment. This conclusion would be comforting, but in all likelihood erroneous. All we have really found is that a frontal assault on O_2, the atmosphere's second most abundant gas, making up 21 volume percent of the atmosphere, is futile. The balance of minor gases, those present at a small fraction of 1 percent, would be more easily upset. In this world of delicately balanced and interconnected parts, even some of the minor components of the atmosphere have significant roles. Their disturbance would have far-reaching consequences, as we shall show in subsequent sections.

We have seen already that small amounts of ozone (O_3) in the upper atmosphere were enough to liberate early life from its confinement deep in the seas. Today, ozone still plays a vital role in protecting living organisms from ultraviolet bombardment. Likewise, low levels of carbon dioxide in the atmosphere affect the temperature and climate of earth by intercepting outgoing, energy-carrying radiation. Here again, a minor gas has far-reaching effects, and we must understand man's impact on its balance.

To obtain a perspective on man's potential for disturbing atmospheric gases, we note that most industrial activity is now based on fossil fuel combustion. Atmospheric disturbances will usually reflect the level of this activity. The most direct and generally the most massive changes in the atmosphere and those that occur directly as the fuel is burned, a process represented simply by

$$C \quad + \quad O_2 \quad \rightarrow \quad CO_2$$

| fossil carbon | atmospheric oxygen | atmospheric carbon dioxide |

Both O_2 and CO_2 are involved in this chemical equation: therefore these two are likely to be disturbed as much as any other atmospheric gas by industrial activity. We found in the last section that the combustion of all the recoverable fossil fuels on earth would deplete about 17×10^{12} (17 trillion) tons of atmospheric oxygen. Thus a tentative ceiling near this value—say 20×10^{12} metric tons—represents the largest magnitude by which man is likely to change any atmospheric gas within a century or two. (Later, a civilization driven by nuclear energy could cause more severe changes.) The 20 trillion (20×10^{12}) ton ceiling is 1 part in 250 by weight of the atmosphere's 5000 trillion (5×10^{15}) metric tons. Table 6-2 shows the approximate weight of each gas in the atmosphere and distinguishes between gases temporarily "safe" from meddling by mankind, existing in excess of 20 trillion tons, and the more "vulnerable" gases, present at less than 20 trillion tons. Water vapor is approximately borderline, present at 14 trillion tons. Water moves in and out of the atmosphere more rapidly than do other gases, and it is thus not so much subject to cumulative changes as are many other gases. However, important alterations of its natural cycle of evaporation and condensation can happen and therefore it is possibly vulnerable. This potential vulnerability is important, because water vapor is a key factor in weather, climate, and man's supply of liquid water.

In summary, the three major gases—nitrogen (N_2), oxygen (O_2), and argon (Ar)—are presently immune to any significant disturbance by man. However, all the other gases of the atmosphere are to some

Table 6-2. MASS OF DIFFERENT GASES IN THE WORLD'S ATMOSPHERE, IN TRILLIONS OF METRIC TONS (10^{12} METRIC TONS) Notice how rapidly the amount diminishes from top to bottom. Most gases are well below the critical disturbance level of 20 billion metric tons.

	Gas or Vapor	Trillions of Metric Tons in Atmosphere
Gases Now Safe from Disturbance by Mankind	Nitrogen (N_2)	3900
	Oxygen (O_2)	1200
	Argon (Ar)	67
Gases Vulnerable to Disturbance by Mankind	Water vapor (H_2O)	14
	Carbon dioxide (CO_2)	2.5
	Neon (Ne)	0.065
	Krypton (Kr)	0.017
	Methane (CH_4)	0.004
	Helium (He)	0.004
	Ozone (O_3)	0.003
	Xenon (Xe)	0.002
	Dinitrogen oxide (N_2O)	0.002
	Carbon monoxide (CO)	0.0006
	Hydrogen (H_2)	0.0002
	Ammonia (NH_3)	0.00002
	Nitrogen dioxide (NO_2)	0.000013
	Nitric oxide (NO)	0.000005
	Sulfur dioxide (SO_2)	0.000002
	Hydrogen sulfide (H_2S)	0.000001

degree susceptible to change by man's activities. We shall now explore how man is changing the global content of some of these gases, and we shall see the consequences that may ensue.

THE GREENHOUSE EFFECT

The climate of earth is strongly affected by the atmospheric level of two gases which are on the list (Table 6-2) of those within man's power to disturb: water vapor (H_2O) and carbon dioxide (CO_2). These two affect climate because they participate in the greenhouse effect.[2] (Water has other roles in climate making which we shall not detail here.) Later in this chapter we shall consider a third climate-changing component of the atmosphere that is affected by man: particulate matter (small particles). Particulate matter was not considered in our earlier discussion of normal atmospheric gases (Table 6-2) because it is a collection of solid or liquid materials with an irregular molecular composition.

The mean temperature of our planet is fixed by a steady-state

---

[2]Ozone (O_3) is a minor participant in the greenhouse effect. However, it has a much more important role in providing an ultraviolet screen. A threat to this role is described on page 208.

balance between the energy received from the sun and an equal quantity of heat energy radiated back into space by the earth. (Other sources of heat, such as industrial production and volcanoes, are at present negligible on a global scale.) Disturbances in either incoming or outgoing energy would upset this balance, and the average temperature of the earth's surface would drift off to a different steady-state value. The resulting changes in world climate could upset food production, create deserts, raise the level of the oceans, establish a new ice age, or cause other monumental disturbances. Human activities could bring about one or more of these changes by several mechanisms. In this section we explore just one—the so-called greenhouse effect.

The carrier for the vast energy transmitted from the sun to earth is sunlight—electromagnetic radiation of wavelength between 0.2 microns (μ) (equal to 2000 angstroms (A)) and 3 μ (30,000 A), with a maximum in the visible spectrum at 0.5 μ (5000 A).[3] Most of this radiation passes unhindered through the atmosphere, except where a cloud cover exists. That is, a cloudless sky is transparent to sunlight, and the sun's radiation beats directly on the surface of the earth, warming it and bathing it in light.

The radiation that sends heat energy from earth back into space occurs in a different region of the electromagnetic spectrum—at wavelengths between 2 μ and 40 μ, with the greatest intensity at about 10 μ. These wavelengths are in the range of infrared radiation, about 20 times longer than the wavelengths (0.2 μ to 3.0 μ) of incoming sunlight. (See Figure 6-5.) And unlike sunlight, infrared radiation cannot travel freely through the earth's mantle of air. Atmospheric water vapor (H_2O) and carbon dioxide (CO_2) both absorb infrared radiation. In this way they act as blankets around earth, hindering the escape of heat into space. Without these blankets, the earth's surface would average an unbelievable $-40°F$ ($-40°C$) instead of the comfortable $60°F$ ($15°C$) actually enjoyed.

If now additional H_2O or CO_2 were added to the atmosphere, the effect would be that of adding a blanket. The earth below, unable to rid itself of energy as easily as before, would heat up. This added blanketing, then, is the essence of the greenhouse effect. (See Figure 6-6.)

The greenhouse effect takes its name from the warmth of greenhouses, a warmth stemming in part from the ease with which warming sunlight enters through the glass panes, and the difficulty met by infrared radiation in escaping out through those same panes with the greenhouse heat. In other words, the glass panes act much as the atmosphere does, allowing the free passage of incoming radiation

---

[3]See Figure 2-7 in Chapter 2 for the place of sunlight in the total electromagnetic spectrum. The conversion factor from microns (μ) to angstroms (A) is 10^4 A/μ; 1 μ is 10^{-4} centimeters (cm) (10^{-4} cm/μ).

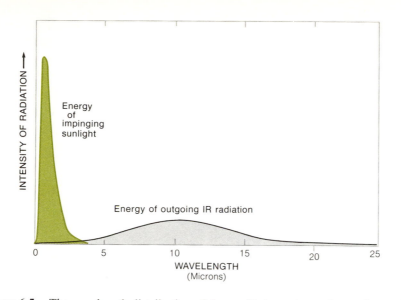

Figure 6-5. The wavelength distribution of the earth's incoming and outgoing radiation. Sunlight occurs mainly in the wavelength range 0.2 to 3 microns (μ); outgoing infrared (IR) in the range of 4 to 40 μ.

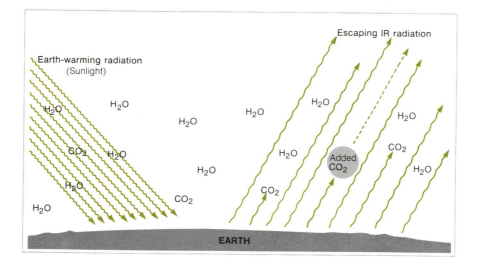

Figure 6-6. The greenhouse effect. Here infrared (IR) radiation is "attempting" to carry away energy brought to earth by sunlight. The escaping energy is captured (absorbed) by H_2O and CO_2. With energy escape hindered, the earth is warmer than it otherwise would be. Any addition of H_2O or CO_2 would cause additional greenhouse heating because even more radiation would be intercepted (as shown by the IR radiation captured by the added CO_2).

but interfering with outgoing radiant energy.[4] There is considerably more water (H_2O) in the atmosphere than carbon dioxide (CO_2) (see Table 6-2), so most of the "greenhouse" heating of the earth's

---

[4]This view of a greenhouse heating is simplistic, because an even larger share of the temperature rise has its origin in the stagnation of the sun-warmed air in the greenhouse.

surface is due to the blanket of H_2O vapor in the atmosphere. However, the H_2O blanket has a large "hole," and in covering part of this hole, CO_2 comes into prominence.

The hole in the H_2O blanket is a blank place in the absorption spectrum of H_2O. It occurs at wavelengths between 7 μ and 14 μ.[5] Carbon dioxide (CO_2) absorbs strongly from 12 μ to 16.3 μ, so it covers the edge of the H_2O hole, the part from 12 μ to 14 μ (Figure 6-7). This absorption makes the hole smaller, reduces heat loss, and thus warms the earth.

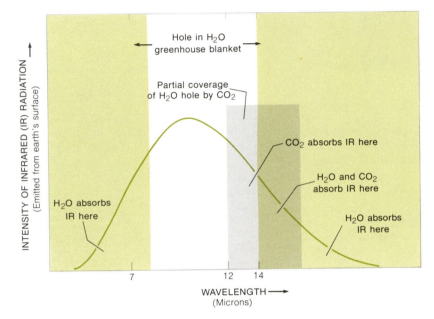

Figure 6-7. Water vapor (H_2O) dominates the atmospheric greenhouse effect, absorbing much outgoing infrared (IR) radiation. The hole in the H_2O absorption spectrum is partially plugged by carbon dioxide (CO_2), giving it, too, a role in the greenhouse warming of earth.

CARBON DIOXIDE, GREENHOUSE, AND CLIMATE

Carbon dioxide (CO_2) is an invisible, odorless gas, nontoxic to living systems. It is unreactive in the atmosphere and persists indefinitely as long as it remains there. However, CO_2 dissolves more plentifully in water than do most atmospheric gases; the oceans hold 50 times

---

[5]The 7 μ to 14 μ hole has two parts. In the central part, 8.5 μ to 11 μ, no absorption occurs and no contribution to greenhouse heating occurs. Near the hole's edges, 7 μ to 8.5 μ and 11 μ to 14 μ, H_2O absorbs weakly, thus retaining a small part of the earth's heat via the greenhouse effect. H_2O does not absorb all infrared radiation equally. Recall (Ch. 2) that electromagnetic radiation is absorbed only at certain wavelengths, where its energy corresponds to the energy of quantum jumps within atoms or molecules. There are no quantum jumps in H_2O molecules corresponding to the energy of electromagnetic radiation between 8.5 μ and 11 μ; hence there is no absorption in that range. Quantum jumps occur infrequently at the fringes of the hole, 7 μ to 8.5 μ and 11 μ to 14 μ, and thus only weak absorption is found in these regions.

more CO_2 than does the atmosphere. In water, CO_2 forms a weak acid, carbonic acid (H_2CO_3), by the reaction

$$CO_2 + H_2O \rightarrow H_2CO_3$$

Part of this acid ionizes to form acidic hydrogen ions (H^+) along with bicarbonate ions (HCO_3^-). In basic solutions, further ionization yields the carbonate ion (CO_3^{2-}). The carbonate ion combines readily with calcium (Ca^{2+}) ions made available by sea creatures to form solid calcium carbonate ($CaCO_3$), the substance of sea shells. Eventually, shells and skeletal material sink to the bottom to become limestone ($CaCO_3$) deposits. In the presence of abundant magnesium ions, carbonate frequently forms seabottom deposits of dolomite ($CaMg(CO_3)_2$) as well.

From geological records it has been surmised that over 10^{17} tons of CO_2 have been spewed into the air by volcanoes since the earth began. This mass is four times greater than that of all gases now in our atmosphere and 40,000 times greater than that of present atmospheric CO_2. Three fourths of this CO_2 has gone into the oceans and seas, there to combine with Ca^{2+} and Mg^{2+} to form limestone and dolomite. Up to one fourth of the original CO_2 has been split up by photosynthesis, the carbon being buried in sediments (only a small fraction being concentrated in the fossil fuels, as we noted earlier) and the oxygen entering the atmosphere to provide our invaluable mantle of O_2.

The two ongoing processes—formation of carbonate rock and burial of organic carbon—have consumed the volcanic output and kept CO_2 at a low level (around 300 ppm) in the atmosphere. The present influx of manmade CO_2—whose occurrence is sudden on a geological time scale—threatens to severely alter the CO_2 balance, and with it, the greenhouse blankets and the temperature of earth.

The CO_2 Greenhouse. Even though H_2O vapor dominates the greenhouse effect and its warming of earth, an increased level of CO_2 would cause some worldwide greenhouse warming by covering part of the infrared absorption hole in the H_2O blanket, as we discussed in the last section. This prospect has caused some alarm because it is feared that the massive quantities of CO_2 being pumped into the air by fossil fuel combustion might alter the atmospheric CO_2 level enough to change climate. To examine this threat, we must evaluate the prospects for substantial changes in the atmosphere's content of CO_2. We proceed by the same method used earlier to assess the danger to world O_2 supply.

Recall that the combustion of fossil carbon produces CO_2 via the oxidation reaction

$$C + O_2 \rightarrow CO_2$$

$$12 \quad\quad 32 \quad\quad 44$$

The mass of CO_2 relative to that of the fossil carbon (C) burned is seen by the mass inventory to be 44/12, or 11/3, which gives the conversion factor

$$\frac{11 \text{ parts CO}_2}{3 \text{ parts C}}$$

The estimated world reserve of fossil carbon amounts to 6.5×10^{12} tons, as noted in Table 6-1. If this carbon were burned entirely, it would yield 24×10^{12} tons of CO_2.[6] This amount is almost 10 times greater than the present mass of CO_2 in the atmosphere, 2.5×10^{12} tons. (See Table 6-1.) It has been found that about half of the CO_2 entering the atmosphere remains there for a substantial period of time, the other half being consumed by photosynthesis and by dissolving in the surface layers of the oceans. Thus the potential increase in atmospheric CO_2 from man's industrial activity is approximately $1/2 \times 24 \times 10^{12}$ tons $CO_2 = 12 \times 10^{12}$ tons CO_2, an amount almost 5 times greater than the present 2.5×10^{12} tons of CO_2 in the atmosphere. From this figure, it is clear that man could conceivably raise the level of atmospheric CO_2 by a factor of about five. While this extreme degree of CO_2 contamination may not be reached for various reasons, an eventual doubling of CO_2 levels is most likely, even by the most conservative of estimates.[7] We shall focus here on the consequences of doubling the CO_2 level in the atmosphere, realizing that the final increase might be considerably greater. An estimate by Eugene K. Peterson indicates that atmospheric CO_2 may double relative to its preindustrial levels (about 2.3 trillion tons) by the year 2020. (See Problem 8.)

CO_2 and Future Climates. Granted that by present trends mankind will double atmospheric CO_2 in the lifetime of many people now on earth, what are the expected consequences? Several attempts have been made to calculate the outcome, but the results are only approximate because of the atmosphere's complexity. The best estimate insofar as temperature alone is concerned is that the mean global temperature would increase by about $4°F$ with a doubling of CO_2.

Global Repercussions. The global repercussions of an increasing temperature are not clear. The atmosphere is a complex, dynamic system in which single, straightforward effects are unlikely to occur by themselves. A number of side effects are possible. At best we can recite only some of the plausible risks.

The 1970 *Report of the Study of Critical Environmental Problems* (SCEP) discusses the still unknown causes and effects of atmospheric

---

[6]This figure is calculated by the following conversion:

$$6.5 \times 10^{12} \text{ tons C} \times \frac{11 \text{ parts CO}_2}{3 \text{ parts C}} = 24 \times 10^{12} \text{ tons CO}_2$$

[7]Atmospheric CO_2 is now increasing at the rate of 0.7 ppm per year. This rate will accelerate as world industry expands. The CO_2 level has already increased to 320 ppm by volume, compared to 290 ppm a century ago.

circulation, temperature, and rainfall. In the past, "ice advances and retreats, widespread drouths, changes of ocean level, and so forth, were accompanied by only slight shifts in the mean circulation pattern and only small changes in the average temperature over large parts of the earth." Is there a sensitive trigger mechanism for these global changes in climate, such as small changes in greenhouse heating? That this is conceivable may be seen from this SCEP passage: "It is likely . . . that small changes in the mean conditions of the atmosphere or the surface [of the earth] can then produce a change from one climatic regime to another."

Among the possibilities is the chance that a small increase in temperature would cause worldwide changes in rainfall pattern. This change would bring chaos to agriculture and water-hungry industry. There is much evidence that precipitation is somehow linked to temperature. The earth has sustained many impressive changes in climate in which temperature and rainfall changed together. Something like 4000 years ago, Europe and North America were both dominated by a warmer and wetter climate. Since then, warm and dry intervals have alternated with cold and wet climates. The change in average temperature during many of these excursions of climate has been only a few degrees, no more than could be expected from a doubling of CO_2 in the atmosphere.

There is also a strong likelihood that heating the earth a few degrees would begin to melt the massive ice caps of Antarctica and Greenland. The complete melting would take centuries. It would ultimately raise the ocean level 200 ft and would flood the great coastal cities of the world. It has been suggested that the redistribution of ocean weight could trigger earthquakes and increase volcanic activity.

Multiplier Effects. An unknown number of "multiplier" effects could conceivably make the consequences of adding CO_2 to the atmosphere worse than anticipated. For instance, the release of added CO_2 because of increased volcanic activity could increase the greenhouse effect significantly. Also, the melting of the ice caps would expose land beneath that would reflect less (and thus absorb more) sunlight than did the cap of ice. This absorption would warm these regions and eventually the entire surface of the earth by atmospheric circulation. This temperature increase would be over and above that stemming directly from greenhouse heating—the repercussions of the basic greenhouse phenomenon would be multiplied.

A multiplication of greenhouse heating would also occur as the temperature increase led to the vaporization of H_2O and dissolved CO_2 from the oceans and even thicker greenhouse blankets were formed. This multiplier effect has been accounted for in many calculations; what is too difficult to account for is the influence of worldwide heating on cloud cover. Clouds (like snow and ice) reflect a great deal of solar energy back into space. A slight change in the average cloud cover of earth can mean enormous changes in the energy budget and thus in the temperature of the earth's surface. It is not at all clear whether a doubling of atmospheric CO_2 would increase or decrease average cloud cover, and to what extent. Yet we are plunging relentlessly ahead toward an environmental show-down on this issue by our unremitting growth in population and energy production. This is Russian Roulette on a grand scale. Our world should be governed by more sensible and less risky principles.

PARTICULATE MATTER AND CLIMATE

Our atmosphere is occupied by an abundance of tiny solid and liquid particles, 20 million tons or more in total. The smaller ones, under one micron (μ),[8] are sometimes termed *aerosols*. Aerosols may remain in the atmosphere for extended periods of time, settling only slowly to earth. The larger particles drift to earth in a few days under the influence of gravity or are brought down in drops of rain. The level of particulate matter in the atmosphere is at a steady state in which the particles reaching earth are replaced by the arrival of an equal quantity of new particles from a variety of sources.

Salt particles enter the atmosphere in abundance as droplets of ocean spray evaporate, leaving the tiny salt particles behind. Volcanic eruptions spew vast quantities of fine dust into the atmosphere. Dust storms and dust from strip mining and land clearing make up a large source of atmospheric particulate matter. Smoke from forest fires and, more recently, from industry and automobiles makes a substantial contribution.

It is clear that natural and man-made sources both contribute large amounts of particulate matter to the global atmosphere. At present, nature appears to have a slight edge, but man is catching up fast through incessant growth. By present trends, man-made sources of particulate matter should exceed nature's sources within a few decades.

Small particles in the atmosphere influence climate. Their role is exactly opposite to that of the greenhouse effect. Recall that the greenhouse blankets cause the earth to warm up because they interfere with outgoing radiation—radiation whose escape cools the earth by draining away energy. Particles, by contrast, interfere with incoming radiation from the sun. Sunlight that would otherwise warm the earth is reflected back into space by dust, and a cooler world results.

Particles also serve as nuclei around which tiny ice particles form. In this way they may encourage cloudiness and cause a further cooling effect.

---

[8] Recall that $1\ \mu = 10^{-4}$ cm.

The degree of cooling caused by atmospheric dust is even more difficult to calculate then the warming caused by the greenhouse effect. Recent calculations by Rasool and Schneider indicate that a fourfold increase in atmospheric particulate matter would decrease the temperature at the earth's surface by about 6°F. Such a temperature drop is believed sufficient to trigger a new ice age. The results of Rasool and Schneider suggest that this factor is much more important than is greenhouse warming, but of course this finding is not yet conclusive.

It is reasonable to ask why we do not simply look at the present trend in worldwide temperatures to see which way our climate is directed. This task is not simple. The earth's average temperature is subject to many fluctuations, which make it difficult to spot trends before being engulfed by them.[9]

Trends of the recent past, of course, are a matter of record. The temperature movements of the last century are shown in Figure 6-8.

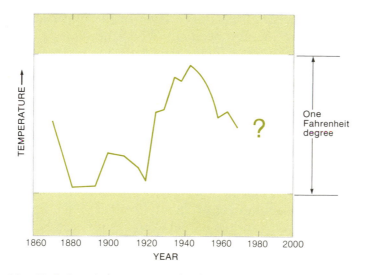

Figure 6-8. Variations in temperature in the Northern Hemisphere for the past century are too erratic to project a definite trend. Man's influence is also uncertain. It has been suggested that greenhouse heating caused the upward swing early in this century and that cooling due to atmospheric particulate matter is causing the present downswing.

The earth warmed up about 1°F in the first half of this century, and it has since been cooling down at the same rate. It is impossible to tell whether the present cooling trend will continue. If this cooling trend is due to the increased addition of particulate matter by man, it will certainly accelerate in the next few decades because of the rapid

---

[9]The difficulty of spotting trends in erratic quantities influenced by multiple factors is exemplified by the stock market, where millions of man-hours have been spent predicting trends in advance, with only spotty success. Incomparably greater efforts have been expended predicting stock market trends than climatic trends—yet the long-term fate of civilization rests on the latter.

growth in population and industrial activity. Rasool and Schneider speculate that the 6° cooling necessary to start an ice age could be reached within the next century.

OZONE AND THE SST

Another minor ingredient of the atmosphere with major importance is ozone (O_3). It was explained on pp. 188–190 that the formation of O_3 in the upper atmosphere 600 million years ago provided a protective screen against deadly ultraviolet (UV) radiation, making life on land possible. Ozone is still there, most of it in a band 10 to 25 miles above sea level, where it reaches a peak concentration of 10 ppm, and it is still protecting living organisms against the damaging molecular disruptions caused by high-energy UV radiation.

There has been widespread concern that the vital ozone shield will be damaged by the flight of supersonic transports (SST's) in the stratosphere. (The stratosphere is a layer of the upper atmosphere, to be described in the next section.) While the American SST has received a setback (perhaps only temporary) by the refusal of Congress to fund its development, the French-British Concorde and a Russian SST will be plying the stratosphere with increasing frequency.

Like most technological developments that threaten global disturbances, the SST is being pressed into use before its full environmental impact is determined. However, some studies have been made which at least indicate the nature of the problem.

It is generally assumed, in accord with Department of Transportation estimates, that by 1985 a fleet of 500 SST's will be flying 7 hours a day at an altitude of about 12 miles. Each of these will consume enormous quantities of fuel—66 tons for each hour of flight of the proposed American SST. And with each hour aloft, approximately 83 tons of water (H_2O), 207 tons of carbon dioxide (CO_2), 3 tons of carbon monoxide (CO), and 3 tons of nitric oxide (NO) will enter the thin air of the stratosphere.

At first the emission of water vapor was of greatest concern to scientists. The stratosphere is so dry (about 3 ppm by weight of H_2O) that these massive injections of H_2O would significantly alter its moisture content in regions of heavy flying—raising it by over 60 percent (to about 5 ppm). This change, it was felt, might do two things. It might increase high-altitude cloudiness, possibly affecting climate and giving the sky a permanent haziness. It was also believed that the increased moisture might deplete ozone (O_3) through a series of chemical reactions. However, careful calculations (reported in 1970) showed that the O_3 screen was not endangered by moisture. With this conclusion in hand, the SST lobby pressed forward vigorously, but it was finally defeated in 1971 under the combined pressure of economic factors and other environmental considerations (noise, excessive fuel consumption, etc.).

The Concorde, the British-French supersonic transport (SST), aims for the stratosphere, where the effect of its abundant chemical emissions is still uncertain. (*Photo by YAN/Rapho Guillumette.*)

As so often happens in matters involving world dynamics, the first simple answers are not necessarily complete. As the American SST project was grinding to a halt, chemistry professor Harold Johnston of the University of California completed calculations which suggested that the ozone shield was indeed threatened by the SST fleet. The chemical culprit was not water (H_2O), but nitric oxide (NO). In a carefully detailed study, Johnston showed that significant amounts of ozone would be removed by the simple chemical reaction

$$NO + O_3 \longrightarrow NO_2 + O_2$$

nitric   ozone   nitrogen   diatomic
oxide   dioxide   oxygen

This reaction consumes NO as well as O_3, so one might normally expect that the damaging NO would soon be used up. Unfortunately, a second reaction occurs which regenerates NO molecules from NO_2, and the cycle can proceed over and over again, consuming great quantities of O_3 with small initial amounts of NO. The second reaction takes place between NO_2 and lone oxygen atoms, shown in the second section to be an integral part of the ozone screen. The reaction is

$$NO_2 \quad + \quad O \longrightarrow NO \quad + \quad O_2$$

nitrogen   oxygen   nitric   diatomic
dioxide   atom   oxide   oxygen

Basing his calculations on the speed of these and other reactions, Johnston estimated that the 500 plane fleet of SST's would cut the ozone content of the upper atmosphere in half, and thus a greatly increased penetration of damaging UV radiation could occur. This analysis is not necessarily the final answer to the SST question, but it is the most complete result we have to date, and it should put us on the alert against quick, consoling analyses of phenomena that are tremendously complex.

ATMOSPHERIC LAYERS AND INVERSIONS

The atmosphere of earth is layered. The bottom layer, next to the earth, is the troposphere. The troposphere, extending about 7 miles up (10 in the tropics) is the home of man. Above it is the stratosphere, extending up to about 30 miles. Next are the mesosphere, a layer occupying a band from 30 to 50 miles up; the ionosphere, at 50 to 350 miles; and finally the exosphere, at 350 to 500 miles. Figure 6-9 shows the first four layers of the atmosphere.

The atmosphere, pulled by gravity, hugs the earth. Its weight is responsible for atmospheric pressure. The pressure and density of air are greatest at sea level. With increasing altitude, the air thins rapidly. (See Figure 6-9.) Its density is approximately halved for each 3.5 miles above sea level. At 30 miles, near the top of the stratosphere, the atmosphere has thinned to about 1/100 of its sea-level density. At 500 miles, the density is less than 1/1,000,000,000 of its value at sea level. Conditions here are very similar to those in the vacuum of outer space.

Because of the rapid thinning of air with increased altitude, the lowest layer, the troposphere, contains most of the air, about 75 percent of the total amount in the atmosphere. The stratosphere, a layer three times thicker, contains only 15 percent of the atmosphere's mass.

Atmospheric Temperatures. The sun beating on the earth keeps the earth's surface warm, and thus the lower troposphere is warmed. The air becomes cooler as we move upward, as we can verify if we drive a car into the mountains or ride a commercial airliner where outside temperatures around $-40°F$ are often announced. At the top of the troposphere, temperatures average $-70°F$.

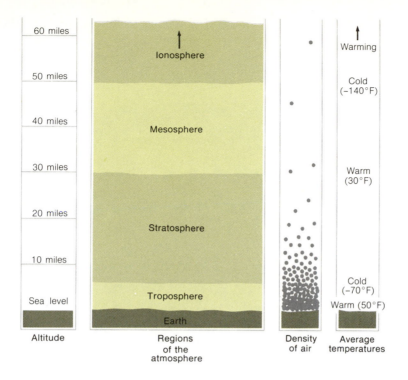

Altitude	Regions of the atmosphere	Density of air	Average temperatures
60 miles	Ionosphere		↑ Warming
50 miles			Cold (−140°F)
40 miles	Mesosphere		
30 miles			Warm (30°F)
20 miles	Stratosphere		
10 miles			Cold (−70°F)
Sea level	Troposphere		Warm (50°F)
	Earth		

Figure 6-9. The four layers of atmosphere up to 60 miles altitude. The density of air is shown schematically on the right. The crowding together of air molecules, which determines density, can be visualized as proportional to the crowding of dots in the diagram.

The rate of decrease of temperature with altitude in the troposphere is called the *lapse rate*. The lapse rate is normally about 17°F for each mile of elevation, or just over 3°F per 1000 ft.

The trend in which cooling occurs with increasing elevation persists throughout the troposphere, but it does not carry over into the stratosphere. In fact, quite the opposite occurs.

Recall that the ozone (O_3) layer in the stratosphere, from 10 to 25 miles high, captures ultraviolet (UV) radiation. The amount captured is about 3 percent of the total incoming radiation from the sun. Absorption of this radiation warms the upper stratosphere to about 30°F. The lower stratosphere, in contact with the cold upper troposphere, is a frigid −70°F. Thus a situation exists in which the atmosphere's lowest layer, the troposphere, becomes cooler with increasing altitude (from 50° to −70°F), and the next layer, the stratosphere, becomes warmer with added height (from −70° to 30°F). These opposing trends in temperature are the most distinctive characteristics of the troposphere and stratosphere, and they have great significance to the environment of man.

Because warm air is less dense than cold air, warm air has a tendency to rise and cold air to sink. Thus in the troposphere, warm air near the ground normally rises into the cold regions above, and the cold air descends.[10] As a result, the troposphere is mixed by these currents of air. (See Figure 6-10.) Among other things, this mixing is responsible for clouds and precipitation, because moisture-laden warm air cannot hold its moisture as it is carried aloft and cooled.

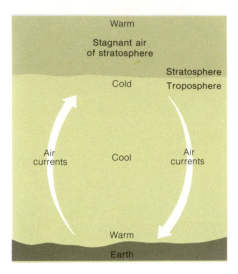

Figure 6-10. Vertical air currents stir the air of the troposphere much of the time. Air in the stratosphere is stagnant.

The stratosphere is quite different. Here the warm air is already above the cool air. There is no tendency for the two to mix, and therefore vertical air currents and precipitation are absent. The stratosphere is a vast, essentially stagnant layer of air encircling our globe (Figure 6-10).

Because vertical air currents and precipitation are absent in the stratosphere, pollutants released there are swept away by neither winds nor rainfall. They stay in place for years.[11] Any steady source of pollution in this region would lead to an accumulation for several years to higher and higher concentrations. For this reason the stratosphere, although vast in extent, is vulnerable to man-made intruders like the SST.

Inversions. In comparison with the monotonous vertical stagnation of the stratosphere, the churning air currents of the troposphere provide enormous variety in the immediate environment of man. Winds, clouds, and precipitation occur in endless forms. However, occasionally these air currents disappear for a time in certain regions.

---

[10]In actual fact the rate at which temperature drops with altitude must exceed some minimum value before vertical air currents arise. This requirement exists because warm air expands and cools as it rises. If the cooling due to expansion leaves it colder than the overlying air, it will sink back to its original position. The critical rate of drop in temperature depends on humidity; it is about 3° F per 1000 ft of altitude, close to the actual lapse rate of the troposphere.

[11]Two years is often cited as the mean residence time of gases in the stratosphere. This time increases with altitude, reaching 3 to 5 years in the upper stratosphere.

This calm leads to another erratic feature of the troposphere that affects man: *inversion layers*. These are layers of stagnant air (like a miniature stratosphere) that appear from time to time in various parts of the troposphere. Most often they occur next to the ground (*ground inversions*), and they can extend upward from a few yards to a few miles.

Ground inversions occur very frequently at night. The reason is simple. Without sunlight, the earth's surface and adjacent air cool rapidly because of radiation energy losses from the ground. Soon the air hugging the cold ground becomes cooler than the air above. This condition, we learned earlier, leads to stagnation: the cold air is so dense that it remains stagnant next to the ground. The inversion will last as long as the temperature remains "inverted"—cool air beneath warm air. Usually the inversion will break up the following day when the ground is heated by sunlight. Then air next to the ground is warmed and the normal circulation of the troposphere is restored. Occasionally, however, meteorological conditions are such that inversions can last for days or weeks.

Inversions and Air Pollution. We have seen that air pollution is now reaching such a magnitude that the global environment is threatened with disturbance. Various pollutants, mixed into and diluted by the vast air mass of the troposphere and the stratosphere, do not reach high levels of concentration, but some of them have unique properties that make them troublesome even at very low concentrations. In many cases the global atmosphere seems much more sensitive to their presence than is the human lung. For instance, nitric oxide (NO) in the stratosphere at about 0.01 ppm appears adequate to slash the ozone shield in half, but the federal standards established for 1975 by the Environment Protection Agency (EPA) will allow us to breathe NO (combined with NO_2) continuously at 0.05 ppm.[12] (This limit would be even less restrictive if it were not for the role played by NO and NO_2 in photochemical air pollution.)

Similarly, particulate matter in the troposphere and stratosphere at 3 micrograms (μg) per cubic meter (m^3) may trigger a new ice age,[13] but continuous inhalation of 75 μg/m^3, a concentration 20 times higher, is regarded as acceptable by federal standards.

From these considerations, it would appear that the global environment would collapse before adverse health effects became

----

[12]This amount is only 5 times higher than 0.01 ppm, a level destructive to the ozone (O_3) shield in the stratosphere. However, since air in the stratosphere is so thin, 0.01 ppm at 15 miles represents an absolute concentration (expressed in atoms per unit volume of air) that is 150 times smaller than that of breathable air.

[13]A fourfold increase in the present particulate content of the atmosphere would consist of $4 \times 20 \times 10^6 = 80 \times 10^6$ metric tons of particles. This amount is 80×10^{18} micrograms (μg) of particles (conversion factor: $(10^6$ g/metric ton$) \times (10^6$ μg/g$) = 10^{12}$ μg/metric ton). If distributed uniformly in the 30×10^{18} cubic meters (m^3) of the combined troposphere and stratosphere, the concentration would be 80×10^{18} μg/30×10^{18} $m^3 = 2.67$ μg/m^3, which rounds to 3μg/m^3. The assumption of uniform distribution is not valid, so this concentration must be recognized as an average.

noticeable. This prediction would undoubtedly be true if there were no inversions. Without inversions, the dirty air we produce would be dissipated quickly into the vast atmosphere, causing few immediate ill effects. However, the frequent stagnation of low-lying air by inversions changes the entire picture. Pollutants poured into such an inversion layer do not become diluted. Instead they accumulate near ground level, mostly over cities, in the breathing space of man. (See Figure 6-11.) Because of inversions, man is forced to inhale his own airborne garbage without the benefit of worldwide dilution, and it sometimes makes him sick.

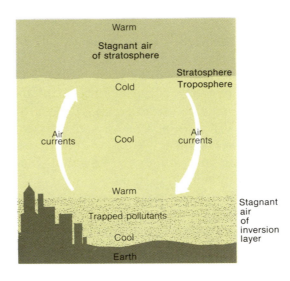

Figure 6-11. The stagnant air of inversions traps pollutants near the ground, in the environment of man.

In the next chapter we shall focus on the local air pollutants released and accumulated by inversions in the lower troposphere—that part of the atmosphere that forms the direct environment of man.

Exercises

1. The area of earth is about 200 million square miles. What weight of air resides over each square mile? What weight of O_2, CO_2, and particulate matter can be found over each square mile? (Find figures in the text for the total mass of these components in the air.)

2. Some scientists think that a substantial contribution to the atmosphere's supply of free oxygen (O_2) was made by the photochemical reaction (induced by electromagnetic radiation—in this case, sunlight) of water:

$$2H_2O \xrightarrow{\text{sunlight}} 2H_2 + O_2$$

   Ordinarily, H_2 and O_2 would recombine by the reverse reaction ($2H_2 + O_2 \rightarrow 2H_2O$), but it is known that molecular hydrogen (H_2) is light enough to escape the gravitational pull of the earth and disappear into space, leaving O_2 behind. How many metric tons of water (H_2O) would be needed to supply the atmosphere's present supply of O_2? Is this amount small, large, or comparable to the 1.5×10^{18} metric tons of H_2O in the oceans?

3. Duncan and Swanson (USGS Circular 523, 1965) estimate that an enormous 2×10^{15} barrels of oil are contained in oil shales throughout the world. This amount is about 2.8×10^{14} metric tons of oil, containing approximately 2.4×10^{14} metric tons of fossil carbon. Most of these deposits are not very rich, and only about 1 part in 5000 of the oil is recoverable under present conditions. If economic means were discovered for recovering all of this oil, would its combustion make a serious dent in the world's O_2 supply? What fraction of atmospheric O_2 would be used up in the process?

4. How much dry plant material (represented by glucose, $C_6H_{12}O_6$) was needed to produce the 8×10^{15} metric tons of organic carbon (C) now buried in sedimentary rocks?

5. Petroleum consists mainly of long hydrocarbon segments, $-CH_2-CH_2-CH_2-CH_2-CH_2-CH_2-$, with various end groups attached. Therefore to good approximation, the formula CH_2 represents petroleum insofar as the approximate two-to-one atomic ratio of hydrogen (H) and carbon (C) is concerned. Hence the combustion of petroleum can be represented by

$$2(CH_2) \;+\; 3O_2 \longrightarrow 2CO_2 \;+\; 2H_2O$$

petroleum	diatomic	carbon	
fragment	oxygen	dioxide	water

   a. What is the weight ratio of oxygen (O_2) to organic carbon (C) consumed? Does it differ from our usual approximation of 8/3, and if so, does it much affect our estimate of O_2 depletion by the burning of fossil fuels?

   b. What is the weight ratio of CO_2 to C? Does it differ from our usual ratio of 11/3, used in the text?

6. Table 6-1 gives world reserves of petroleum carbon (C) as 0.2×10^{12} tons. Much of it will be burned in internal combustion engines. Without

pollution controls, 10 percent of the fossil carbon burned in an internal combustion engine is converted to carbon monoxide (CO), the other 90 percent ending up mainly as CO_2. If the entire world reserve of petroleum were burned in internal-combustion engines without pollution control, how many tons of CO would be produced? Could this amount substantially alter the global level of CO shown in Table 6-2? (Carbon monoxide is removed from the atmosphere in a time estimated as two to three years, apparently by soil microorganisms. Therefore only a fraction, two to three years' worth, of the above total CO would remain in the atmosphere. This change will reduce our calculated level of manmade CO, but it should not alter our conclusions.)

7. Over the ages, great deposits of limestone ($CaCO_3$) have accumulated on earth as a result of a reaction with abundant volcanic carbon dioxide (CO_2). The next reaction is (see pp. 202–206 for the steps in this reaction)

$$CO_2 + H_2O + Ca^{2+} \longrightarrow CaCO_3 + 2H^+$$

Assuming that 7×10^{16} tons of CO_2 followed this chemical route, how much limestone would exist on earth? (More than this amount of CO_2 was emitted, but much of it combined with magnesium to form dolomite and considerable quantities entered sedimentary rocks as reduced carbon.) Crude measurements have shown that there are about 2×10^{17} tons of $CaCO_3$ in the crust of the earth. Your answer should be in this vicinity.

8. About 6 billion tons of carbon in fossil fuels (mainly coal and oil) are burned per year throughout the world. How many tons of carbon dioxide does this process add to the atmosphere per year? Considering that the atmosphere now contains about 2.5 trillion tons of CO_2, in how many years will the CO_2 content double, assuming the same rate of combustion continues? (The rapid increase in the rate of combustion will reduce the doubling time, but it will be partially offset by the fact that only about half of the CO_2 remains in the atmosphere, the remainder going into the oceans and green plants. Considering all these factors, it has been estimated that the preindustrial CO_2 content will double by the year 2020 (Eugene K. Peterson, *Environment*, April 1970, p. 32). It is already up 10 percent.

9. Each year approximately 2 billion barrels of motor fuel (mainly gasoline) are burned in the United States alone. Each barrel contains about 250 lb. Gasoline is actually a mixture of many kinds of molecules, but we can simplify the problem by assuming that all of it is heptane (C_7H_{16}). The combustion of heptane, if we ignore the slight amount of CO produced, can be represented by

$$C_7H_{16} + 11O_2 \longrightarrow 7CO_2 + 8H_2O$$

What is the approximate mass of carbon dioxide released yearly into the atmosphere from this source?

10. The Kaiparowits power plant proposed for construction in southern Utah would be the world's largest, producing 5 million kilowatts of electrical power and burning about 50,000 tons of coal in a single day. Simplifying this problem by assuming that exactly 48,000 tons of fossil carbon (C) are consumed each day by the reaction $C + O_2 \rightarrow CO_2$, calculate the approximate daily and yearly tonnage of carbon dioxide

added to the atmosphere by this plant alone. How much O_2 is consumed per year?

11. Suppose it were determined beyond scientific doubt that emissions of CO_2, if continued at present rates, would have greenhouse repercussions (creation of deserts, melting of the ice caps) catastrophic to man by the year 2000. Develop a brief scenario for the world's response to such a threat. You might wish to discuss factors such as political cooperation versus national goals, population growth, the drive of underdeveloped nations to industrialize, the growing standard of living in industrial countries, the role of inefficient but convenient electrical energy, the probability (and desirability) of replacing fossil-fuel combustion by nuclear power, and similar considerations.

12. How much nitric oxide (NO) would be emitted into the stratosphere in two years (the average length of time that gases remain in the stratosphere) by 500 SST's each flying 2500 hours per year and emitting 3 tons of NO per hour of flight? (This result will overestimate NO production somewhat, because one third of the projected fleet will be smaller planes, emitting about 1.5 tons of NO per hour. Later many more SST's may be flying, so NO levels could well exceed those calculated here.)

13. (Refer to Problem 12.) The first step in the cycle by which nitric oxide (NO) destroys ozone (O_3) was given in the text as

$$NO + O_3 \longrightarrow NO_2 + O_2$$

How much NO_2 is produced by this reaction, assuming that all the NO calculated in the previous problem reacts in this way?

14. (Refer to Problem 13.) About 15 percent of the mass of the atmosphere is in the stratosphere. Therefore 15 percent of the nitrogen dioxide (NO_2) listed in Table 6-2 is normally in the stratosphere. How many tons is this amount? Compare this figure to the amount produced in two years by an SST fleet (Problem 13) and thereby deduce whether the SST fleet can significantly change the normal level of NO_2 in the global stratosphere.

15. The total amount of gas in the stratosphere is about 750 trillion tons. Water (H_2O) exists in the stratosphere at 3 ppm by weight. How many tons of water are normally in the stratosphere? By what percentage would a fleet of 500 SST's alter the global level of stratospheric water in two years' time? (Use the emission levels given in the text.) (It is sometimes assumed that SST-released gases will be confined to a giant swath of stratosphere in the Northern Hemisphere where the SST's are most heavily used, consisting of 10 percent of the total stratosphere. The percentage increase in this swath would then be 10 times higher than the value calculated above for the entire stratosphere.)

16. SST fuel is like most petroleum-derived hydrocarbon mixtures, having approximately 2 hydrogen atoms for each carbon atom. The fuel can therefore be represented by the empirical chemical formula CH_2. It burns by the chemical reaction

$$2CH_2 + 3O_2 \longrightarrow 2CO_2 + 2H_2O$$

From this equation, verify that the quantities given in the section on ozone and the SST for CO_2 and H_2O emissions are indeed reasonable, and not someone's efforts to disguise the true magnitude of SST effluents. (The quantities in that section were obtained from page 72 of the SCEP report cited at the end of this chapter.)

Glossary

Aerosol   suspension in air of particles of diameter less than 1μ.

Dissociation   the breaking apart of a molecule, usually into ions or radicals.

Exosphere   the layer of the atmosphere from 350 to 500 miles high.

Fermentation   a reaction by which a cell can obtain energy from sugar without oxidation.

Greenhouse effect   the retention of heat on earth because of water vapor and carbon dioxide in the atmosphere.

Inversion layer   layer of stagnant air in the troposphere that occurs when wind currents disappear. In a *ground inversion*, the stagnant layer is next to the ground.

Ionosphere   the layer of the atmosphere from 50 to 350 miles high.

Lapse rate   rate of decrease of temperature with altitude in the troposphere.

Mesosphere   the layer of the atmosphere from 30 to 50 miles high.

Oxygen-mediating enzymes   molecules in cells which sidetrack oxygen to prevent this byproduct of photosynthesis from indiscriminately oxidizing cell material.

Photosynthesis   chemical reaction occurring in cells that produces sugars and oxygen from carbon dioxide and water using the energy of sunlight.

Respiration   chemical reaction in cells that produces energy from the reaction of glucose with oxygen.

Stratosphere   the layer of the atmosphere from 7 to 30 miles high.

Troposphere   the layer of the atmosphere from ground level to about 7 miles high.

Additional Reading

Starred selections are most suitable for the nontechnical reader.

*Brown, Theodore L., *Energy and the Environment*. Columbus, Ohio: Merrill, 1971.

*Chandler, T. J., *The Air Around Us*. Garden City, N.Y.: The Natural History Press, 1969.

*Peterson, Eugene K., "The Atmosphere: A Clouded Horizon," *Environment*, April 1970, p. 32.

Man's Impact on the Global Environment, A Report of the Study of Critical Environmental Problems (SCEP). Cambridge, Mass.: MIT Press, 1970.

Riehl, Herbert, *Introduction to the Atmosphere*. New York: McGraw-Hill, 1965.

Scientific American, Vol. 223, No. 3 (September 1970). Issue on the biosphere.

Scorer, Richard, *Air Pollution*. Oxford: Pergamon Press, 1968.

HYDROGEN, CARBON, NITROGEN, OXYGEN, SULFUR The five elements highlighted above, in a diversity of chemical forms, have a profound impact on urban air quality.

7 AIR POLLUTION IN THE ENVIRONMENT OF MAN

AIR POLLUTION IN THE HABITAT OF MAN

Man's exposure to atmospheric air is more intimate and continuous than is his exposure to any other segment of the environment. Man is constantly surrounded by and bathed in air, and his sensitive lung tissue is exposed to about 20,000 liters (approximately 20,000 quarts) of inhaled air each day.

Lungs are designed for exchanging gases between air and the bloodstream; oxygen (O_2) continuously enters the bloodstream so that it can be circulated to all parts of the body, and spent carbon dioxide (CO_2) is expelled into the air from the bloodstream. The lungs' facility for exchanging O_2 and CO_2 carries over to other materials, including many toxic gases and particles in the air. Thus the lungs provide a portal by which atmospheric pollutants can enter man. In addition, the lungs themselves can be damaged by many contaminants: lung diseases such as lung cancer, emphysema, and chronic bronchitis are among the most common afflictions found in regions having high levels of air pollution.

Man's place in the atmosphere (as opposed to the place of air in man) is limited. Except for infrequent excursions in airplanes and skyscrapers, man lives out his existence in the lowest few hundred feet of air. Even within this narrow band of atmosphere, his range is severely restrained: for the most part it includes only the air immediately above the urban and suburban complexes of earth. In these limited patches of atmosphere that constitute man's home, air pollutants of the most varied kind are released in astounding amounts. Whether they are carried away or remain to afflict man depends on the characteristics of the atmosphere in the individual regions. Atmospheric characteristics are classified and described in the study of meteorology, a brief but crucial part of which we now describe.

The Meteorology of Air Pollution. Meteorology is the science of the atmosphere, its wind and weather. Its bearing on air pollution is critical because pollutants are carried by the winds and air currents or left in place if the air is stagnant, and they are frequently swept out of the air by precipitation.

The bottom 7 miles of our atmosphere—the troposphere—is a layer in which three quarters of the earth's air is concentrated.[1] Here, air is churned frequently by vertical air currents. These currents normally serve to disperse ground-level pollutants, to mix them in the vast atmosphere where their effects are less obnoxious. The churning occurs as a result of the characteristic temperature distribution of the lower atmosphere—warm at ground level and cold above. Warm air has a tendency to rise and cold air to sink, so that regions of the lower atmosphere gradually mix together by the vertical air currents. This mixing aids the removal of pollutants from local environments.

Occasionally, cold air gets layered beneath the warm, and no churning occurs. The stagnant region in which temperature is so inverted is called an *inversion* or *inversion layer*. Its most significant feature is its vertical stagnation (although horizontal air movement can occur). Pollutants released within an inversion are trapped there until the inversion breaks up. Most serious are the *ground inversions* that occur next to the earth's surface—in the environment of man— for these inversions accumulate man's pollutants and cause unduly high human exposure.

In Figure 7-1 a map is shown which indicates the general frequency of inversions over different parts of the United States. It shows inversions to be common over California and nearby states and in a region surrounding western North Carolina. The northwest states and southern Florida are least affected. Local areas may depart from the overall trend—valley areas are notoriously high in inversion frequency, and mountain chains (like those east of Los Angeles) can trap air and enhance inversion formation.

With or without inversions, winds blowing along the ground also have a considerable influence on pollution. Persistent horizontal winds tend to carry pollutants to outlying regions (sometimes to other cities). An area like the Los Angeles basin, plagued with light winds as well as frequent inversions, is a natural candidate for severe air-pollution incidents.

All meteorological factors together—inversion frequency and height, winds, precipitation—combine to make some regions more

---

[1] Chapter 6 describes the topics of the next few paragraphs more fully.

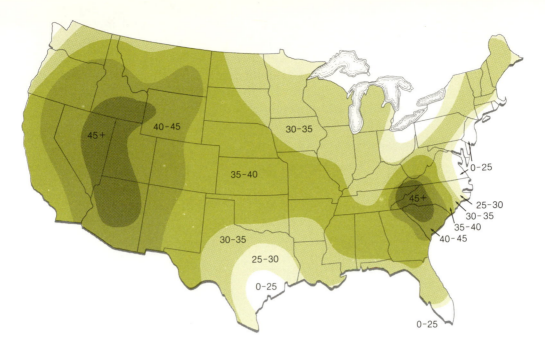

Figure 7-1. Map showing the fraction of time that inversions occur in different regions of the United States. [From Charles R. Hosler, "Low-Level Inversion Frequency in the Contiguous United States," *Monthly Weather Review, 89,* 319 (1961).]

susceptible than others to attack by pollution. Figure 7-2 is a map reflecting how all these factors combine to give different meteorological potentials for air pollution. Again, there are local exceptions where unusual meteorological conditions prevail.

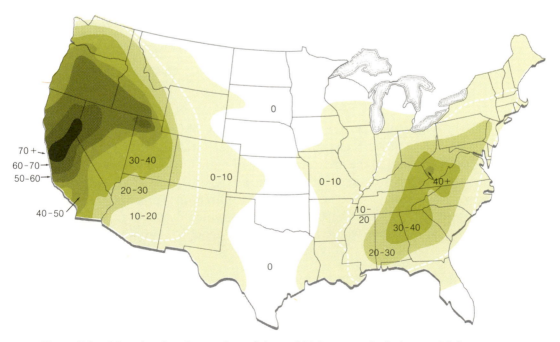

Figure 7-2. Map showing the number of days of high meteorological potential for air pollution in a five-year period. (From Office of Air Programs Publication No. AP-101.)

Various meteorological factors also have a profound impact on the localized exposure of people living downwind from a tall smoke-stack. Some examples shown in Figure 7-3 indicate that inhabitants on the ground may escape exposure on some days, but when different meteorological conditions prevail, they will bathe in concentrated fumes. The crucial role of meteorology applies on a smaller scale too—fixing one's exposure to nearby automobile exhaust or even to a companion's cigarette smoke.

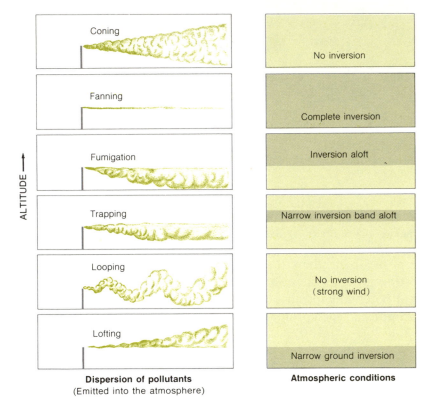

Figure 7-3. Various forms assumed by smoke plumes, depending on meteorological conditions.

Composition of Air. In the previous chapter (Table 6-2) the total mass of each gas in the atmosphere was identified. Here our focus narrows to the breathing space of man, and our concern is more with the local concentration of atmospheric gases than with their worldwide mass. For reference, we compile in Table 7-1 the composition of dry "background" air, that which exists at ground levels in remote regions where man's local influence is negligible.[2]

---

[2]The concentrations listed in Table 7-1 are parts per million (ppm) by volume, not by weight. (Percentage figures, too, are volume percents.) Concentrations in air are almost always expressed this way, while concentrations in water or food are usually expressed as ppm by weight.

In gases, each molecule, no matter what type it is, takes up almost exactly the same space. Thus ppm by volume is equivalent to ppm in the number of molecules present. For example, 320 ppm by volume of carbon dioxide (CO_2) in air means that there are 320 molecules of CO_2

Table 7-1. CONCENTRATION OF ATMOSPHERIC GASES IN CLEAN, DRY AIR AT GROUND LEVEL

(Clean air also contains small quantities of natural dust particles.)

Gas	Concentration, ppm, by Volume	Concentration, Percent, by Volume
Nitrogen (N_2)	780,000	78.09
Oxygen (O_2)	209,500	20.95
Argon (Ar)	9,300	0.93
Carbon dioxide (CO_2)	320	0.032
Neon	18	0.0018
Helium (He)	5.2	0.00052
Methane (CH_4)	1.5	0.00015
Krypton (Kr)	1.0	0.0001
Hydrogen (H_2)	0.5	0.00005
Dinitrogen oxide (N_2O)	0.2	0.00002
Carbon monoxide (CO)	0.1	0.00001
Xenon (Xe)	0.08	0.000008
Ozone (O_3)	0.02	0.000002
Ammonia (NH_3)	0.006	0.0000006
Nitrogen dioxide (NO_2)	0.001	0.0000001
Nitric oxide (NO)	0.0006	0.00000006
Sulfur dioxide (SO_2)	0.0002	0.00000002
Hydrogen sulfide (H_2S)	0.0002	0.00000002

Clean, unpolluted air has a rather constant composition, with only small variations around the world except for water vapor and a few minor gases. The amount of water vapor depends most on temperature because, as we have said (Chapter 5), vapor pressure increases significantly with temperature. Air holds little water vapor if it is cold or if it was once cold but was not able to pick up more water as it warmed up. Significant amounts of water vapor can be captured only by warm air near a source of water. Oceans are the main source of atmospheric water, but lakes, forests, croplands, and snowfields can all supply large amounts of water to the atmosphere. Depending on water sources available for evaporation and upon temperature, the atmosphere's water content can vary from 5 percent in the humid tropics to under 0.01 percent at the frigid poles. Typically, the water load is 1 to 3 percent.

If we remove the variable water load from clean air, the dry air left behind is 99 percent nitrogen and oxygen (Table 7-1). Well beneath these in concentrations are numerous minor gases, some of them having significant global roles despite their modest abundance, as shown in Chapter 6. And while the concentrations shown in Table 7-1 are observed only far from civilization, some residue of man's activities inevitably appears even there. Carbon dioxide (CO_2),

---

for each million total molecules. (Since different molecules weigh different amounts, ppm by weight is ordinarily different from ppm by volume or by number of molecules. Thus CO_2, whose molecular weight of 44 is above average (28), constitutes about 500 ppm by weight of the atmosphere, larger than the 320 ppm by volume or by molecules.)

for instance, was 290 ppm before the industrial revolution, 10 percent beneath its present level of 320 ppm. However, all the gases listed do occur naturally to some degree, and we may assume that Table 7-1 represents fairly well what the atmosphere was like before it received so much of man's airborne refuse.

THE IMPACT OF AIR POLLUTION

Air pollution is a worldwide problem now receiving worldwide attention. It reaches its greatest magnitude in regions that combine a high population density with industrialization and affluence—for instance, western Europe and the northeastern United States. It is impelled to even greater heights in those specific locations plagued by frequent inversions—Los Angeles and London, for example. During prolonged inversions, air pollution can cause immediate sickness and death—such episodes occurred in London in 1952; in Donora, Pennsylvania, in 1948; and in the Meuse Valley in Belgium in 1930. And evidence is mounting that chronic (long-term) effects are even more destructive to health in all urban regions.

Air pollutants of many chemical forms exist, and they arise from many sources. The principle pollutants and their major sources in the United States are shown in Table 7-2. A more detailed breakdown for each major air pollutant will be shown in subsequent sections.

Table 7-2. EMISSIONS IN MILLIONS OF TONS OF THE PRINCIPAL AIR POLLUTANTS IN THE UNITED STATES[a]

Source	Carbon Monoxide (CO)	Oxides of Sulfur (SO_2, etc.)	Partic- ulates	Oxides of Nitrogen (NO_2, etc.)	Hydro- carbons (HC)
Transportation	63.8	0.8	1.2	8.1	16.6
Fuel combustion in stationary sources	1.9	24.4	8.9	10.0	0.7
Industrial processes	9.7	7.3	7.5	0.2	4.6
Solid-waste disposal	7.8	0.1	1.1	0.6	1.6
Miscellaneous	16.9	0.6	9.6	1.7	8.5
Total	100.1	33.2	28.3	20.6	32.0

[a]NAPCA Publication No. AP-73, 1970; 1968 emissions.

Many pollutants beyond the five listed in the table are emitted into the air in variable amounts. Some, like pesticides, radioactivity, and metals, will be described in separate chapters. Others, like hydrogen fluoride (HF) and certain industrial chemicals, have a smaller overall impact (although the local impact is occasionally severe). There is insufficient space to describe all of them here.

Carbon monoxide (CO) is emitted in greater quantities than is any other pollutant. However, we cannot judge the seriousness of a pollutant by its tonnage emissions. This mistake is frequently made, in ignorance of the diverse factors that typically enter questions of environmental impact. One factor is a pollutant's health impact: industrial standards for sulfur dioxide (SO_2) are 20 times more stringent than those for CO (federal air pollution standards reflect a comparable discrimination against SO_2). This fact implies that 1 ton of SO_2 may constitute a threat to health as great as 20 tons of CO.

There are further complications in judging impact. Chronic effects for the various pollutants may be serious in altogether different proportions than the acute (short-term) effects. Human susceptibility may vary from person to person. Property damage is still another factor. The point of release also should be considered—street-level CO from automobile exhaust causes a proportionately greater exposure than does SO_2 emitted from a tall power-plant stack. Furthermore, the seriousness of various pollutants differs from city to city, and from one location to another within an urban complex. Because of these variations, we cannot generalize on the relative impact of pollutants on individuals in a society. In overall influence, SO_2 (with its sulfur oxide relatives) and particulate matter have sometimes been ranked as most serious, but such a ranking must be considered in the light of all the reservations we have noted.

The threat of air pollution is as difficult to judge by sight as by abstract generalizations. Most air pollutants are invisible. Of the major pollutants, only nitrogen dioxide (NO_2) and particulate matter are visible. The rest can inflict damage on the clearest of days.

Air Pollution and Human Health. Our foremost concern with air pollution is its effect on human health. Property damage must be relegated to a secondary (although still not negligible) place of importance. So, too, must broad ecological considerations, for air pollution is relatively slight outside the dense urban complexes of man.

Damage to health occurs as pollutant molecules interact unfavorably with the intricate molecules and fluids of the human body. Living systems are so complex that the detailed chemistry of these interactions is unclear in all but a few exceptional cases. (For one exception see the third section of this chapter, in which the well understood role of carbon monoxide (CO) is described.) In large part we must rely on direct observation to uncover the effect of pollution on health. Because of human variability, studies of long-term, slowly acting (*chronic*) effects are extremely difficult to make, but the short-term or *acute* effects are in some cases immediately apparent. Nowhere are they more obvious than in pollution episodes.

Pollution *episodes* are incidents of abnormally high air-pollution levels—usually encouraged by inversions. During such incidents, eyes frequently water and breathing becomes more labored. Athletic performance drops. Certain groups of people suffer exceptionally. Older people, asthmatics, and young children are most sensitive. Many heart patients die in the first day of episodes; lung patients several days later. The 1952 air-pollution episode in London claimed 5000 lives and caused extensive illness.

Recent studies, in which sophisticated statistical methods were used, suggest that air pollution causes long-term effects as well. Economists Lester B. Lave and Eugene P. Seskin have found evidence that a general 50 percent reduction in air pollution would add 3 to 5 years to the life expectancy of newborn babies in the United States. Such a reduction would lower the deathrate significantly for some important diseases: 25 percent for deaths from lung cancer, 50 percent for bronchitis, 25 percent for respiratory diseases in general, and 20 percent for cardiovascular disease.

The Economics of Air Pollution. Air pollution has serious economic repercussions. Human health itself has an economic component by virtue of medical costs and workdays lost. In the health studies just cited, Lave and Seskin estimate that the nation would save $2 billion annually by a 50 percent abatement of air pollution. By another estimate, the total health toll is $6 billion. Added to this figure is damage to crops and materials at $5 billion and depression of property values in polluted areas by about $5 billion. A 1972 report from the Environmental Protection Agency (EPA) places the total annual cost of air pollution near $14 billion. Pollution cleanup would cost less—about $12 billion. Thus the battle against air pollution would be worthwhile on economic grounds alone. And economic tools can be used in aiding the battle.

The concept of taxing pollution rather than regulating it has gained ground in recent years. A pollution tax in proportion to emissions would provide economic incentive for cleanup, particularly by the big polluters. This scheme has been labeled a "license to pollute," but it is now recognized that a high enough tax could indeed achieve a vast improvement in air quality with fewer arbitary regulations. This reasoning has led to wide endorsement for a tax on sulfur emissions—amounting to 10 to 15 cents per pound of sulfur released.

Federal Legislation. This nation is now under federal mandate to improve its air quality by 1975. Under provisions of the Clean Air Act of 1970, the Environmental Protection Agency (EPA) established in 1971 national air-quality standards for the most hazardous air pollutants (Table 7-3). The states are required to set and enforce limits on emissions from existing sources so that these standards will be reached by 1975. The EPA must set limits on new sources and on automobiles. This is to be done in such a way that the clean air in outlying areas suffers no substantial degradation in quality. Each state was asked to submit by January 31, 1972—after a public hearing—its plan for meeting the federal regulations. State emergency plans were requested to deal with pollution episodes. All plans required approval by the EPA.

The EPA listed six general control techniques for achieving federal

Table 7-3. FEDERAL AIR QUALITY STANDARDS TO BE REACHED BY 1975[a]

Pollutant Class	Period	Federal Air Standards
Carbon monoxide (CO)	8 hour average	9 ppm
	1 hour average	35 ppm
Oxides of sulfur (SO_x)	Annual average	0.03 ppm
	24 hour average	0.14 ppm
Particulates	Annual average	75 $\mu g/m^3$
	24 hour average	260 $\mu g/m^3$
Oxides of nitrogen (NO_x)	Annual average	0.05 ppm
Hydrocarbons (HC)	3 hour average	0.24 ppm
Oxidants	1 hour average	0.08 ppm

[a]These figures are maximum permitted levels of each type of pollutant for the period indicated. The values listed are primary standards, necessary to protect human health. Secondary standards, based on overall environmental effects, are more strict for particulates and sulfur oxides. (Oxidants are secondary pollutants described later in this chapter.)

standards. We shall present specific examples in subsequent sections.

1. Fuel Substitution. This method may entail replacing a high-sulfur fuel with a low-sulfur fuel or the use of nuclear energy in place of energy from fossil fuels.

2. Process Changes. Changes are required in equipment and maintenance of a kind that will reduce air pollution.

3. Gas-Cleaning Equipment. Special equipment can be used to reduce pollutants in emitted gases.

4. Process Relocation and Use of Tall Stacks. These techniques will not reduce emissions, but may remove them from the immediate environment of man.

5. Source Shutdown. The shutdown of obsolete plants would reduce air pollution.

6. Changed Transportation Policies. Patterns of highway construction and the development of mass transit systems can aid the battle for clean air. Inspection stations to detect individual violators can be established.

7. Land-Use Policies. These policies would ensure the optimal use of land to produce the least environmental impact.

With such control techniques and strong federal legislation, one would expect the growth in air pollution to grind to a halt. However, many experts feel that nothing can stop the rising tide of air pollution as long as growth continues. Federal air-pollution engineers have testified that sulfur oxides and nitrogen oxides will increase somewhat even if control efforts are successful. Clearly, the giant coal-fired power plants now planned for the Southwest will cause a significant degradation in the quality of the now clean air above the Grand Canyon and other national parks. Furthermore, neither federal regulations nor most types of control technology are directed at ridding the air of fine particles, those most damaging to health and visibility.

Until clean air is an achieved fact, it should not be assumed that the mere stroke of a legislative pen will reduce the human impact of dirty air.

CARBON MONOXIDE

Carbon monoxide (CO) in the atmosphere is in large part a product of the automobile's internal-combustion engine. Without pollution controls, this engine emits an average of 2.9 lb of CO for each gallon of gasoline burned, or nearly 1 ton of CO for each 2 tons of fuels consumed. A worldwide gasoline consumption in the neighborhood of 400 million tons is therefore associated with about 200 million tons of CO spewed into the atmosphere yearly. Table 7-4 shows how this figure compares with CO emissions from other sources. Clearly, the internal-combustion engine is a large factor in worldwide CO emissions; it is the overwhelming factor in urban environments.

Table 7-4. ESTIMATED YEARLY WORLDWIDE EMISSIONS OF CARBON MONOXIDE BY MAJOR SOURCES, BOTH MAN-MADE AND NATURAL[a]

Source	Millions of Tons Emitted per Year
Gasoline (internal combustion engine)	193
Coal	12
Noncommercial fuel	44
Incineration	25
Forest fires	11
Marine organisms	4
Oxidation of plant terpenes	60
Total	349

[a]Data from E. Robinson and R. C. Robbins, *Sources, Abundance, and Fate of Gaseous Atmospheric Pollutants Supplement* (Menlo Park, Calif.: Stanford Research Institute, 1969).

The greatest uncertainty in Table 7-4 is concerned with the level of CO emissions by marine organisms. No significant biological role for CO has been unraveled, although some organisms give off CO. The floats on kelp (seaweed) have been found with CO concentrations as high as 800 ppm. More significant are the microscopic ocean siphonophores which expel bubbles containing up to 80 percent CO. They may contribute amounts of CO to the atmosphere far larger than the 4 million tons estimated for marine sources in Table 7-4.

Reactions and Natural Removal of Carbon Monoxide. Carbon monoxide (CO) appears to be chemically inert insofar as reactions with most atmospheric constitutents are concerned; it therefore undergoes few chemical reactions in air. Nonetheless, there is a natural tendency for CO to react with oxygen (O_2), thereby becoming oxidized to carbon dioxide (CO_2):

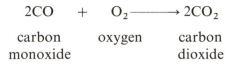

$$2CO \quad + \quad O_2 \longrightarrow 2CO_2$$

carbon oxygen carbon
monoxide dioxide

This oxidation (like other spontaneous reactions) is impelled by thermodynamic factors, most notably by the release of energy during the reaction. However, the reaction proceeds very slowly in air because of the very large activation energy (Chapter 5) needed to make the reaction go. As a consequence, the oxidation of CO has no apparent significance in the atmosphere. It has overwhelming significance in the world as a whole, however, for if destruction by oxidation did not occur somewhere, CO would accumulate to higher and higher levels, eventually forming a toxic blanket around earth. It appears now that this crucial oxidation step occurs in the soil. Several soil microorganisms, using their enzyme catalysts to make an otherwise sluggish reaction clip along, apparently remove CO by oxidation at the same rate that it is injected into the atmosphere. The average CO molecule apparently remains in the atmosphere about 2.5 years (some recent evidence suggests 0.1 years) before its capture and conversion by these soil species. In this way, CO levels in the general atmosphere are kept down to a tolerable steady-state level of 0.1 ppm (Table 7-1).

It would be tragic if man introduced into the environment a quantity of some toxic agent sufficient to destroy the CO-eating microorganisms. While such a colossal blunder is unlikely, we need occasional reminding that our safe existence hinges on many creatures, large and small, and that consideration for their welfare may in the end make possible our own survival.

Effects of Carbon Monoxide. Carbon monoxide (CO) is a gas without color or odor. It is quite inert chemically, as mentioned, and therefore causes little property damage. The one activity that gives CO notoriety is its strong inclination to combine with the hemoglobin of blood.

Hemoglobin is an iron-containing protein that carries vital oxygen (O_2) to body tissues. Hemoglobin (here shown by symbol Hb) picks up O_2 in the lungs by the process

$$Hb \quad + \quad O_2 \longrightarrow O_2Hb \quad \text{(in the lungs)}$$

hemoglobin oxygen oxyhemoglobin

This association is reversible, and the bound O_2 is released once HbO_2 reaches the tissues

$$O_2Hb \longrightarrow Hb \quad + \quad O_2 \quad \text{(released in tissue)}$$

oxyhemoglobin hemoglobin oxygen

The free Hb then returns to the lungs for a new supply of oxygen.

If carbon monoxide (CO) is present in the lungs, it will take the place of O_2 by combining with hemoglobin in the blood. In general, hemoglobin will associate with few substances other than O_2, but CO is a good chemical imitation of O_2 because it is diatomic, stable, oxygen containing, and of comparable size to O_2. Thus it can replace O_2 by combining with hemoglobin to form carboxyhemoglobin

$$Hb \quad + \quad CO \longrightarrow COHb$$

hemoglobin   carbon monoxide   carboxyhemoglobin

This process ties up free Hb in the blood, immobilizing this essential carrier of O_2 so that the body is deprived of oxygen. This association, too, can be reversed to yield free Hb again, but only with difficulty. Carbon monoxide has a much greater chemical affinity for Hb than does O_2: when CO and O_2 are present at equal concentrations, CO ties up about 220 times more Hb than does O_2, leading to almost complete oxygen starvation and sure death. Fortunately, the concentration of CO in the lungs stays well below the O_2 level. But with a 220-fold advantage, even low-level CO can immobilize enough Hb to cause a dangerous shortage of oxygen. Figure 7-4 shows the actual percentage of hemoglobin immobilized by different atmospheric concentrations of CO.

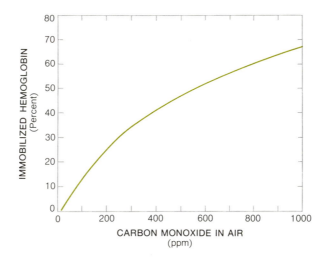

Figure 7-4. Percentage of hemoglobin immobilized by various carbon monoxide (CO) levels in the air. This percentage represents the reduction in the O_2-carrying capacity of the blood.

Because of O_2 deprivation, 1000 ppm CO will kill quickly and 250 ppm will cause a loss of consciousness (Figure 7-3 shows the percentage of hemoglobin immobilized for each CO level given). The upper limit for industrial exposure to healthy workers (the *threshold limit value* or TLV) is 100 ppm. At this level, most people experience dizziness, headache, and lassitude. Concentrations as low as 10 ppm begin to have an effect on the central nervous system:

there is an impairment in the discrimination of time intervals and brightness and in visual acuity. Some apparent effect on performance has been noted down to 5 ppm.

By comparison, federal measurements in five major cities have shown an average (not maximum) concentration of over 7 ppm CO. Taxis in New York contain 30 ppm CO average and 50 ppm CO maximum. Levels as high as 50 to 100 ppm are found in crowded streets during rush hour; higher levels occur in automobile tunnels. These levels go up and down with automobile traffic, as shown in Figure 7-5.

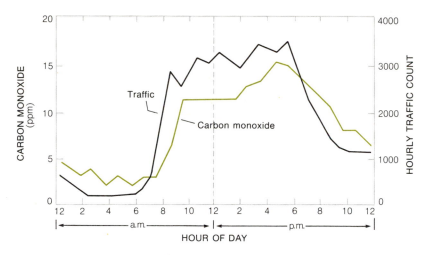

Figure 7-5. Carbon monoxide levels in midtown Manhattan go up and down with traffic flow. [From K. L. Johnson, L. H. Dworetzky, and A. N. Heller, *Science, 160,* 67 (1968).]

Observed CO concentrations are clearly high enough to invoke a negative response in man. Since CO is concentrated in streets and tunnels, it is most likely to affect people when they are driving and need sound judgment and quick reflexes.

Recent statistical studies of deathrates in Los Angeles by Alfred C. Hexter and John R. Goldsmith suggest that ordinary city levels of CO may cause death as well as sickness. Up to eleven deaths per day may be attributed to CO in Los Angeles County alone. Another study showed a possible connection between CO and heart disease. More work is needed at this stage to define the specific contribution of CO to death and chronic illness.

Control of Carbon Monoxide Pollution. The full oxidation (combustion) of carbon-containing organic materials yields carbon dioxide (CO_2):

$$C \quad + \quad O_2 \longrightarrow CO_2 \quad \text{(full oxidation)}$$

carbon in   oxygen   carbon
organic   dioxide
materials

For reasons we shall state, oxidation can be less than complete, which leads to the production of carbon monoxide (CO):

$$2C \;+\; O_2 \longrightarrow 2CO \quad \text{(underoxidation)}$$

carbon   oxygen   carbon
monoxide

Such underoxidation (or incomplete combustion) may occur because there is insufficient O_2 for full oxidation, or because reactions occur so fast that oxidation cannot be completed. Both reasons for underoxidation plague the internal combustion engine. Maximum power is achieved with rich fuel mixtures (rich in hydrocarbon fuel, deficient in O_2), and the rapid engine cycling does not allow time for all of the O_2 present to react. Hence CO is indeed an abundant contaminant in automobile exhaust.

There are two approaches in present efforts to gain control over automobile CO emissions: first, a reduction in the amount of CO formed by obtaining more complete combustion in the first place; and second, the elimination of leftover CO by the completion of its oxidation to CO_2. Changes in carburetion that provide more O_2 (air) aid the first approach, and various reactors (both catalytic and thermal) in the exhaust stream contribute to the second. Unfortunately, the internal combustion engine does not emit CO alone: it is also adding oxides of nitrogen, hydrocarbons, lead compounds, and numerous minor components to the air. This engine is such a versatile polluter that it is not by any means sufficient to turn off the spigot for one pollutant alone. Yet controlling all the pollutants is a formidable task requiring compromises and sacrifices. A subsequent section will be devoted to the unique problem that is encountered in reducing simultaneously all the pollutants from this remarkable and ubiquitous pollution machine.

SULFUR COMPOUNDS

Sulfur dioxide (SO_2) and related sulfur (S) compounds are judged by many experts to be the most serious class of air pollutants worldwide. However, they do not stem from man's industry alone. The presence of SO_2 in unpolluted air (0.002 ppm, Table 7-1) reflects several natural sources of airborne sulfur. Volcanoes emit both SO_2 and hydrogen sulfide (H_2S); the anaerobic (without air) decomposition of dead plant material releases abundant H_2S; and evaporating ocean spray leaves tiny particles of sulfate salts such as sodium sulfate (Na_2SO_4) in the air.

On a global basis, sulfur emissions from man-made sources now rival those from nature. These man-made sources are concentrated in local regions of the earth where population density and the potential for damage are greatest.

Sulfur dioxide (SO_2) is man's most abundant sulfur pollutant. Small quantities of sulfur trioxide (SO_3) and hydrogen sulfide (H_2S) are also emitted. One source alone—the burning of coal—is responsible for over half of all man-made sulfur emissions. All coal contains sulfur, usually in amounts between 0.5 and 5 percent. With burning, this sulfur is oxidized to sulfur dioxide:

$$S \quad + \quad O_2 \quad \longrightarrow \quad SO_2$$

sulfur in coal   oxygen   sulfur dioxide

The SO_2 escapes up smokestacks into the atmosphere. (Later we show how to calculate the amount of SO_2 emitted by coal-burning power plants.) There it joins SO_2 and other sulfur compounds from petroleum combustion and refining and from metal smelters. (See Chapter 5.) Table 7-5 shows the approximate amounts of SO_2 emitted by each source, both in the United States and worldwide.

Table 7-5. MAJOR SOURCES OF SULFUR DIOXIDE (SO_2) POLLUTION IN THE UNITED STATES AND WORLDWIDE

| Source | Millions of Tons Emitted per Year | |
	USA[a]	World[b]
Coal—combustion	20.7	102
Petroleum—combustion and refining	7.2	28.5
Smelting [over 80 percent worldwide from copper (Cu) smelters]	3.9	15.7
Miscellaneous	1.4	—
Totals	33.2	146

[a]Data for 1968 from National Air Pollution Control Administration Publication No. AP-73, 1970.

[b]Data for mid-1960's from E. Robinson and R. C. Robbins, *Sources, Abundance, and Fate of Gaseous Atmospheric Pollutants Supplement*. Menlo Park, Calif.: Stanford Research Institute, 1969.

These amounts are growing rapidly because of expanding population and industrialization. Figure 7-6 (p. 236) shows worldwide growth in man-made SO_2 pollution since 1860 and the trend expected from now to the year 2000 if pollution control is not applied. This figure shows that levels of SO_2 will approximately double by the turn of the century unless growth is slowed or valid pollution-control methods are found and used. The likelihood of effective pollution control will be discussed shortly.

Reactions and Natural Removal of Sulfur Compounds. Sulfur compounds in the atmosphere are linked together by a series of chemical reactions. These reactions convert sulfur compounds into the gaseous forms (SO_2, H_2S) that are responsible for getting sulfur airborne in the first place, and subsequent chemical reactions put them in a chemical form in which they can be removed from the atmosphere. In this way, sulfur compounds are like many other environmental substances: they cycle through air, oceans, soil, and

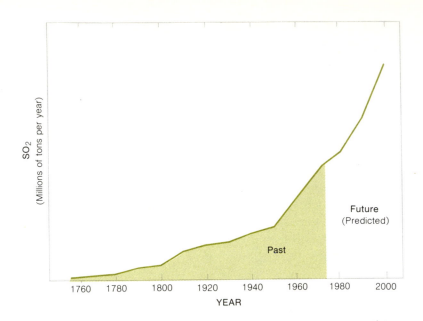

Figure 7-6. Man's emissions of sulfur dioxide, past and future. The emissions predicted from now to the year 2000 are based on a lack of significant pollution control, and are therefore possibly too high. [Data from E. Robinson and R. C. Robbins, *Sources, Abundance, and Fate of Gaseous Atmospheric Pollutants Supplement* (Menlo Park, Calif.: Stanford Research Institute, 1969).]

living systems, constantly interacting with other substances in an endless variety of ways. Here we shall outline only those features of this complex cycle that have a bearing on atmospheric sulfur pollution.

In chemical terms, the atmosphere of earth is an oxidizing environment because of its high oxygen (O_2) content. (See Chapter 6.) Not surprisingly, then, the most important reactions that happen to atmospheric sulfur compounds are oxidation reactions. First the enormous quantities of hydrogen sulfide (H_2S) released to the atmosphere by decaying plants are converted to sulfur dioxide by oxygen.[3]

$$2H_2S \quad + \quad 3O_2 \longrightarrow 2SO_2 \quad + \quad 2H_2O$$

| hydrogen sulfide | oxygen | sulfur dioxide | water |

This reaction occurs within a few hours of the time when H_2S is released, a fortunate circumstance since it prevents the buildup of toxic H_2S in the atmosphere.

The loss of H_2S is accompanied by a gain in SO_2, as the above

---

[3] Chemists recognize that many reactions like those shown in this and subsequent sections take place in several chemical steps, some only poorly understood. For instance, atmospheric O_2 may react with H_2S by first forming atomic oxygen (O) or ozone (O_3). The exact chemical pathway will depend on the presence of moisture and particles that may act as catalysts and on sunlight. However, after the reaction is complete and all the changes are tallied up, the net loss and gain of molecules is reflected accurately by the chemical reactions shown.

reaction shows. While SO_2 is also toxic, its production by this reaction is scattered around the world, and SO_2 thus fails to reach high concentration in any one place. A quite different role is played by man-made SO_2, which sometimes reaches suffocating levels, particularly during some pollution episodes.

Sulfur dioxide, no matter what its origin, does not remain in the atmosphere for long. It, too, is consumed by oxygen, this time producing sulfur trioxide:

$$2SO_2 \ + \ O_2 \longrightarrow 2SO_3$$

sulfur   oxygen   sulfur
dioxide   trioxide

It normally takes several days for this reaction to consume newly formed SO_2, but the time may be shortened to hours if the air contains abundant moisture or impurities that act as catalysts.

The sulfur trioxide (SO_3) formed in the process reacts almost immediately with water (H_2O) molecules in the atmosphere to form sulfuric acid (H_2SO_4)

$$SO_3 \ + \ H_2O \longrightarrow H_2SO_4$$

sulfur   water   sulfuric
trioxide   vapor   acid

The H_2SO_4 molecules attract more water, soon forming small droplets of concentrated sulfuric acid in the air. This aerosol mist is largely responsible for the cloudy appearance of sulfur-containing air, including the air in smokestack plumes from power plants and smelters. Some of the sulfuric acid in the droplets combines with ammonia (NH_3) existing naturally in air (Table 7-1) to form ammonium sulfate $(NH_4)_2SO_4$.

$$H_2SO_4 \ + \ 2NH_3 \longrightarrow (NH_4)_2SO_4$$

sulfuric   ammonia   ammonium
acid   sulfate

Finally, after a time that averages about one week, both sulfuric acid droplets and small ammonium sulfate particles are washed out of the atmosphere by falling rain. Some SO_2 is also captured and brought to earth by rain, and additional amounts are absorbed on plant leaves. In this way, all emitted sulfur compounds are returned eventually to earth and the atmospheric sulfur cycle is completed (Figure 7-7, p. 238). But during their residence in the atmosphere, sulfur emissions are a nuisance and a threat to health and property. We turn to these problems now.

Sulfur Pollutants and Their Effect. Both hydrogen sulfide (H_2S) and sulfur dioxide (SO_2) are colorless and therefore invisible gases. But though concealed from sight, they do not escape our

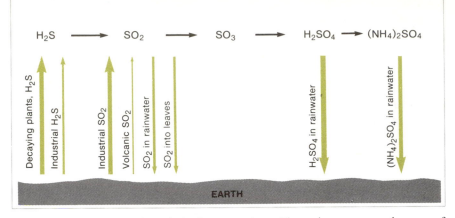

Figure 7-7. The sulfur (S) cycle in the atmosphere. The major sources and means of removal are shown. The wide arrows indicate the most significant ways in which sulfur is transferred between earth and atmosphere. The horizontal arrows above indicate chemical reactions (unbalanced as shown) which lead from one sulfur compound to another.

sense of smell. Both have a strong odor, H_2S smelling like rotten eggs (and sometimes called rotten-egg gas) and SO_2 having a suffocating, irritating odor. And both are extremely toxic, causing death at about 500 ppm. Fortunately, H_2S is not emitted in the great quantities that are typical of released SO_2, and it is therefore a lesser pollution hazard than is SO_2.

Sulfur trioxide (SO_3) is a colorless, toxic liquid with a choking odor. Sulfuric acid (H_2SO_4), droplets of which form rapidly as SO_3 reacts with atmospheric water, is a strong, corrosive acid. Ammonium sulfate (($NH_4)_2SO_4$) is a rather innocuous, saltlike material, somewhat acidic when dissolved in water.

The effects of sulfur pollutants on health are complicated by interactions between the various sulfur compounds and nonsulfur pollutants in urban air. Many facets of the health problem are still unclear. Laboratory studies on animals suggest that SO_2 contamination at levels as high as 5 ppm has no adverse effect on health. But SO_2 in the atmosphere is never the lone contaminant. The threat of SO_2 appears to be greatly enhanced by aerosols in the air—either solid particles or droplets of H_2SO_4. Altogether, SO_2 and its relatives are regarded by many as the most damaging and irritating of all the contaminants in polluted air.

The worst pollution episodes have all had high concentrations of SO_2 in combination with an abundance of particles. Londoners in their 1952 disaster found the SO_2 concentration rising to 1.3 ppm. Donora levels reached 2 ppm, while in the Meuse Valley episode, SO_2 reached an astounding peak of 38 ppm.

Sulfur dioxide has its greatest impact on the respiratory system. Immediate symptoms include coughing, chest pains, shortness of breath, and constriction of the airways. In the long run, SO_2 is thought to be a contributing factor in the rising incidence of emphysema and bronchitis. Two recent studies have raised another specter—that SO_2 and the derivatives of SO_2 formed in body fluids may cause genetic injury through damage to the system's DNA.

Sulfur dioxide and its relatives are among the most corrosive gases affecting property and plant life. Under the influence of atmospheric SO_2, metals corrode, fabrics age, and leather becomes stiff and cracked. Paper becomes brittle, and thus priceless library collections are endangered. Limestone and marble (both being calcium carbonate, $CaCO_3$) in buildings and monuments are chemically converted to $CaSO_4$, which crumbles away. Cleopatra's Needle has corroded more in London since 1878 than it did in thirty centuries in the Egyptian desert. Many irreplaceable stone monuments from ancient cultures (including the Parthenon in Athens) are threatened by seemingly endless releases of SO_2.

Sulfur dioxide emissions may also be among the few that have a broad ecological impact beyond the fringes of urban centers. The eventual products of atmospheric SO_2—sulfuric acid (H_2SO_4) and ammonium sulfate ($(NH_4)_2SO_4$)—are both acidic. When they are captured in raindrops and brought to earth in rainfall, their acidic characteristics affect the land. Rain and snow in the northeastern United States are now up to 100 times more acidic than natural rainfall, with acidic pH's ranging from 3.5 to 5. (See Chapter 8 for a definition of pH.) A similar fallout in Scandinavia, born in the industrial emissions of Western Europe and England and carried north in currents of air, is turning lakes and rivers acidic, endangering species and threatening the ecological balance of the aquatic communities. This unfortunate incident reemphasizes two broad environmental concepts, the first being that pollution can be both interstate and international in impact. The second significant concept is that air and water pollution do not exist in separate compartments. In a real sense, there are finely woven interconnections that relate all corners of the intricate environment of earth, and continued damage to one may mean eventual destruction of another.

Control of Sulfur Pollution. More schemes have been developed for controlling sulfur pollution than for any other class of air contaminant. Despite this fact, these methods are not fully effective; they are not adequately proved for the wide range of industrial conditions that exist; and they are costly in economic terms.

In this section we shall confine our attention to the control of sulfur dioxide (SO_2), the major source of sulfur in urban air. This substance, like most air pollutants, is a gas. Control is commonly directed at immobilizing SO_2, usually by chemical reactions that convert the sulfur of SO_2 to some relatively nonvolatile form. In Chapter 5 in the section on chemical equations, we described how a great deal of the SO_2 produced in copper (Cu) smelters is converted to liquid sulfuric acid (H_2SO_4), a useful industrial chemical, by means of the reaction

$$2SO_2 \ + \ O_2 \ + \ 2H_2O \longrightarrow 2H_2SO_4$$

sulfur   oxygen   water   sulfuric
dioxide   acid

Why not convert all SO_2 emissions to H_2SO_4? The conversion to commercial grade H_2SO_4 is fully effective only with streams of gas containing high concentrations of SO_2—above 3.5 percent. Only part of the smelter gas (that from the converter furnaces) contains SO_2 at these high levels; most of the remainder is near 1 percent SO_2 and is discarded to the atmosphere. As a consequence, almost half of the smelter-produced SO_2 is released to the air even with maximum pollution control. For many smelters, control is negligible. Smelters that belch around 500 tons of SO_2 into the air daily are not uncommon in the western states. (See Chapter 5.) Many of these smelters exist in unpopulated regions, and their human impact is therefore blunted.

Coal-burning power plants are less fortunately situated. They must be clustered in regions where a heavy demand for electricity exists—populated regions like the northeast States. In addition, there are far more power plants than smelters, each releasing roughly as much SO_2 as does a smelter—several hundred tons per day. These factors conspire to give coal-burning power plants the highest rank for exposing human populations to sulfur compounds. These plants will be discussed in a wider context later.

Unfortunately, the SO_2 emitted by power plants is too dilute (about 0.2 percent) for ready conversion to sulfuric acid. Many other chemical reactions have been proposed to remove dilute SO_2 from the waste gases of power plants, smelters, and other industries. Some of these methods are summarized in Table 7-6. A few have been tried, with moderate promise. Each process looks simple on paper, but there are many technical problems related to the large volumes of gas requiring treatment, high dilution, impurities, and side effects. To cope with these problems, engineering designs of elaborate form have been conceived. Catalysts, flow chambers of certain shapes, and special conditions of temperature are needed to make the reactions proceed quickly and selectively. Additional reactions and processes are frequently needed to dispose of the enormous quantities of byproducts or to recycle the chemical substance reacting with SO_2. In some cases, the byproducts are marketable (for example, pure sulfur, S), or can be converted into marketable form (S, H_2SO_4, $(NH_4)_2SO_4$). Frequently the byproducts must be discarded, creating a sizeable solid waste problem.[4]

---

[4]The limestone injection process, a promising method shown in Table 7-6, produces a solid waste that is mostly calcium sulfate, or gypsum ($CaSO_4$). A mass analysis gives the

Table 7-6. PROPOSED CHEMICAL PROCESSES FOR RIDDING STACK GASES OF SULFUR DIOXIDE[a]

Process	Chemical Substance Reacting with SO_2	Chemical Form of Sulfur after Reaction[b]	Balanced Chemical Reaction
Claus	H_2S	S (solid)	$2H_2S + SO_2 \rightarrow 3S + 2H_2O$
U. of Massachusetts	CO	S (solid)	$2CO + SO_2 \rightarrow S + 2CO_2$
Parsons	H_2	H_2S (removable gas)	$3H_2 + SO_2 \rightarrow H_2S + 2H_2O$
Limestone (dolomite) injection	$CaCO_3$	$CaSO_3$ (solid)	$CaCO_3 + SO_2 \rightarrow CaSO_3 + CO_2$
		$CaSO_4$ (solid)	$2CaCO_3 + 2SO_2 + O_2 \rightarrow 2CaSO_4 + 2CO_2$
Manganese oxide	MnO	$MnSO_3$ (solid)	$MnO + SO_2 \rightarrow MnSO_3$
Ammonia scrubbing	NH_3	NH_3	$2NH_3 + SO_2 + H_2O \rightarrow 2(NH_4)_2SO_3$
Alkaline scrubbing	NaOH	$NaHSO_3$ (solution)	$NaOH + SO_2 \rightarrow NaHSO_3$
Sodium sulfite scrubbing	Na_2SO_3	$NaHSO_3$ (solution)	$Na_2SO_3 + SO_2 + H_2O \rightarrow 2NaHSO_3$
Potassium formate	KOOCH	$K_2S_2O_3$ (solution)	$2KOOCH + 2SO_2 \rightarrow K_2S_2O_3 + 2CO_2 + H_2O$

[a]The chemical reactions shown here often constitute only a first stage of the total sulfur (S) removal process, subsequent reactions being employed to complete the removal and recycle some of the reacting substances.

[b]Now subject to removal.

Other promising efforts at SO_2 removal have centered around the improvement (by use of vanadium pentoxide (V_2O_5) catalyst) of the conversion to sulfuric acid (H_2SO_4) so that dilute SO_2 gases can be treated; adsorption of SO_2 on char; scrubbing by seawater; and treatment with alkalized alumina (Al_2O_3). Approaches based on using or producing low-sulfur coals will be discussed in the section "Major Air-Pollution Sources—Coal-Burning Electric Plants," pages 257–263.

Because of the immense technical problems, none of the above methods has proved entirely satisfactory. A 1970 report from the National Research Council (NRC) shows the seriousness of the situation: even with a full commitment to SO_2 control, three to ten years would pass before a reasonable control technology would be widely available. A pattern of increasing SO_2 emissions throughout the 1970's is therefore possible because of the rapidly expanding use of coal-produced electrical energy and other sulfur-emitting industries. Without a major commitment by government and industry, SO_2 levels may increase throughout the rest of this century, following the threatening upward trend shown in Figure 7-6.

relationship of the amount of $CaSO_4$ to the amount of SO_2 captured

Reaction:	$2CaCO_3$	$+ 2SO_2$	$+ O_2 \rightarrow$	$2CaSO_4$	$+ 2CO_2$
Mass analysis:	200	128	64	272	88

showing that the ratio is 272 parts $CaSO_4$/128 parts SO_2, or just over 2 to 1. Thus the removal of 500 tons of SO_2 per day from a large power plant or smelter would create over 1000 tons each day of solid $CaSO_4$, which would need to be hauled away and discarded. Some of it might be washed into nearby soil and water by impinging rainfall. Similar problems exist with dolomite ($MgCa(CO_3)_2$) injection, where magnesium atoms replace some calcium atoms in each step above.

PARTICLES

Small particles in enormous quantities occupy the earth's atmosphere. Some of the particles are solid and some are droplets of liquid. Most are composed of millions of atoms and molecules associated with one another. Their chemical composition and chemical properties are highly variable. Some will dissolve in water, some will react chemically with other atmospheric gases, and some will catalyze important atmospheric reactions. Some contain complex, carcinogenic (cancer-producing) substances, and others contain toxic metals. Altogether, the chemistry of atmospheric particles is enormously variable and complex, with much yet to be understood.

Both natural and man-made particles occupy our atmosphere. Natural particles are somewhat more abundant on a global basis (Chapter 6), but particles produced by man overwhelm the urban air. They come from many sources—coal-burning power plants, steel mills, cement plants, and gasoline (internal-combustion) engines are prominent examples. Table 7-7 shows a breakdown of sources within the United States.

As usual, we must be careful when we interpret tonnage-emissions tables. While such tables present the simplest overall picture, the environmental impact of particles hinges on far more than particle weight. Health effects, for example, depend strongly on both the chemical nature of the particles and particle size. Clearly, particles of radioactive plutonium oxide (PuO_2), toxic in the extreme, are a greater threat to health than are particles of inert silica (SiO_2) having the same mass. And weight for weight, small particles are far more dangerous than large. To begin with, it takes a far greater number of small particles than of large ones to provide a given weight.[5] Combined with their greater numbers per unit of mass, small particles are an additional hazard because they penetrate into the lungs more deeply than do large ones. (The implications of this problem will be discussed later.) It is unfortunate indeed that both emission inventories and federal air standards are based on mass alone. For instance, federal air-quality standards to be reached by 1975 for particulate matter (Table 7-3) allow a year-long average of 60 micrograms in each cubic meter (m^3) of air. But 60 micrograms could conceivably contain anywhere from 10^5 to 10^{14} particles, a billionfold variation. Therefore while official standards can be met by ridding air of its large, weighty particles, true progress cannot be recorded until the hordes of small particles (and those most chemically toxic) are reduced. We shall examine prospects for this effort shortly.

The size of particles depends on their source. Particles larger than 10 microns (μ) are produced mechanically, through grinding, erosion,

---

[5] One gram of 10 micron (μ) diameter particles of typical density contains about one billion (10^9) particles. One gram of 1 μ particles contains 1000 times more particles, about one trillion (10^{12}). In general, the number of particles will increase in proportion to (1/particle diameter)^3.

Table 7-7. BREAKDOWN OF PARTICULATE EMISSION SOURCES IN THE UNITED STATES[a]

Source	Emissions, Millions of Tons per Year	Percent of Total
Transportation	1.2	4.3
Motor vehicles	0.8	2.8
Gasoline	0.5	1.8
Diesel	0.3	1.0
Railroads	0.2	0.7
Vessels	0.1	0.4
Nonhighway use of motor fuels	0.1	0.4
Fuel combustion in stationary sources	8.9	31.4
Coal	8.2	29.0
Fuel oil	0.3	1.0
Natural gas	0.2	0.7
Wood	0.2	0.7
Industrial processes	7.5	26.5
Iron and steel	1.9	6.8
Grey-iron foundries	0.2	0.3
Other metals	0.1	0.6
Cement	0.9	3.1
Stone, sand, rock, etc.	0.9	3.1
Coal cleaning	0.2	0.7
Phosphate rock	0.2	0.7
Lime	0.4	1.6
Asphalt batching	0.5	1.9
Other mineral products	0.2	0.6
Oil refineries	0.1	0.4
Other chemical industries	0.1	0.3
Grain handling and storage	0.8	2.8
Pulp and paper	0.7	2.5
Flour and feed milling	0.3	1.1
Solid-waste disposal	1.1	3.9
Miscellaneous	9.6	33.9
Forest fires	6.7	23.7
Structural fires	0.1	0.4
Coal refuse burning	0.1	1.4
Agricultural burning	2.4	8.4
Total	28.3	100.0

[a]NAPCA Publication No. AP-73, 1970; 1968 emissions.

or spraying. The size range from 1 μ to 10 μ may be produced mechanically or may originate as a dust or ash. Particles under 1 μ (aerosols) more frequently come from condensation of nonvolatile molecules together as they cool (as in smoke or auto exhaust) or as a solvent evaporates. Figure 7-8 (p. 244) shows the typical size range for particles of various origins.

Natural Removal of Particulate Matter. As we said in the last chapter, particles in the air eventually reach earth by being swept out with rain or settled out by gravity. Aerosol particles, under 1 μ in diameter, are influenced very little by gravity, but they coagulate

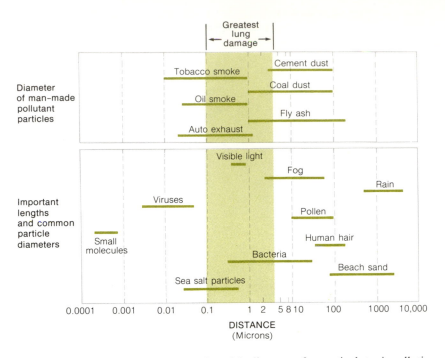

Figure 7-8. The most frequent range of particle diameters for particulate air pollution. These diameters are compared with other significant sizes in the lower half of the diagram.

to form large particles which do then settle to earth. Particles over 10 μ settle rapidly, usually reaching the ground within a day or two. Few particles remain in the lower atmosphere for more than one or two weeks.

Effects of Airborne Particles. The most obvious effect of atmospheric particulate matter is dirtiness—the visible dirtiness of air and the dirty grime that settles from air onto buildings, sidewalks, and clothing. In industrial areas, over 1000 tons of grime may be deposited over a square mile each year. This grime is more than simply unpleasant. It promotes the corrosion of metals and the deterioration of canvas awnings and other fabrics. It damages paint and vegetation and can cause short circuiting in high-voltage lines.

Particulate pollution is visible, in contrast to most of the air pollution caused by individual molecules. Most particles in air have a diameter within a factor of ten of the wavelength of light (about 0.5 μ, Chapter 2) and so can scatter light to cause the typical "smoggy" appearance of urban air. Single molecules, by contrast, are too small to scatter light. However, a few of them (most notably nitrogen dioxide, NO_2) can discolor the air by absorbing certain wavelengths of light.

Actual particle counts in urban air show why visibility is frequently poor. Los Angeles air has been found to contain up to 100 billion particles per cubic meter. Other cities are not far behind.

The effect of particles on human health is our greatest concern.

A ground inversion in St. Louis, Missouri, enshrouds the city in smog, but the Gateway Arch reaches through to clear air above. (*Photo by St. Louis Post-Dispatch.*)

Inhaled particles larger than 10 μ are lodged in the nostrils. Particles in the size range from 5 μ to 10 μ are captured by the mucous lining of the upper airway; they are carried to the throat and swallowed. Particles under 2 μ in size are the greatest threat: they penetrate the deeper structures of the lung, including the alveolar sacs, where no protective mucous blanket exists. Some of these particles may be retained in the lungs, although the extremely small particles, under 0.1 μ, are usually swept out with exhaled air.

Particles reaching the deeper parts of the lung may bring with them toxic pollutants that have adsorbed on their surfaces. It has been suggested that particles enhance the damage to lungs caused by sulfur dioxide (SO_2) because they carry SO_2 to deep regions of the lung that are not otherwise reached. Particles also catalyze the conversion of SO_2 to more corrosive SO_3 and sulfuric acid. Such enhancements of one pollutant's (for example, SO_2's) effect by another (particles) is called *synergism*. Synergism is common in all areas of environmental pollution.

Inhaled particles may be toxic in their own right. Soot particles are known to contain benzopyrene, the organic, cancer-producing substance described in Chapter 4. Lead-containing particles from automobile exhaust may transmit toxic lead (Pb) to the lung and thence to the bloodstream. (The lead problem will be discussed more completely in Chapter 9.) Other examples are known, but particles are so complicated by the enormous variety of substances they contain that many of their toxic constituents are probably still undiscovered.

Control of Particulate Pollution.   Methods for controlling particles are based largely on techniques for reducing particle formation in the first place (prevention) or on mechanical or electrical collection once they are formed (cure). Only in unusual cases can chemical reaction be used to destroy particles.[6]

The simplest mechanical collector is a large settling chamber, to which incoming gases release large particles, above $40\ \mu$. Particles down to $5\ \mu$ can be removed by a cyclone collector, which spins the gas in a violent spiral until the particles are thrown to the outside, where they can be collected. Scrubbers (or washers) use a spray of water to wash out particles, Most of these methods fail to remove particles smaller than $5\ \mu$. but one, the Venturi scrubber, is effective to sizes below $1\ \mu$. Fabric filters or "baghouses" allow dirty air to circulate into and through fabric bags. The fabric filters remove nearly all particles from the existing air down to sizes between $10\ \mu$ and $1\ \mu$, and occasionally lower, depending on the nature of the fabric.

The electrostatic precipitator is perhaps the most useful collector of particulate matter because it works well with small particles. Most particles, large and small, carry or can be given a small amount of static charge. A high voltage (between 12,000 and 100,000 volts) between metal plates will draw these particles to the plates by simple electrostatic attraction. Once deposited on the plates, the particulate matter can be simply washed or shaken off and then discarded. The fraction of particles removed from dirty air by this method ranges between 95 and 99.5 percent.

Although the "prevention" approach to particle control is dominated by the foregoing "cures," there is one major exception. The

---

[6]Chemical reaction usually fails to consume particles because the molecular composition of the particulate matter is too diverse for a uniform reaction to occur, and even if reacted, the particle is still usually a particle, changed only in chemical nature. The oxidation (by burning) of organic particles is the only prominent exception, in this case because gases are formed.

combustion of fuels, garbage, and other organic matter is often incomplete, releasing unburned organic materials as smoke and soot particles. However, in a well-designed furnace, the organic materials are completely oxidized to carbon dioxide (CO_2) and water (H_2O), leaving no sooty residue. Such furnaces are especially important in waste incineration, as anyone will recognize after witnessing the dense smoke from burning garbage dumps or piles of discarded tires.

In summary, the variety of approaches to particle control promises to alleviate much of the impact of particulate pollution, provided proper equipment is bought and installed. The main problem remaining is that of small particles. (See Figure 7-9.) Only a few of the collection systems will trap these particles, and even then at less than 100 percent. In addition, many small particles come from sources that are difficult to control, including the automobile. Worst of all, many small particles form in the atmosphere itself and are therefore impossible to trap. The most striking example is found in photochemical pollution, where tiny particles grow in the air, nurtured on a constant outpouring of oxides of nitrogen and hydrocarbons from many assorted enterprises. Control of the particles must strike at these two chemical roots. In the next section we discuss the first of the two—the oxides of nitrogen.

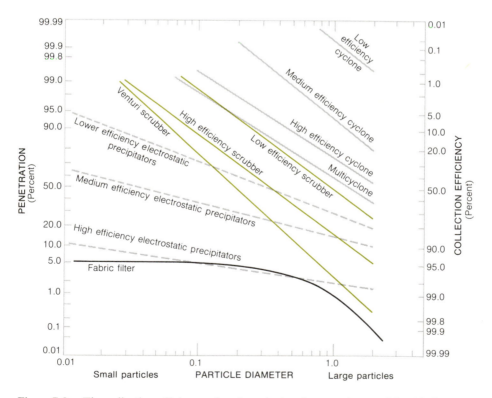

Figure 7-9. The collection efficiency of various devices for trapping particles. Notice that all collectors lose efficiency in trapping small particles. (From *Particulate Pollutant System Study*, Volume II, Midwest Research Institute, 1971.)

OXIDES OF NITROGEN

There are at least five kinds of small nitrogen-containing gas molecules in the atmosphere: N_2, NO, NO_2, N_2O, and NH_3. Diatomic nitrogen (N_2) dominates all others in quantity, making up 78 percent of the total atmosphere (Table 7-1). However, N_2 is colorless, odorless, nontoxic, and most often chemically unreactive; its chief environmental role is to provide a large reservoir of nitrogen from which other nitrogen-containing compounds (including some essential to life) originate. Among these other compounds, dinitrogen oxide (N_2O) is most abundant in the natural atmosphere (0.2 ppm) but not significant as a man-made pollutant; ammonia (NH_3) is occasionally an air pollutant; and nitric oxide (NO) and nitrogen dioxide (NO_2) are major air pollutants. In this section we concentrate on the latter two.

Whenever air is raised to a very high temperature, NO is produced (Chapter 5) by the chemical reaction

$$N_2 \quad + \quad O_2 \longrightarrow 2NO$$

diatomic	diatomic	nitric
nitrogen	oxygen	oxide

Such heating occurs in flames, and therefore most NO comes from combustion of fuels—coal, gasoline, oil. Worldwide emissions, tabulated by fuel source, are shown in Table 7-8. (Natural sources—lightning and biological processes—account for even greater quantities in the atmosphere as a whole.) All emission values are reported

Table 7-8. WORLDWIDE URBAN EMISSIONS OF NITROGEN OXIDES IN MILLIONS OF TONS[a]

Fuel	Source Type	NO_2 Emission
Coal	Power generation	12.2
	Industrial	13.7
	Heating	1.0
Petroleum	Refineries	0.7
	Gasoline	7.5
	Kerosene	1.3
	Fuel oil	3.6
	Residual oil	9.2
Natural gas	Power generation	0.6
	Industrial	1.1
	Heating	0.4
Others	Incineration	0.5
	Wood fuel	0.3
	Forest fire	0.8
Total		52.9

[a]From E. Robinson and R. C. Robbins, *Sources, Abundance, and Fate of Gaseous Atmospheric Pollutants Supplement* (Menlo Park, Calif.: Stanford Research Institute, 1969).

in terms of NO_2, the chemical form to which NO invariably becomes oxidized in the atmosphere. (Since NO and NO_2 exist side by side, they are sometimes represented by the collective symbol NO_x.)

Reactions and Natural Removal of the Oxides of Nitrogen. Once formed by combustion, NO persists in the atmosphere an average of four days (occasionally down to a few hours in dirty urban air) before it is oxidized. The oxidation is probably accomplished through atmospheric ozone (O_3), perhaps by the very same reaction in which the O_3 of the stratospheric UV screen is threatened by NO from the SST (Chapter 6):

$$NO \quad + \quad O_3 \longrightarrow NO_2 \quad + \quad O_2$$

| nitric oxide | ozone | nitrogen dioxide | diatomic oxygen |

The resultant NO_2 is partly absorbed by vegetation, water, and soil, and partly converted to nitric acid by a hydrolysis reaction with airborne water:

$$3NO_2 \quad + \quad H_2O \longrightarrow 2HNO_3 \quad + \quad NO$$

| nitrogen dioxide | water | nitric acid | nitric oxide |

Nitric acid may be brought down directly in rain, or it may first combine with ammonia (NH_3) to form ammonium nitrate (NH_4NO_3). In either case the nitrogen is eventually washed to earth, where it becomes a nutrient for plants. At this stage it enters the biological phase of the famed nitrogen cycle. This story is interesting in itself, but we must leave it to focus on the unique effects of NO and NO_2 while they reside in the air.

Effects of NO and NO_2. While NO is a colorless gas, NO_2 is unusual among simple air pollutants because it is colored, giving any air that contains it a brownish-yellow appearance.[7] An NO_2 level of 1 ppm, sometimes found in urban centers, would reduce visibility to a distance of about 10 miles.

The compound NO is relatively inert and only moderately toxic. Like carbon monoxide, NO can combine with hemoglobin (forming metheglobin), thus reducing oxygen transport. However, this effect does not appear to be important at normal urban concentrations of NO, which are usually less than 1 ppm.

In contrast to NO, NO_2 is a highly toxic and corrosive gas. It causes respiratory irritation and distress. Long-term effects are not yet clear. At urban levels, commonly under 1 ppm, the direct effects of NO_2 are only moderately significant. However, NO_2 is a building

---

[7] The color of NO_2 is due to its absorption of light in the visible spectrum. It absorbs mainly blue light: the light left over, from the other end of the spectrum (red-orange-yellow), is responsible for the brown-yellow color.

block in photochemical air pollution (next section), and because of this role it has an importance of the first rank.

Control of Pollution by NO and NO_2. The oxides of nitrogen (NO_x) are among the most difficult pollutants to bring into check. As long as the combustion of fuels with air continues to be such a prominent part of civilized existence, NO and NO_2 will be produced. However, certain changes in the design and operation of combustion devices will help to limit them. The major changes are based on the following chemical principles.

It was shown in Chapter 5 that the reaction $N_2 + O_2 \rightarrow 2NO$ produces very little NO at room temperature. The chemical equilibrium in air heated to flame temperatures is shifted to produce much more NO, up to 3.5 percent. Figure 7-10 shows how the percentage of NO increases as air is heated through the normal range of combustion temperatures, 2400° to 4500°F.[8] It shows also that air from which most of the O_2 has been removed produces less NO than does

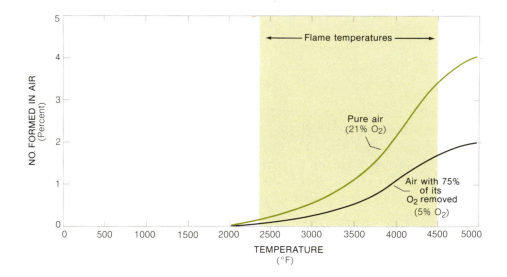

Figure 7-10. Percentage of nitric oxide (NO) formed from air (at equilibrium) by the reaction $N_2 + O_2 \rightarrow 2NO$. [Data from P. R. Ammann and R. S. Timmins, *A. I. Ch. E. Journal, 12,* 956 (1966).]

---

[8] We might ask why the NO concentration does not fall again as the combustion gases cool off, thus eliminating the problem of NO from hot flames. The reason is found in the high activation energy for the reaction $N_2 + O_2 \rightarrow 2NO$ and the reverse reaction $2NO \rightarrow N_2 + O_2$. A high activation energy (Chapter 5) makes these reactions so sluggish at normal temperatures that equilibrium concentrations cannot be reached even in weeks. At high temperatures, such as those existing in a flame, reactions go much faster and equilibrium is reached promptly. Thus the large equilibrium concentration of NO is reached quickly in a flame, but as the combustion gases cool, the reactions slow down and become "frozen," unable to provide the low NO concentration that one would expect at equilibrium in a cool gas. Only if a catalyst could be found to lower activation energy would NO approach its low equilibrium concentration in cooled combustion gases. Some progress has been made in developing such a catalyst; it may eventually be a useful tool for reducing NO pollution.

ordinary air when heated.[9] We can take advantage of these two trends to reduce NO pollution, but not without sacrifices.

First of all, we can reduce flame temperatures by diluting the normal fuel-air mixture with additional cool air, with the recirculated products of combustion, or with other nonflammable gases. This method reduces NO production, but the cooled flame also generates less useful energy. High temperatures, not low, are generally sought in energy conversion because the efficiency of conversion of heat to useful (mechanical or electrical) energy increases with temperature.[10] This rule is true for power plants as well as for automobiles and diesel locomotives. Therefore reducing flame temperatures is an unpalatable (but sometimes necessary) means for reducing NO.

The other approach suggested by Figure 7-10 for cutting flame-produced NO is to choke off the supply of oxygen (O_2). To some degree, this occurs naturally in combustion because the organic fuel consumes and thus depletes O_2:

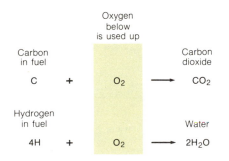

With an increase in the fuel-air ratio, more O_2 can be depleted. This approach indeed reduces NO, but for some reason that is not fully understood, the reduction is less than that anticipated. Even worse, insufficient oxygen leads to the incomplete combustion of fuel, which encourages the formation of carbon monoxide (CO) and hydrocarbons (general symbol, HC). Hence this approach, if pushed too far, simply trades one pollutant (NO) for several others (CO and HC). In a later section we shall discuss this dilemma, a major obstacle when we try to build a nonpolluting automobile engine.

Since it is inherently difficult to prevent NO from forming in

---

[9] Since O_2 is one of the basic constituents forming NO in the reaction $N_2 + O_2 \rightarrow 2NO$, a reduction in O_2 supply will starve the production of NO. In mathematical terms, the equilibrium relationship among the various gases is given by

$$C_{NO}^2 = K C_{O_2} C_{N_2}$$

where K is an equilibrium constant and C_{NO}, C_{O_2}, and C_{N_2} are the concentrations of NO, O_2 and N_2, respectively. This equation shows precisely how C_{NO} rises with oxygen concentration.

[10] Recall Chapter 1, in which the maximum efficiency (fraction of heat energy converted to useful energy) of any machine was given by efficiency = $(T_{hot} - T_{cold})/T_{hot}$. This efficiency increases as the hot temperature of the flame, T_{hot}, rises well above that of the cooling environment, T_{cold}. (Temperatures must be expressed in absolute temperature units.)

combustion, a second line of attack is to seek a cure for the affliction—to try to clean up the NO in the spent combustion gases. (We focus on NO because it is the root of the problem—most NO_2 is formed by the oxidation of NO in the atmosphere.) Approaches have been proposed or tried that were based on scrubbing the combustion gases with a spray of liquid or by adsorption on special solid materials. The large volume of combustion gases and the difficulty of disposing of the wastes have been discouraging.

More promising are methods for the chemical reduction of NO. Recall that NO is formed by oxidation of nitrogen (N_2) gas, $N_2 + O_2 \rightarrow 2NO$. If an oxygen-hungry (reducing) gas could be mixed with NO in the combustion effluent, perhaps the oxygen atom could be taken back, reducing NO to harmless N_2. Many reducing gases have been proposed for various situations. Some of these gases and their chemical reactions with NO are shown below:

$$2CO + 2NO \longrightarrow N_2 + 2CO_2$$

carbon
monoxide

$$4NH_3 + 6NO \longrightarrow 5N_2 + 6H_2O$$
ammonia

$$CH_4 + 4NO \longrightarrow 2N_2 + CO_2 + 2H_2O$$
methane

Two problems have not yet been entirely resolved: the selection of catalysts and temperature conditions suitable for large volumes of gas, and the avoidance of undesirable side reactions. For instance, if the combustion gases contain sulfur dioxide (SO_2), the use of carbon monoxide (CO) can produce toxic carbonyl sulfide (COS) through the reaction

$$3CO \quad + \quad SO_2 \longrightarrow 2CO_2 \quad + \quad COS$$

| carbon | sulfur | carbon | carbonyl |
| monoxide | dioxide | dioxide | sulfide |

In short, many problems still plague efforts to control the oxides of nitrogen. Combustion sources are so varied that no general solution is in sight. We must hope that individual approaches to the many sources will bring progress in the years ahead.

HYDROCARBONS AND PHOTOCHEMICAL OXIDANTS

More hydrocarbons (general symbol, HC) are emitted by natural sources than by man. Methane (CH_4), the most abundant hydrocarbon in the atmosphere (1.55 ppm background), is produced worldwide in the anaerobic decomposition of dead plant material. Methane is joined by a class of hydrocarbons of more intricate molecular structure known as terpenes.[11] These substances emanate from plants. Terpene molecules bond together (polymerize) to form tiny aerosol particles that are visible as the "blue haze" over many forested areas.

Only 15 percent or so of atmospheric hydrocarbons come from the activity of man. Their human impact is out of proportion to their abundance because, as with other man-made pollutants, they are released mostly in urban air.

Hydrocarbons by themselves cause little damage. However, in the presence of light, a small fraction of hydrocarbons in the air react chemically with oxygen atoms, ozone, and oxides of nitrogen to form photochemical oxidants. These toxic and irritating compounds are known as secondary pollutants because no source emits them directly. Instead they form by chemical reactions between emitted compounds—the *primary pollutants*.

Most airborne hydrocarbons, including methane (CH_4), have little tendency to react in the atmosphere, and they have only slight human significance. The ones that do react are called *reactive hydrocarbons*: they must be singled out for control if we are to decrease photochemical oxidants. Table 7-9 (p. 254) shows the detailed worldwide sources of reactive hydrocarbons. Emissions are compared with total hydrocarbon output for comparison. The outstanding entry in this table is gasoline (used mainly in the internal-combustion engines of cars) which leads all other sources in both total emissions and the percentage of emissions that are reactive. Next most serious is incineration, used to dispose of solid wastes.

Reactions and Natural Removal of Hydrocarbons. The eventual fate of all atmospheric hydrocarbons is oxidation to carbon dioxide (CO_2) and water (H_2O). In this process, each carbon atom or hydrogen atom becomes bonded exclusively to atoms of oxygen from the atmosphere. The bonding with oxygen atoms occurs in stages, and the intermediate substances produced at some of the stages in sunlight are undesirable photochemical oxidants.

Photochemical oxidants are defined as oxidizing substances produced under the influence of light.[12] Whether photochemical oxidants

---

[11] A simple but rather typical terpene is limonene, found commonly in citrus fruit and in the needles of some pines:

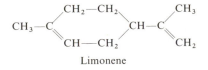

Limonene

[12] Recall from Chapter 5 that many reactions are promoted by electromagnetic radiation (light). Light is absorbed by molecules, and its energy aids in rupturing the molecules' bonds so that the reaction can proceed.

Source	Emissions		
	Total Hydrocarbons	Percent Reactive	Reactive Hydrocarbons
Coal			
Power generation	0.2	15	0.03
Industrial	0.7	15	0.10
Heating	2.0	15	0.30
Petroleum			
Refineries	6.3	14	0.88
Gasoline	34	44	15
Kerosene and oils	0.4	18	0.06
Evaporation and transfer loss	7.8	20	1.6
Other			
Solvent use	10	15	1.5
Incinerators	25	30	7.5
Wood fuel	0.7	15	0.10
Forest fires	1.2	21	0.25
Totals:	88		27

[a]From E. Robinson and R. C. Robbins, *Sources, Abundance, and Fate of Gaseous Atmospheric Pollutants Supplement* (Menlo Park, Calif.: Stanford Research Institute, 1969).

are formed from hydrocarbons depends in large part on the intensity of sunlight and the chemical nature of the hydrocarbons. Remember (Chapter 4) that hydrocarbons consist of an enormous number and variety of organic molecules, all composed of hydrogen and carbon. The chemical behavior of these molecules depends upon the way in which the various atoms are bonded together. Specifically, it is found that the alkanes or paraffin hydrocarbons, in which all atoms are joined by single bonds, do not react very readily to form photochemical oxidants. The olefins (alkenes), in which some carbon atoms are joined to one another by double bonds, are the major reactive hydrocarbons, producing photochemical oxidants rapidly in the presence of sunlight and the oxides of nitrogen.

The reactions that occur as olefins are oxidized in polluted air are varied and complex, with a pattern not yet fully understood.[13] In fact, the importance of these photochemical oxidation processes in urban air was not even recognized until 1951, when A. J. Haagen-Smit demonstrated that the irritating "smog" of Los Angeles was mostly of photochemical origin. This type of pollution is now referred to as *photochemical smog*.

A key step in the formation of photochemical smog is the splitting

---

[13] These processes are very complex. T. A. Hecht and J. H. Seinfeld [*Environmental Science and Technology, 6,* 47 (1972)] have outlined a scheme of 81 distinct chemical reactions that may occur in the photochemical oxidation of propylene (C_3H_6) alone. Fifteen of them are suggested to have central importance.

apart of nitrogen dioxide (NO_2) molecules with the aid of sunlight to yield free oxygen atoms.[14]

$$NO_2 \xrightarrow{\text{sunlight}} NO + O$$

nitrogen   nitric   oxygen
dioxide   oxide   atom

Oxygen atoms are extremely reactive (Chapter 6, Section 2); many of them react with O_2 to form ozone (O_3), $O_2 + O \rightarrow O_3$. Other O atoms along with O_3 attach reactive hydrocarbons by way of a number of chemical processes. Among the varied products are many eye-burning compounds, including formaldehyde, acrolein, and the notorious photochemical oxidant peroxyacetyl nitrate, commonly known as PAN:

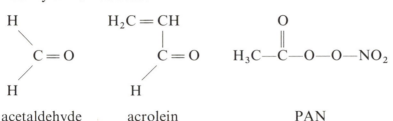

acetaldehyde   acrolein   PAN

More ozone is produced by the incessant chemical reactions, and ozone levels soar. Various molecules in this complex atmospheric soup bond together, forming larger molecules and eventually tiny particles. These particles and the brown colored NO_2 give the air a· dirty, oppressive appearance.

Eventually the hydrocarbons are oxidized completely to CO_2 and H_2O. Until that end occurs, many compounds having only partial oxygen attachment (like the three pictured above) are both irritating and toxic to human populations.

Effects of Hydrocarbons and Photochemical Oxidants. A great variety of hydrocarbons exist, ranging in size and complexity from simple methane (CH_4) to large polymer molecules and polynuclear aromatic compounds having hundreds of atoms. The smaller molecules form gases; with increasing size, the molecules are pulled together into liquids and waxlike solids by the growing intermolecular forces. Most hydrocarbons, particularly the small, gaseous ones, are nontoxic. A few complex hydrocarbons in the polynuclear aromatic class (such as benzopyrene) are intensely carcinogenic, as we saw in Chapter 5. Still, taken together, the direct impact of all hydrocarbons in the atmosphere is small compared to that of secondary photochemical pollutants they give birth to, described next.

Ozone (O_3), a colorless gas, is a powerful oxidizing agent. The extra oxygen atom it acquires in the reaction $O_2 + O \rightarrow O_3$ transfers

---

[14] This photochemical reaction utilizes only the energetic component of sunlight that has a wavelength less than 4000 A (0.4 μ). See Figure 2-7.

with ease to many other substances, including most organic materials. For this reason, O_3 attacks rubber, causing it to harden and crack. Just as easily, it oxidizes key organic molecules in living systems, causing sickness and death. Ironically, its power to destroy germs long made ozone in the air a symbol of good health. Unfortunately, its destructive powers are much broader. Rats exposed to as little as 1 ppm for 8 hours a day over a year-long period develop bronchitis and fibrosis. Exposure to 1 ppm for 1 hour has been found to decrease man's breathing capacity; a few ppm cause sickness and death. Because of its severe toxicity, maximum industrial exposure (the TLV value) is a slight 0.1 ppm for O_3, 50 times smaller than the value for SO_2. Yet with considerable regularity (about once monthly) ozone concentrations in the Los Angeles basin exceed 0.4 ppm.

Peroxyacetyl nitrate (PAN) is both toxic and irritating. Concentrations down to a few parts per billion cause eye irritation. Plants are particularly sensitive: a fraction of 1 ppm causes extensive damage to vegetation. Crop losses in the millions of dollars have occurred in Los Angeles County and elsewhere because of this and other photochemical oxidants. In Los Angeles, PAN concentrations frequently exceed 0.05 ppm.

Control of Photochemical Oxidants.   Control over photochemical oxidants can be gained only by striking at their chemical roots: hydrocarbons and oxides of nitrogen (NO_x). We saw the difficulties in reducing NO_x emissions in the last section. Here we deal with hydrocarbon controls.

Table 7-9 shows that most reactive hydrocarbons originate in combustion, the two leading sources being the gasoline (internal-combustion) engine and incinerators. The fundamental role of combustion is always to burn organic materials (including hydrocarbons), not to allow their escape. Hence hydrocarbon emissions reflect faulty combustion devices—those with incomplete fuel combustion. The obvious cure is to modify combustion technology so that we have more complete combustion. Several approaches are possible, but none is faultless.

To try to reach total combustion, we can use larger quantities of oxygen—that is, increase the air-fuel ratio. This approach works, but at the same time it increases the rate of formation of undesirable nitric oxide (NO), which requires oxygen for its formation (see last section). Thus with combustion processes in general, a compromise must be reached, for too much oxygen encourages oxides of nitrogen, and too little encourages hydrocarbons (and carbon monoxide). A partial solution to this dilemma for incinerators and other furnaces is two-stage combustion, in which excess oxygen is not added until a second, cooler stage is reached. The lower temperature of this stage discourages the formation of NO despite the presence of excess oxygen.

Other methods of hydrocarbon control are used for many lesser sources. For automobiles, the use of catalysts promises to help complete the oxidation of hydrocarbons. (The particular problem of automobile emissions will be discussed separately in the last section.) In summary, reasonable hydrocarbon control is within man's grasp, the principal hurdles being the cost of control and the concomitant increase in NO emissions as control is used.

MAJOR AIR-POLLUTION SOURCES— COAL-BURNING ELECTRIC PLANTS

Air pollution arises in sources of the greatest variety. Many industrial activities contribute, as do certain domestic and agricultural practices. The basic processes that lead to air pollution are equally varied—combustion is the overwhelming villain, but ore smelting, grinding and mixing, tilling of the land, and evaporation of solvents and fuels all contribute substantially.

Of the specific sources of air pollution that plague civilization, two are outstanding: coal-fired electric generating plants and gasoline-driven internal-combustion engines in automobiles. Both derive energy from combustion. Each emits pollutants of considerable chemical variety in vast quantities. Each presents an extreme challenge to cleanup technology. We shall deal with each in turn, starting here with electric generating plants that derive electrical energy from the combustion of coal.

Coal-burning power plants dot the countryside, each emitting large quantities of combustion gases from smokestacks (see photo). These gases contain three major pollutants: sulfur dioxide, particulates, and oxides of nitrogen. Alongside these are trace quantities of hydrocarbons and carbon monoxide, of heavy metals like mercury

Crystal River power station smoke plume. (*Photo by Roy O. McCaldin.*)

and cadmium, and of radioactive uranium and thorium isotopes. (See Chapter 11.) Most of them can be traced directly to the basic fuel—coal.

The Problem with Coal. Coal is largely the organic remains of ancient plant life. Its origin can be traced back 60 to 300 million years, a time generally dominated by low swamps and luxurious plant growth.

Swamp water often is stagnant, rich in organic material, and lacking in oxygen. Vegetation that falls into such water is protected from total decay by its oxygen-free (anaerobic) condition: plant debris can therefore accumulate in deep bottom layers. Traces of minerals in the plants become part of the deposits; more inorganic material is added by windblown dusts and mineral-rich water. Eventually such deposits are covered by sediments, are subjected to great pressures and in some cases elevated temperatures, and are converted to seams of coal by the working of intricate chemical reactions extending over millions of years. The coal is laced with chemical elements brought in by the vegetation itself, by water, and by wind. The resulting composition of coal of different types (ranks) is suggested by Table 7-10.

The broad spectrum of chemical elements in coal creates a considerable pollution problem, for many of these elements enter the air as undesirable contaminants when the coal is burned. Foremost among the unwanted elements is sulfur, a natural constituent both of plant life and of inorganic minerals deposited from water. With combustion, the sulfur (S) is oxidized to gaseous sulfur dioxide (SO_2), $S + O_2 \rightarrow SO_2$. The sulfur dioxide escapes up the smokestack to become a major pollutant for man. Quantities of SO_2 as great as 500 tons or more per day are emitted by large power plants.[15]

A second major contaminant in coal consists of the various minerals, most of which become inorganic ash upon combustion. The ash content of coal is commonly 5 to 10 percent, as shown in

---

[15] A 1000 megawatt plant will consume about 10,000 tons of coal per day. If the coal contains 2.5 percent S (2.5 parts S/100 parts coal), average for power generation in the U. S., the total S burned per day is 10,000 tons coal × (2.5 parts S)/(100 parts coal) = 250 tons S. Most of this S forms SO_2 (a small percentage becomes sulfur trioxide and inorganic sulfate), the amount being calculated by an atomic mass analysis of the underlying reaction which converts S to SO_2.

	sulfur in coal	oxygen	sulfur dioxide
	S +	O_2 →	SO_2
Mass inventory:	(32.1)	(32.0)	(64.1)

This equation shows that the ratio of SO_2 to S is given by the conversion factor 64.1 parts SO_2/32.1 parts S, or very close to 2 parts SO_2/1 part S. Therefore 250 tons of sulfur forms SO_2 in the amount

$$250 \text{ tons S} \times \frac{2 \text{ parts } SO_2}{1 \text{ part S}} = 500 \text{ tons } SO_2$$

This quantity of sulfur dioxide is produced each day. It mixes with the combustion gases and escapes almost totally into the atmosphere.

Table 7-10. THE PERCENTAGE ELEMENT-BY-ELEMENT COMPOSITION
OF TYPICAL COALS IN THE UNITED STATES[a]

Rank	State	Inorganic Ash[a]	Sulfur	Hydrogen	Carbon	Nitrogen	Oxygen
Anthracite	Pa.	9.0	0.6	3.4	79.8	1.0	6.2
Semianthracite	Ark.	10.1	2.2	3.7	78.3	1.7	4.0
Bituminous coal							
Low volatile	Md.	12.3	3.1	4.5	74.5	1.4	4.2
Medium volatile	Ala.	9.9	0.8	4.9	76.7	1.5	6.2
High volatile B	Ohio	3.8	3.0	5.7	72.2	1.3	14.0
Subbituminous coal	Wash.	10.9	0.6	6.2	57.5	1.5	23.4
Lignite	N. Dak.	5.2	0.4	6.9	41.2	0.7	45.6

[a]The inorganic ash is the residue remaining after combustion, and it contains by itself many chemical elements, including oxygen (O), silicon (Si), aluminum (Al), iron (Fe), and others. (See Table 7-11.) (Data from U.S. Bureau of Mines Information Circular No. 769, 1954.)

Table 7-10. The major constituents in ash are shown in Table 7-11. There are many minor components of ash (not shown), some of them toxic. For instance, coal contains 1 to 2 ppm of the slightly radio-active isotopes uranium-238 and thorium-232. These isotopes concentrate to about 10 ppm in the ash with combustion. Some heavy metals are found at comparable levels. Most of them concentrate in the ash, but mercury (Hg) is so volatile that it escapes directly into the combustion gases.

Electric-Power Generation. Enormous quantities of coal—nearly 250 million tons a year in this country alone—are burned to produce power. A large power plant consumes 20,000 tons or more of coal a day. The giant Kaiparowits plant, scheduled for construction in southern Utah just 75 miles north of the Grand Canyon, will gobble 50,000 tons in a single day. Everything is large scale in these plants. The size of the coal-burning furnace and accessories dwarfs man, as is suggested by Figure 7-11.

The salient features of a power plant are simple (Figure 7-11, p. 260). Coal is pulverized and blown into giant furnaces for ignition. Most of the carbon of each tiny coal fragment is burned to CO_2; hydrogen is burned to H_2O; and sulfur is burned to SO_2. A slight amount of carbon becomes carbon monoxide (CO), various hydro-

Table 7-11. CHEMICAL COMPOSITION OF INORGANIC ASH[a]

Chemical Substance	Range of Weight Percent
Silica (SiO_2)	34–38
Alumina (Al_2O_3)	17–31
Iron oxide (Fe_2O_3)	6–26
Calcium oxide (CaO)	1–10
Magnesium oxide (MgO)	0.5–2
Sulfur trioxide (SO_3)	0.2–4
Unburned carbon (C)	1.5–20

[a]From L. J. Minick, *Am. Soc. Testing Mater. Proc., 54,* 1129 (1954).

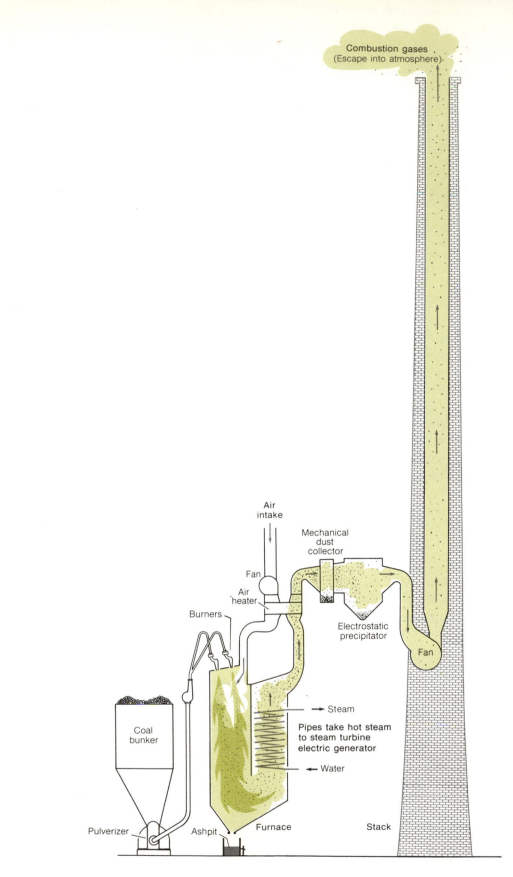

Figure 7-11. The principal features of a coal-burning electric generating plant.

carbons, and unburned specks of carbon. The inorganic or rocklike minerals melt under the intense heat, forming microscopic droplets. These droplets solidify, incorporate the unburned carbon specks, and become tiny particles of fly ash. Some settle into the ashpit (Figure 7-11), but others are carried away with the combustion gases.

At this stage the hot furnace gases contain many components: N_2 and O_2 from air; CO_2, H_2O, and SO_2 from fuel combustion; NO from the high-temperature reaction between N_2 and O_2; abundant fly-ash particles; and traces of CO, SO_3, NO_2, metallic vapors, hydrocarbons, and other organic compounds.

Next the hot combustion gases exit over banks of water-containing pipes, designed to absorb heat and create steam to drive giant electric generators. The gases, having given a great part of their intense heat energy to the steam, are now useless wastes and are discarded through a smokestack. By present practice, all the various gases and vapors in the combustion products, some of them major pollutants, enter the atmosphere undiminished. Only the solid particles of fly ash are controlled, often by the use of several types of collectors, including electrostatic precipitators. Collection efficiency ranges from 60 to 99 percent on a weight basis.

Even though the trapping of particles is common practice, power plants are still a major source of particulate matter in the air (Table 7-7). The reason is simply that the amount of inorganic ash from coal is so tremendous that even with a high-percentage collection, enormous quantities of particles remain.

Control of the other pollutants is no simpler. The enormous volumes of gas and soot from which pollutants must be removed presents a formidable challenge. A large power plant emits 10^8 cubic meters (m^3) of combustion gases per day. This amount is roughly the volume of air in 100,000 houses. To thoroughly clean so much effluent in a day is presently impossible.

Many partial solutions to power-plant pollution are being instituted. Improved collectors are able to trap more particles. Special furnace designs will reduce oxides of nitrogen. And increasingly tall stacks—some now 1000 ft high—release the remaining pollutants further from human lungs.

Sulfur dioxide (SO_2) is perhaps the most serious challenge. The difficulty of removing SO_2 from enormous volumes of waste gases has prompted efforts of a different kind: the reduction of the sulfur (S) content of coal. Coal fields in some states yield low-sulfur coals (1 percent and below, compared to a more common sulfur content of between 2 and 3 percent). Figure 7-12 (p. 262) shows how the average sulfur content of coal varies from state to state. Many communities are requiring the use of coal having 1 percent or less sulfur to alleviate sulfur emissions into the atmosphere.

Because supplies of low-sulfur coal are limited, expensive, and often not available in nearby regions, means have been sought for

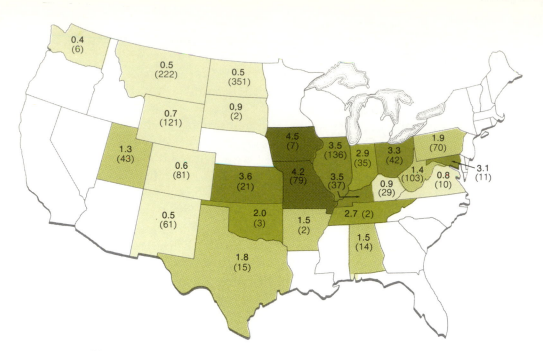

Figure 7-12. The average percentage sulfur content of coal reserves in those states having substantial coal deposits. The total coal reserves in each state, expressed in billions of tons, are shown below the percentage figure in parenthesis. (Since coal in any state is not uniform, some coal in each state is available with a sulfur content less than the average shown.) (Compiled from data in Bureau of Mines Information Circular 8312, 1966.)

removing sulfur from coal. About half the sulfur, that contained in discrete inorganic particles of iron pyrite (FeS_2), is perhaps removable; the rest is organic sulfur (sulfur atoms chemically bonded to carbon atoms) for which removal prospects are dim.

In the long run, it might be possible to convert coal rather completely to a gaseous fuel similar to natural gas, mostly methane (CH_4). Sulfur could then be removed in the gas-forming stage. Such *coal gasification* shows long-range promise for sulfur removal. It may also alleviate the growing shortage of natural gas around the world.

In the meantime the insatiable industrial and public demand for power is doubling every 10 years. Both coal-fired and nuclear plants (Chapter 11) are being constructed to meet the demand. Many experts believe that growth will outpace effective pollution control, which would lead to a net increase in air pollution in coming years.

Some plants are being built in remote locations to minimize their impact. Most controversial are six giant plants in the largely unoccupied southwestern United States (Figure 7-13), designed to supply Los Angeles, Phoenix, and other growing urban centers with future power. These plants, some completed (see photo) and others still in planning, are beginning to pour pollutants across some of the most magnificent landscape on earth, including Grand Canyon

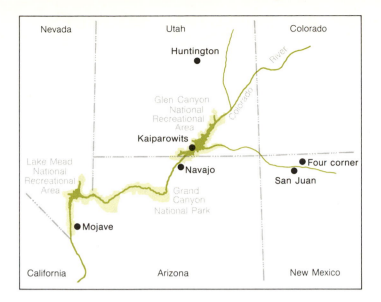

Figure 7-13. The location of six plants in the Southwest, now under development to serve growing urban centers. Of these plants, Kaiparowits, not yet built, is designed to be the world's largest.

National Park and an almost endless number of other spectacular redrock canyons, pinnacles, and landforms. The central issue of the debate is whether the clean air of this national showplace is to be sacrificed in the name of unremitting growth, or whether growth can perhaps be moderated to maintain environmental quality. On a finite earth, growth must eventually yield. But if growth is doggedly pursued to its utter limit, many of the earth's beautiful places will be scarred, and life at best will become a more dreary experience.

The Four Corners Power Plant in northwestern New Mexico pours smoke into the desert air. (*Photo by William Marling.*)

MAJOR AIR-POLLUTION SOURCES—
THE AUTOMOBILE'S GASOLINE ENGINE

Nearly 100 million motor vehicles are plying the highways of this country, half the total for the world. They are carrying passengers a trillion miles a year, enough to go to the moon and back 2 million times. To do so, they are consuming 100 billion gallons of gasoline, and through combustion they are converting it and air into more than 60 million tons of carbon monoxide (CO) and 6 million tons of nitric oxide (NO). They are spewing out 500 million pounds of lead (Pb), an important gasoline additive. And they are emitting 16 million tons of hydrocarbons so diverse in nature that all have not been identified. Of the more than 200 detected, some are powerful carcinogens like benzopyrene.

No way exists to disperse these pollutants through tall stacks; they are discharged at street levels where cars and people intermingle most (see photo).

Engines. The *internal-combustion* engine is one in which combustion gases drive the moving parts to produce power. Important internal-combustion engines are the gasoline engine, diesel engine, and gas turbine engine. The gasoline engine, sometimes termed the spark-ignition engine, is used in most automobiles and light trucks. The bulkier diesel engine powers most heavy trucks. And the expensive but relatively lightweight gas turbine engine propels large aircraft.

Cars and people jumbled together on city streets lead to serious pollution exposure. (*Photo by Jan Lukas/Rapho-Guillumette.*)

External combustion engines are those in which outside combustion heats a contained fluid, which then drives the engine. If the fluid is water-steam, the engine is called a steam engine. However, other driving fluids can be used: the new Lear engine uses a mixture of water and other chemicals and some Datsuns will employ trichlorotrifluoroethane, a halocarbon chemical.

It is most common to run an external combustion engine on a unique sets of strokes known as the Rankine cycle. Therefore this form of external combustion engine is frequently referred to as the *Rankine* engine.

The Gasoline Engine. Of the above engines, the gasoline engine, which powers most automobiles, creates by far the largest amount of vehicular pollutants. The gasoline engine works by using a piston to compress a mixture of gasoline vapor and air in a cylinder to a pressure about 10 times that of the atmosphere.[16] A spark ignites the mixture and the pressure soars to 30 or 40 times atmospheric pressure as the hot gases expand. These gases drive the piston out of the cylinder, forcing an attached drive shaft to rotate, which then transmits power to the wheels.

The burning of the fuel-air mixture after spark ignition is sudden and intense. Temperatures reach 4500°F almost instantly, and an abundance of nitric oxide (NO) is produced (see pp. 248–252). As the piston is thrust out of the cylinder, the combustion gases cool in less than a hundredth of a second. The heating-cooling cycle occurs so rapidly that much of the carbon of the fuel is unable to complete its oxidation to carbon dioxide (CO_2). A large part of this carbon remains as carbon monoxide (CO), and some is left in the form of hydrocarbons (HC), aldehydes (RCHO), and other organic substances. Additional hydrocarbons are unburned near the cylinder wall because the metal conducts heat away from this region, retarding combustion.

We see that high pollution levels from a gasoline engine are encouraged by the fundamental characteristics of that engine. The high temperature promotes the excess oxidation (overoxidation) of nitrogen to NO; the rapid cycling leads to the incomplete combustion (underoxidation) of fuel. The simultaneous existence in automobile exhaust of both overoxidized contaminants like NO and underoxidized contaminants like CO and HC makes it almost impossible to clean up the exhaust with any single chemical treatment.

Even then, emissions could be cut by various adjustments in fuel, fuel-air ratios, and spark timing if gasoline engines were run at a steady pace. They are not. In a typical day a car's engine is started, stopped, accelerated, idled, and decelerated many times. Each mode of this erratic use requires different and conflicting adjustments, which makes an overall tuning that would be appropriate for all

---

[16]This multiplication of pressure is essentially equal to the compression ratio, discussed further along. Compression ratios for the gasoline engine vary from about 6 to 11.

occasions impossible. This compromise is the first of many needed in controlling automobile pollution.

An example of typical changes in exhaust pollution levels with altered driving conditions is shown in Table 7-12. On the whole, acceleration and deceleration increase emissions.

Table 7-12. CHANGES IN EXHAUST POLLUTION LEVELS WITH DRIVING MODE FOR AUTOMOBILES WITHOUT POLLUTION CONTROL

Exhaust Component	Driving Mode			
	Idle	Cruise	Acceleration	Deceleration
Carbon monoxide (CO)	52,000 (5.2%)	8,000 (0.8%)	52,000 (5.2%)	42,000 (4.2%)
Hydrocarbons (HC)	750	300	400	4,000
Nitric oxide (NO)	30	1,500	3,000	60

It should be clear by now that the contrast between combustion in the gasoline engine and that in coal-fired power plants is extreme. The burning of coal is slow, steady, and fairly complete, with pollutants arising largely from impurities in the fuel itself. Combustion in the gasoline engine is rapid and erratic, characteristics that by themselves favor the formation of pollutants.

Clearly, the mood of the nation favors cleaning up major pollution sources like the gasoline automobile engine. This mood has been translated by the Clean Air Act of 1970 (see pp. 228–230) into stringent federal requirements for 1975 automobile emissions. Carbon monoxide and hydrocarbon levels in 1975 model cars, and nitric oxide levels in 1976 cars, are to be reduced 90 percent from those for 1970 models. Next we shall discuss means for achieving these objectives.

Improving the Gasoline Engine. The gasoline engine involves many intricate compromises, some pitting environmental quality against engine performance. In the past the latter was paramount; the rather sudden need to consider environmental standards as well has caused some chaos as the giant auto industry gropes for a new set of compromises.

The fuel/air ratio is one example of compromise. With too much air, excess oxygen (O_2) is around to form nitric oxide (NO); conversely, a mixture rich in fuel suffers incomplete combustion and produces carbon monoxide (CO) and hydrocarbons (HC).[17] The

---

[17] A chemically balanced mixture with exactly enough oxygen to consume all the fuel is a *stoichiometric mixture*. It can be calculated as follows. We represent gasoline by a typical hydrocarbon, octane (C_8H_{18}). Its complete combustion occurs by the reaction

$$\underset{\text{octane}}{2C_8H_{18}} + \underset{\text{oxygen}}{25O_2} \rightarrow \underset{\substack{\text{carbon} \\ \text{dioxide}}}{16CO_2} + \underset{\text{water}}{18H_2O}$$

Partial mass inventory: 260   800

The O_2/fuel weight ratio is $(800/262) = 3.08$, a quantity that can be expressed as 3.08 parts O_2/1 part fuel. However, air is mainly nitrogen, whose weight must be added in to get the air/fuel

fuel-rich mixture is better for smooth engine performance, but in recent years, leaner mixing ratios have been introduced to reduce CO and HC. But at best, as Figure 7-14 shows, substantial pollution occurs at all mixing ratios; compromise simply trades one group of pollutants and a degree of favorable engine performance for another pollutant. Hence modification in the fuel/air ratio will not totally solve the problem, and we must look further.

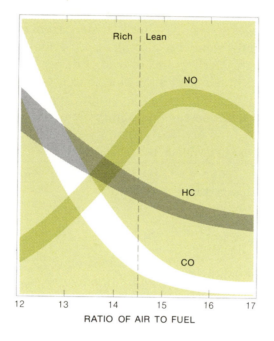

Figure 7-14. Relative emissions of NO, HC, and CO as a function of air/fuel ratio in the gasoline engine.

Another variable is related to fuel quality and the gasoline engine's compression ratio. The *compression ratio* is the factor by which the fuel-air mixture is compressed before spark ignition. As any gas is compressed it gets hot, and the more it is compressed the hotter it gets. Hence high compression ratios heat the fuel-air mixture before combustion; they also cause increased temperatures in the burned gases because the temperature starts at a high level. The increased temperature improves engine efficiency (more miles per gallon of gasoline), but it also increases emissions of NO. Not surprisingly, compression ratios in the 1960's inched up to about 11 to gain

---

ratio in place of the O_2/fuel ratio. From Table 7-1, O_2 is only 21 percent of the atmosphere; the ratio of air to O_2 mass is therefore approximately 100 parts air/21 parts O_2. Multiplication gives

$$\frac{3.08 \text{ parts } O_2}{1 \text{ part fuel}} \times \frac{100 \text{ parts air}}{21 \text{ parts } O_2} = \frac{14.7 \text{ parts air}}{1 \text{ part fuel}}$$

The air/fuel ratio is therefore close to 14 or 15, depending on the precise composition of the gasoline.

performance; currently, they are down to 8.5 for environmental compromise.

There is a further complication. Fuel-air mixtures preheated by undue compression tend to ignite spontaneously, before normal spark ignition, causing an engine "ping" and reduced performance. To combat this problem, high-octane fuels must be used. These fuels have additional environmental impact.

The octane number is a measure of a fuel's ability to resist early ignition. The octane number increases with both chain branching and aromatic content in the gasoline's hydrocarbons, as shown by Table 7-13. Octane number is also increased about 10 points by the addition of a mere 2 or 3 grams of lead (Pb) per gallon of gasoline. (The lead is added as a lead alkyl compound, commonly tetraethyl lead, $Pb(C_2H_5)_4$.) A 10 point gain in octane number, from 90 to 100, permits a substantial increase in compression ratio, from 8 to 11.

Table 7-13. **OCTANE NUMBERS OF DIFFERENT FUEL CONSTITUENTS**[a]

Fuel	Octane Number
Straight chain hydrocarbons:	
butane	94
hexane	25
heptane[b]	0
octane	−19
Branched chain hydrocarbons:	
isooctane[b]	100
Aromatic hydrocarbons	
benzene	106
toluene	120
ethyl benzene	107
xylene (para)	116

[a] The straight chain hydrocarbons are most abundant in natural petroleum; the others must be obtained largely by refining. Aromatic hydrocarbons are added to mostly gasoline at a level of 20 to 25 percent to increase octane number.

[b] These two hydrocarbons are arbitrarily assigned octane numbers of 100 and 0. They establish the scale for all other octane values.

High-octane hydrocarbons are expensive, but the small amount of lead needed for octane improvement is cheap, just over a half cent per gallon. Therefore lead was an integral part of the previous decades' trend toward increased octane numbers, compression ratios, combustion temperatures, and engine efficiencies. However, this overall trend has been stopped and reversed because of three environmental considerations. First, high temperatures create more NO, as already noted. Second, toxic lead (Pb) compounds are exhausted into the air to the extent of about 500 million lb per year (see Chapter 9 for the implications of this problem). Third, lead in exhaust gases destroys the activity of catalysts in reactors that are designed to clean up certain pollutants, as we shall see.

We are thus brought to realize that the intertwining of economic factors, engine performance, and variable driving conditions makes the design of engines immensely complicated. When multiple environmental requirements also must be considered, the complexity of the matter sends shock waves of doubt and controversy through all segments of our unique society-on-wheels.

Pollution Control Devices. Engine adjustments will reduce but not eliminate emissions. Nor will they meet stringent federal standards for 1975. Thus other means are being enlisted in the battle to reduce air pollution from automobiles. Prominent among them are certain control devices that deal with the exhausted combustion gases. They are based on principles that we outlined in the earlier sections on individual pollutants, so they need only brief coverage.

Recall that carbon monoxide (CO) and hydrocarbons (HC) are products of incomplete combustion. Both are rendered harmless when their oxidation is completed. This end is best accomplished by passing exhaust gases through a special chamber where oxidation reactions are encouraged. This chamber, termed a *reactor*, is being developed in two different forms. In the first, the *thermal reactor*, air is added to the exhaust and both are heated to a high enough temperature to complete combustion. In the second, the *catalytic reactor*, exhaust gases are mixed with air and passed over a solid catalyst, which takes the place of high temperature in promoting final oxidation. The catalyst is inactivated by lead (Pb) and therefore requires unleaded gasoline—a prime reason for getting lead out of automobile fuels.

Nitric oxide is more difficult to treat, but a combination of techniques may reduce its level to meet federal requirements. The approaches are the traditional ones for nitric oxide control: they involve either a decrease in combustion temperatures to suppress NO formation or the use of a chemical reducing agent to destroy NO in a separate reactor. In one method, part of the exhaust is simply recirculated back through the engine to dilute and cool the burning fuel mixture.

Finally, a small fraction of a car's hydrocarbon emissions come from parts other than the exhaust. Figure 7-15 (p. 270) illustrates this point. Direct evaporation from the gas tank and carburetor account for about 15 percent of total hydrocarbons, and another 20 percent escape the wrong way past the pistons to become crankcase blowby. The latter problem is handled by *positive crankcase ventilation* (PCV), which simply leads these vapors back through the engine cycle for combustion.

Other Engines. The gasoline spark-ignition engine is a natural polluter. Devices to cure it are much like patches on a bad suit: they do not cover all the defective fabric, and worse, the patches can themselves fall into disrepair. Thus there is a search for alternate engines.

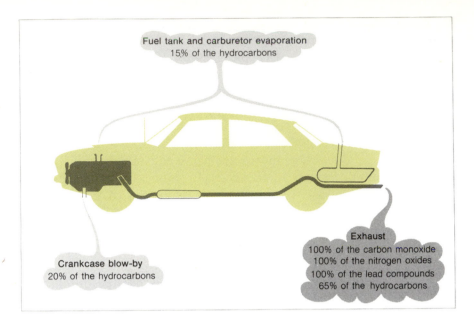

Figure 7-15. An overview of automobile air pollution.

The Rankine or steam engine is intrinsically cleaner because its steady and continuous flame (used to heat the steam) in excess air oxidizes CO and hydrocarbons far better than does the intermittent burning of the gasoline engine. Without compression, temperatures are lower and NO formation is also limited. There are no pre-ignition problems requiring lead additives and causing attendant lead pollution. One new Rankine engine is claimed to have emissions of only 1 ppm CO, no detectable hydrocarbons, and under 0.01 ppm NO. These emissions are thousands of times lower than emissions from the gasoline engine, shown in Table 7-12. Several promising Rankine engines have been developed for propelling cars; most former doubts about bulkiness and slow startup have been dispelled by the new designs.

The gas turbine is another low-pollution engine showing promise. This lightweight engine (hence used in airplanes) requires extreme precision in its construction: it is thus costly to build. In the long run, it shows considerable promise.

Rotary engines of the Wankel type are becoming important in automotive circles. This special form of the internal-combustion engine shows more promise for simplicity of manufacture than for low pollution. It tends to emit more hydrocarbons than does the conventional piston engine, and less NO.

Pollutionless, battery-driven cars have been widely discussed. However, they lack power and range. Much worse, battery recharging would require great quantities of electricity, so by this approach the pollution problem would be merely transferred from the automobile to the electric generating plant.

Exercises

1. Inversions appear to be an unfortunate accident of nature. They trap pollutants over urban areas, where considerable health and property damage occurs. But inversions have a positive side: citizens who must breathe their own pollution are strongly inclined to clean it up. In other words, inversions provide an incentive for each community to control its own emissions. Consequently, emissions from Los Angeles, Pittsburgh, and some other cities have been slashed. Without inversions, local damage would be slight, emissions would expand, and worldwide effects would grow. In this light, discuss and compare a hypothetical world of no inversions with the real world of frequent inversions.

2. Explain the various plume shapes of Figure 7-3. (Keep in mind that vertical currents will stir smoke up and down where no inversion exists, but very little vertical stirring occurs within an inversion.)

3. A ton of coal typically costs a power plant $6. At 3 percent sulfur (S), the ton of coal contains 60 lb of sulfur. If this S were all emitted into the air and a tax of 15 cents/lb of sulfur applied, what would the company's tax bill be relative to its fuel bill? What daily tax would be imposed on a large power plant consuming 10,000 tons of coal per day? Would this tax be likely to prove an economic deterrent to sulfur pollution?

4. What amount of hydrogen sulfide (H_2S) would be needed to convert a smelter's daily 300 tons of sulfur dioxide (SO_2) to sulfur (S) using the Claus process outlined in Table 7-6?

5. How much sulfur (S) would be produced each day while 280 tons of carbon monoxide (CO) were used to clean up a plant's SO_2 pollution by the University of Massachusetts process outlined in Table 7-6?

6. Union Carbide Corporation has announced a new process for removing the element fluorine (F) from industrial emissions (*Chemical & Engineering News*, February 14, 1972). The fluorine-containing pollutants are reacted with charcoal (C) at high temperatures to form inert carbon tetrafluoride (CF_4). It is claimed that only 1 lb of charcoal will remove 6.3 lb of fluorine. Do you believe this claim? Why?

7. An average person breathes 20,000 liters of air per day. How many particles does someone in Los Angles inhale on a day when photochemical smog is responsible for 10^{11} particles/m^3 of air?

8. Charleston, West Virginia, has had concentrations of particulate matter in air as high as 960 $\mu g/m^3$ (observed in 1960). How many milligrams of this matter would be inhaled in a day by a person breathing 20,000 liters of air?

9. The strong carcinogen benzpyrene occurs at an average level of 0.006 $\mu g/m^3$ of city air. How much benzpyrene will be inhaled in a 1-year period, considering that a man breathes 20,000 liters of air per day? (Only part of the inhaled benzpyrene will be retained in the lungs; the rest will be removed with particles that are cleared from the lungs.)

10. The 20,000 liters of air entering our lungs each day contain about 5×10^{26} molecules. How many of them are sulfur dioxide (SO_2) molecules on a day when SO_2 averages 0.14 ppm in the atmosphere, the federal limit for the 24-hour average SO_2 concentration?

11. The proposed Kaiparowits power plant in southern Utah will consume about 50,000 tons of coal a day, more than any other plant in the world. Fortunately, the coal for this facility is low in sulfur, about 0.5 percent. How much sulfur dioxide (SO_2) will this plant produce per day?

12. (Refer to Problem 11). If the SO_2 from the Kaiparowits plant could be cleaned up by the limestone injection process (Table 7-6), how much gypsum ($CaSO_4$) would accumulate as solid waste per year? (See footnote 4.)

13. One of the six large Southwest power plants mentioned in the text is the Navajo plant in northern Arizona. When completed in 1976, this plant will burn 24,000 tons of coal per day containing 8 percent inorganic ash. Modern electrostatic precipitators will capture most of the fly ash, although the exact amount is not certain. A plausible estimate is 99 percent capture, much better than the national average. Therefore 1 percent of the fly ash will remain as fine particles in the emitted combustion gases. How many tons of particulate matter per day will the Navajo plant emit to the atmosphere under these circumstances? How many tons of captured fly ash will become solid waste?

14. The Navajo power plant (see Problem 13) is one of the most controversial in the Southwest, partly because its enormous requirements for fuel will be met by stripping coal from Navajo and Hopi tribal lands, and partly because it is closest of all the plants to the Grand Canyon. The Environmental Statement for this plant cites a growing demand for electricity and points out that other power sources cannot meet the immediate demand: nuclear plants could not be licensed soon enough, natural gas is becoming scarce, oil is expensive and higher in sulfur content than the coal to be used, hydroelectric power sites have been taken, and alternate power sources have not been developed. Nowhere does the Environmental Statement suggest curtailment of growth in power production as an alternate direction. List arguments favoring and opposing this alternative. How would growth limitation be achieved?

15. Your warm bath may involve up to 300 lb of tap water heated 40°F by a hot-water heater. The combustion energy from about 1 lb of fossil fuel is required to supply this heat. If the heater is electrical, this combustion occurs at the power plant, but the amount of coal burned there is more like 2.5 lb because such plants are at best only 40 percent efficient. If the 2.5 lb of coal burned for your bath contain 3 percent sulfur (S), how much sulfur dioxide (SO_2) pollution is your bath responsible for? (The baths taken over a year's time by a city full of people add up to a substantial quantity.) Should anyone else share the blame for your bath's pollution? What, if anything, should be done?

16. The stack gases from a power plant typically contain sulfur dioxide (SO_2) at concentrations of 1500 ppm by volume. How many cubic meters (m^3) of pure SO_2 could be obtained daily from a power plant emitting 10^8 m^3 of combustion gases? The density of SO_2 under average atmospheric conditions is 2.5 kg/m^3. Using this number, calculate how many metric tons of SO_2 effluent are contained in the

stack gases. (Does this amount make sense in view of other evidence in this chapter on SO_2 emissions from power plants?)

17. Most sulfur dioxide (SO_2) in the atmosphere is oxidized to sulfur trioxide (SO_3), then hydrolyzed to sulfuric acid (H_2SO_4), as noted in the text. The overall reaction yielding H_2SO_4 can be represented by $2SO_2 + O_2 + 2H_2O \rightarrow 2H_2SO_4$. How much sulfuric acid is produced per year from the effluents of a plant or smelter emitting 500 tons of SO_2 daily?

18. (Refer to Problem 17.) To an approximation, rain is made acidic (pH about 4) with 0.01 g of sulfuric acid in one liter (1000 g) of water, a ratio of 10^{-5} parts $H_2SO_4/1$ part H_2O. About how many tons of rain can be made acidic by the yearly quantity of sulfuric acid that comes from plant SO_2 emissions, calculated in the previous problem?

19. (Refer to Problem 18.) A typical yearly rainfall of 28 in. brings 2 million tons of rainwater to each square mile (mi^2) of the earth's surface. How many square miles of earth would the acidic rain of Problem 18 cover? To a very rough approximation (summarizing the last three problems), this figure is the land area that can be steadily inundated by acidic rain (pH close to 4) as a result of a facility emitting 500 tons of SO_2 per day. You can see from this result why acidic rain is becoming a widespread problem in the United States and Europe.

20. Federal legislation has been proposed to tax sulfur-containing emissions at rates up to 15 cents/lb of sulfur (S) in heavily polluted regions (those not meeting the federal primary air-quality standards). Is this tax high enough to provide a deterrent to the sulfur dioxide (SO_2) emitted from copper smelters, which comes primarily from the smelting of chalcopyrite ($CuFeS_2$) (explained in Chapter 5)?

$$2CuFeS_2 + 5O_2 \rightarrow 2Cu + 2FeO + 4SO_2$$

The copper (Cu) produced markets for about 50 cents/lb. (Ask yourself this question: How much tax would be levied on the sulfur contained in the SO_2 produced with each pound of Cu? What percentage would this tax be of the revenue from the pound of Cu?)

21. A recent publication (Bureau of Mines Technical Progress Report 43, September 1971) pegs the average mercury (Hg) content of coal at 0.2 ppm. If we assume that all mercury escapes through the smokestack as vapor, how much mercury will be released to the atmosphere each year from a large power plant burning 10,000 tons of coal of average Hg content daily?

22. The rush-hour traffic commuter, responsible for peak pollution levels, is subsidized heavily by extra lanes of roads and highways that he alone needs and by all the extra maintenance, policing, and tax exemption that go with those lanes. William Vickrey of Columbia's Department of Economics has stated in *Science* (March 31, 1972, p. 1417) that if a mass-transit rider (who produces very little pollution) were subsidized equally, he not only would ride free but would be paid a bonus. Can you think of any other examples where pollution is encouraged by the misdirection of taxes or subsidies?

23. Lead is added to gasoline at an average level of 2.5 g/gal. How many pounds of lead go into the 100 billion gallons of gasoline consumed in this country each year? Does this number explain the approximate magnitude of lead emissions cited in the text?

24. Gasoline consists of a variety of hydrocarbons, each requiring a slightly different air/fuel ratio for complete combustion. What is the air/fuel ratio needed for the total combustion of benzene (C_6H_6), a major aromatic compound in gasoline?

25. The typical car without pollution controls emits about 5 g of nitric oxide (NO) per mile. With automobiles in this country traveling close to one trillion miles a year, estimate the tons (either metric or common) of NO emitted by automobiles annually.

26. (Refer to Problem 25.) The total NO emitted by automobiles is soon converted to nitrogen dioxide (NO_2) by the reaction $2NO + O_2 \rightarrow 2NO_2$. How much NO_2 are automobile emissions responsible for each year? If you worked the problem properly, your answer should account for most of the oxides of nitrogen from transportation sources listed in Table 7-2. Did you succeed? (Simple methods like this one are used to arrive at total air-pollution emissions of the kind in Table 7-2.)

27. Natural gas, mainly methane (CH_4), has a density of about 0.2 lb/ft^3. How many cubic feet (ft^3) of natural gas would be needed each day to clean the nitric oxide (NO) from a stack emitting 200,000 lb of NO daily? The cleanup reaction is $CH_4 + 4NO \rightarrow 2N_2 + CO_2 + 2H_2O$, (see p. 252). At a typical industrial cost of $0.50 for 1000 ft^3, what would the company pay for natural gas each day for this cleanup? (This amount is not the sole cost—capital expenditures for this unproved process could be quite high.)

Glossary

Acute   short term.
Anaerobic   condition in which no free oxygen is present.
Chronic   long term.
Coal gasification   conversion of coal to gaseous form.
Compression ratio   factor by which the fuel-air mixture is compressed in an engine.
External-combustion engine   engine in which an outside heat source heats a contained fluid, which then drives the engine.
Gas turbine engine   a possible future low-pollution vehicular engine.
Internal-combustion engine   engine in which enclosed combustion gases drive the moving parts to produce power.
Meteorology   the science of the atmosphere and its weather.
Photochemical oxidants   toxic and irritating compounds formed by the reaction of hydrocarbons with ozone, oxygen, and oxides of nitrogen. Also called *secondary pollutants* because no source emits them directly.
Photochemical smog   air pollution resulting from reactions in the atmosphere that produce photochemical oxidants.
Pollution episode   incident of abnormally high air pollution.

Positive crankcase ventilation (PVC) the process of leading crankcase blowby (uncombusted gasoline) back to the engine cycle for combustion.

Primary pollutants polluting compounds directly emitted from sources that produce them.

Rankine engine steam engine.

Reactor chamber in which carbon monoxide and hydrocarbons in engine exhausts are fully oxidized to render them harmless. A thermal reactor heats the exhaust. A catalytic reactor uses a catalyst instead of heat to cause oxidation.

Rotary (Wankel) engine a type of internal-combustion engine.

Synergism enhancement of the effect of one pollutant by the presence of another pollutant.

Threshold limit value (TLV) the upper limit for industrial exposure to a given substance for healthy workers.

Additional Reading

Starred selections are most suitable for the nontechnical reader.

*Air Conservation, A Report of the Air Conservation Commission of the American Association for the Advancement of Science. Baltimore, Md.: Horn-Shafer, 1970.

Atkisson, Arthur, and Richard S. Gaines, Eds., Development of Air Quality Standards. Columbus, Ohio: Merrill, 1970.

*Cleaning Our Environment, The Chemical Basis for Action, A Report by the Subcommittee on Environmental Improvement, Committee on Chemistry and Public Affairs, American Chemical Society, Washington, D. C., 1969.

*Giddings, J. Calvin, and Manus B. Monroe, Eds., Our Chemical Environment. San Francisco: Canfield, 1972.

Hagevik, George H., Decision-Making in Air Pollution Control. New York: Praeger, 1970.

Purdom, P. Walton, Ed., Environmental Health. New York: Academic, 1971, Chapter 5.

Scorer, Richard, Air Pollution. New York: Pergamon, 1968.

Starkman, Ernest S., Ed., Combustion-Generated Air Pollution. New York: Plenum, 1971.

Stern, Arthur C., Ed., Air Pollution, 2nd ed. Vols. I, II, and III. New York: Academic, 1968.

Strauss, Werner, Ed., Air Pollution Control, Part One. New York: Wiley, 1971.

*Treshow, Michael, Whatever Happened to Fresh Air? Salt Lake City, Utah: University of Utah Press, 1971.

*Wise, William, Killer Smog. New York: Audubon/Ballantine, 1970.

1 H																	2 He
3 Li	4 Be											5 B	6 C	7 N	8 O	9 F	10 Ne
11 Na	12 Mg											13 Al	14 Si	15 P	16 S	17 Cl	18 Ar
19 K	20 Ca	21 Sc	22 Ti	23 V	24 Cr	25 Mn	26 Fe	27 Co	28 Ni	29 Cu	30 Zn	31 Ga	32 Ge	33 As	34 Se	35 Br	36 Kr
37 Rb	38 Sr	39 Y	40 Zr	41 Nb	42 Mo	43 Tc	44 Ru	45 Rh	46 Pd	47 Ag	48 Cd	49 In	50 Sn	51 Sb	52 Te	53 I	54 Xe
55 Cs	56 Ba	57 La	72 Hf	73 Ta	74 W	75 Re	76 Os	77 Ir	78 Pt	79 Au	80 Hg	81 Tl	82 Pb	83 Bi	84 Po	85 At	86 Rn
87 Fr	88 Ra	89 Ac	104 Ku	105 Ha	106	107											

HYDROGEN, NITROGEN, OXYGEN, SODIUM, MAGNESIUM, PHOSPHORUS, SULFUR, CHLORINE, CALCIUM Nine elements found to varying degrees in the waters of earth conspire to influence water quality, water use, and the ability of water to support life.

8 OUR WATER ENVIRONMENT

ENVIRONMENTAL ROLE OF WATER

Spaceship earth is an oasis of two marvelous chemicals: oxygen (O_2) and water (H_2O). In Chapter 6 we detailed the profound role of O_2 in shaping the environment of earth. This chapter focuses on the chemical compound H_2O, which is even more significant to the environment than O_2.

Life on earth began in the water of ancient seas. The intricate chemistry of life was molded around this watery medium, and the evolution to more complex, land-living creatures seemed only to increase the vital biological role of water. In man, water forms the essential medium in which the chemical reactions of tiny cell factories proceed; it supports the integrity of cell walls so his cells do not collapse; it transports blood; it forms a pool for his digestion; and it holds and helps transport the electrically charged ions that generate nerve signals and make the human brain possible.

Our physical environment hinges on water no less than does our biological environment. As a vapor, water absorbs radiation to influence the heat balance and temperature of our environment, and it brings moisture to the continents. As a liquid, water erodes and shapes our land, transports and concentrates minerals, and moderates climate. As a solid, ice, water gouges glacial valleys and lakes, pulverizes rock by expanding when it freezes and thereby creates soil, and stores winter moisture.

In listing the services provided by water, we should not ignore its profound esthetic role. Free-flowing rivers and blue lakes are things of magnetic beauty, attracting us to look and to wonder; we find inspiration and renewal as we admire and enjoy the primordial liquid that has shaped us and our planet.

Water plays another, more material role for man. It carries away his wastes. As the most abundant liquid on earth, water runs steadily to sea along a vast network of rivers. This moving water can be a receptacle for sewage; it can be used to rinse away grime or toxic chemicals; or it can remove waste heat from the boilers of civilization. Natural water can purify some of these wastes; it can dilute others until they are harmless; and it can carry the rest out of sight. But with the incessant growth in world population and industry, one city's wastes become the next city's drinking water. Natural purification breaks down, fish die, odors form, diseases are transmitted, and the central medium of all life becomes a garbage-filled carrier of stinking death.

In this chapter we shall describe the nature, availability, use, and abuse of natural water. We shall detail the underlying chemical processes that have degraded water under the onslaught of man. And, finally, we shall show why water inherits—from the arrangement and electrical distribution of atoms in the H_2O molecule—the characteristics that make its environmental role so profound.

The beauty of water has inspired man through the ages. (*Photo by Klink/Photofind.*)

OUR WATER ENVIRONMENT

Polluted water is an eyesore, a hazard to health, and a killer of fish and wildlife. (*Photo by Richard F. Conrat/Photofind.*)

THE EARTH'S WATER RESOURCES

The total water on earth is an enormous 1.5×10^{18} metric tons, a quantity 300 times larger than the weight of the entire atmosphere (Chapter 6). If this water were distributed equally among all human beings now alive, each person would possess a staggering 380 million metric tons of water. He could use 15,000 metric tons each day of his life and not run out.

How, then, can man worry about water? To start with, most water is inaccessible—it is in oceans, ice caps, and underground *aquifers* (water-bearing beds), and some is even in the air. Only a small fraction is on the earth's surface and directly accessible to man. The relative amount of water of each type is illustrated in Figure 8-1, which shows the depth each would reach if spread as a uniform blanket over the earth's entire surface.

The surface water, while only 0.02 percent of the total, represents a substantial 2 ft blanket of water around the earth. This slight fraction of the world's water would be enough for man's needs if it were well distributed and clean. Since it is neither, water becomes one of the primary concerns of man.

Much thought has been given to tapping the four less accessible

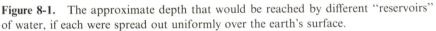

Figure 8-1. The approximate depth that would be reached by different "reservoirs" of water, if each were spread out uniformly over the earth's surface.

water reservoirs of Figure 8-1. Ocean water can be used by coastal cities if its salt is removed. Desalting has been achieved by distillation and reverse osmosis—two of the separation methods illustrated in Chapter 1. However, the desalting (desalination) of ocean water is now too expensive to be used widely. The distant ice caps seem less promising still, but even here serious proposals have been made to tow their giant icebergs to port cities, or to shoot massive snow-balls along tubes to the populated continents!

Underground water is already used extensively throughout the world, through wells and springs. Unfortunately, underground reservoirs are renewed only slowly by natural seepage, and steady use is depleting some of them. Arizona farmers in Pinal County must dig 250 ft for well water that was within 50 ft of the surface in 1910.

Water vapor, the other inaccessible reservoir, seems barely worth considering. It holds a mere 1 in blanket of water. This means that if all moisture were wrung from the sky and the clouds as rain, the depth of rainfall would average only 1 in (more, of course, in some areas, less in others). But this 1 in is renewed time and again, as new water evaporates from oceans and the old falls as rain and snow. Hence truly enormous quantities pass overhead in a year's time. Attempts are now underway to "milk the clouds," so that some of the airborne water can be brought to needy locations on earth. The ecological and human consequences of such *weather modification* projects are uncertain.

Whether or not we succeed in directing the sky's water, it is already invaluable to all land-based life, for it is the sole source of water for the lands of earth. Water vapor, carried by world air currents, is part of an endless cycle—the *hydrological cycle* or *water cycle*—that brings water to land and then returns it to the oceans along rivers (Figure 8-2). This cycle is powered by the energy of the sun, which in a single day evaporates over a trillion metric tons of ocean water. When this water condenses to a liquid and falls as rain, it is fresh water—free of salt—because ocean evaporation is a distillation process (Chapter 1) that separates and purifies ocean water by leaving its nonvolatile salts behind.

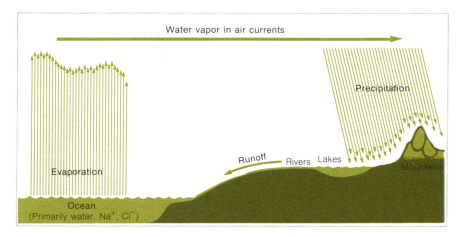

Figure 8-2. The hydrological cycle, which brings water from ocean to land.

The rainwater from the churning currents of the hydrological cycle falls to earth in an irregular pattern. Most of the rain falls back into the oceans. What rain reaches the land is concentrated in certain belts of high rainfall, while vast areas of earth receive barely enough to sustain sparse, desert plant life. Figure 8-3 shows the distribution of dry and wet areas. This uneven distribution has prompted great water projects, including dams and river diversions, and more recently the transfer of river water over enormous distances from wet to dry river basins. Most controversial has been the California Water Plan, by which the abundant waters of the northern California Sierra mountains have been diverted to thirsty southern California for agriculture and expanding cities—such as Los Angeles. This project has scarred the land with canals and dams and has badly mutilated some of America's most scenic, free-flowing rivers. Here is another example in which growth has taken its toll in the quality of life. Some water plans are so ambitious that they may have broad-

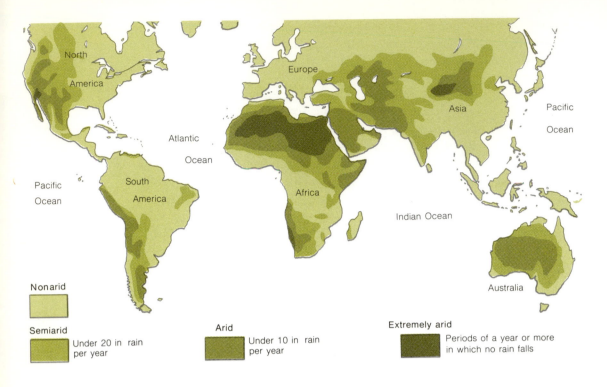

Nonarid

Semiarid | Under 20 in rain per year

Arid | Under 10 in rain per year

Extremely arid | Periods of a year or more in which no rain falls

Figure 8-3. Worldwide variations in precipitation.

scale implications to climate and ecology, too. Notable examples are plans to divert northern-flowing Russian rivers to the industrial south and proposals to bring water from the Yukon basin of Alaska to the dry southwest United States.

THE USES OF WATER

The most essential use of water is simply drinking, to sustain the body fluids necessary for life. Water for drinking should be of the highest quality, free of contaminants that cause ill effects (see pp. 321–323). But while the demands for quality are high, the quantity of water used for drinking is relatively low. Less than 1 part in 4000 of the water used in the United States is employed as drinking water. Here we shall deal with the question of what happens to the 3999 parts of water out of 4000 that are *not* used to quench the thirst of man.

Each day a water flow averaging 4.7 billion metric tons courses down the river channels of the United States.[1] This amount is 1250 billion gal of water a day. Water supplies in this country are commonly

---

[1] The 4.7×10^9 metric tons of stream flow is about 30 percent of the daily precipitation falling on the United States. The other 70 percent returns to the atmosphere through evaporation.

measured in gallons,[2] so we shall use this number of gallons (1250 × 10⁹) as a key reference for the fresh water available for daily use in the United States.

The use of water in this country now exceeds 400 billion gal, one third of the supply. When we consider the unequal distribution of both natural water and water demand, this use of water puts a severe strain on water supplies in some areas, particularly in the industrialized East, where demands are so great, and the arid Southwest, where supplies are so limited. The problems are destined to get worse because the demand for water is growing rapidly. Figure 8-4 shows the explosive growth in water use in the United States. Total use is expected to exceed stream flow shortly after the turn of the century.

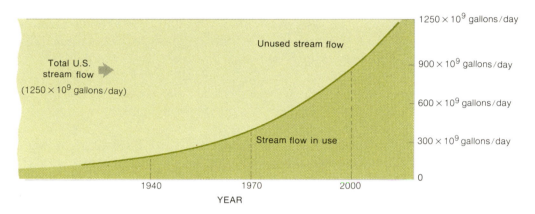

Figure 8-4. The rapid growth in the use of water in the United States. By present trends, use will exceed stream flow early in the next century.

The 400 billion gal used daily represents almost 2000 gal for each man, woman, and child in the United States. What activities consume so much water? Less than a tenth of each person's water share is piped into the home, and an even tinier share is used for drinking, as noted earlier. Most of an individual's share of water goes to agriculture to grow his food; to industry to manufacture his material goods; and to power-generating facilities where his electricity is produced. The amount used in each of these categories is shown in Table 8-1. Water used for hydroelectric power and recreation is not tabulated.

To get a perspective on water use and supply, we must distinguish between the *use* and *consumption* of water. Water that is used is not necessarily lost. It may be used to carry away heat or to wash an industrial product, but it is still water with a potential for use again. Consumed water, on the other hand, is defined as water that is

---

[2] A metric ton (1000 kg or 2200 lb) of water consists of 265 gal; a short ton (2000 lb) contains 240 gal. Other equivalents between common units for water are 8.33 lb of water = 1 gal; 7.50 gal = 1 ft³; and 1 acre ft = 43,560 ft³.

Table 8-1. THE DAILY USE OF WATER IN THE UNITED STATES, BY MAJOR CATEGORY

Nature of Use	Gallons Used
Irrigation	180 billion
Industry	80 billion
Power generation	110 billion
Municipal water	30 billion

permanently lost from man's accessible store of water, either by evaporation or by incorporation in a product. The distinction, then, comes down to this: used water can in theory be recycled, while consumed water cannot. In this light, the growth of water use shown in Figure 8-4 is not so frightening, because recycling and reuse can give us two or three uses of water before it is gone.

The number of times water can be recycled depends on the fraction lost with each use—the ratio of consumption to use. Overall, this ratio is about one fourth, meaning that four cycles are the most we could hope to achieve in stretching our total water supply. However, this ratio varies from one use to another, being one tenth or less for many industries and power generation, but reaching an excessive three fifths (60 percent) for irrigation.

The steel-producing industry is not unusual. Each ton of finished steel requires the use of about 50,000 gal (210 tons) of water. Only 1000 gal of this are consumed, a mere 2 percent of the water used.

Such huge demands for water stem from requirements that arise at nearly every stage of steelmaking. The raw materials—ore and coal—are mined and washed with the aid of water. Hot coke is quenched (cooled) by water after its manufacture. Crystallized coal chemicals are washed with water. Dust is controlled by water. Blast furnaces, open-hearth furnaces, and electric-arc furnaces are cooled by water. Jets of water are used to remove scale from hot steel. These and many other functions are typical of the industrial uses of water. In fact, most industrial uses, like those we have mentioned, fall into two major categories: cleaning and cooling. With such use, something is always added to the water—either impurities or heat. Each affects water quality, and each must be removed—or tolerated— if the water is to be recycled and used again.

In summary, water is used extensively to bring man his food and fibers; his necessities and luxuries. This is best illustrated in Table 8-2, which shows the amount of water used in producing items we take for granted in our modern world.

WATER—PROPERTIES AND NATURAL IMPURITIES

Water is a liquid with unique characteristics that trace ultimately to the highly polar water molecule, H_2O. Here we shall review briefly

Table 8-2. WATER USED IN THE PRODUCTION OF VARIOUS GOODS

Item Produced	Water Used (gal)
A Sunday newspaper	200
Wheat for a loaf of bread	300
A tankful of gasoline	400
1 lb of beef	4000
An automobile	50,000
1 ton of alfalfa	200,000
1 ton of synthetic rubber	600,000

the important properties of this molecule; more complete information on the origin of water's properties can be found in Chapter 3, "Chemical Bonds and Intermolecular Forces."

Both of the hydrogen atoms of H_2O are attached by chemical bonds to the same side of the oxygen atom, which gives H_2O a bent configuration (Figure 8-5). The oxygen atom's thirst for electrons attracts a residue of negative electron charge to the oxygen side, leaving a partial positive charge with the hydrogen atoms on the opposite side. Because of this charge separation, the water molecule is highly polar.

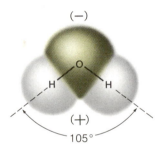

Figure 8-5.   The water molecule.

Intermolecular forces between polar molecules are inherently strong, but water exceeds expectations because it is one of the few simple molecules that form hydrogen bonds—the strongest of intermolecular forces (see Chapter 3). Water molecules therefore cling to one another with unusual tenacity.

The Nature of Water.   Most molecules with only two to five atoms are so weakly attracted to one another that they break free to form a gas at normal temperatures. Methane (CH_4), for instance, is a gas down to $-161°C$. If water followed the typical pattern, there would be no liquid oceans, rivers, or lakes—only a dense atmosphere of vaporized water. As it is, hydrogen bonds hold water molecules together as a liquid in the range of $0°$ to $100°C$ ($32°$ to $212°F$). Hydrogen bonds, then, literally hold the oceans together. But they do much more.

The large amount of energy needed to break hydrogen bonds

makes water an excellent energy absorber.[3] By soaking up energy from the sun, large bodies of water prevent temperatures from soaring. They effectively moderate extremes of climate. Also evaporating perspiration cools the body—a most important natural tool for the regulation of body temperature. In the same way, water is an excellent medium (coolant) for carrying away industrial heat, a prominent use of water mentioned above. The energy-absorbing properties of water, added to its natural abundance, make the occurrence of water extremely convenient for man's energy-centered existence.

Other significant features of water that are related to its hydrogen bonds are its high surface tension, viscosity, heat conductivity, and the expansion of solid water (ice) to a volume larger than that of the liquid. These features, by fixing the way in which water flows, erodes, freezes, and forms waves and spray, have played a major role in shaping the environment of earth—and the nature of man.

The Natural Impurity of Water. Water is never entirely pure. It inevitably carries traces of other substances—various organic compounds, particles, gases, minerals, and ions. This fact should not be surprising, for we learned in Chapter 1 that absolute chemical purity is ruled out by the small size of molecules and their near-infinite numbers, along with the thermodynamic tendency for mixtures to form out of molecules. Therefore we focus on the concept of impurity concentrations. Here we shall look at the status of natural waters, and later the circumstances most likely to lead to concentration levels that threaten the use and enjoyment of water, ecological balance, and human health.

Since water is a polar substance, it tends to dissolve other polar substances, in accord with the chemical principle that "like dissolves like." Therefore likeness in polarity is a useful guideline to impurity concentrations in water. It is to be used as follows: if several substances are present in equal amounts in the surroundings of water, the most polar substance will tend to dissolve in the water to the highest concentration, the least polar substance will dissolve least (but never as low as zero concentration), and the substances of intermediate polarity will dissolve to intermediate levels. Recall the precaution stated in Chapter 5, where we said that there are many exceptions to this rule—it is a guideline only. But it is a guideline invaluable in helping us understand why water contains its various impurities at their observed levels.

Organic Materials in Water. Organic matter is perhaps water's most significant impurity. Organic substances have a wide range of

---

[3] Water absorbs considerable energy when it is vaporized, which requires that all hydrogen bonds must be broken. Water also soaks up a greater than average amount of energy when it is heated, because some hydrogen bonds are broken as the temperature increases and the molecules take on an increasingly loose molecular arrangement. When water vapor condenses back to a liquid and cools, energy is released as hydrogen bonds form again. The yield of energy is exactly equal to the amount absorbed in the reverse process. Thus water moderates extremes of cold weather as well as hot.

polarities and thus a considerable range of solubilities in water. Hydrocarbons are nonpolar and thus dissolve very sparingly in water. We noted in Chapter 1 that oil spills, composed primarily of hydrocarbons, tend to spread across the surface of the water because their hydrocarbon molecules are unable to penetrate (dissolve) appreciably into the water. However, a very slight penetration does occur: enough, unfortunately, to disturb the life cycle of many aquatic organisms.

Organic molecules containing polar groups are inclined to dissolve at higher concentrations than hydrocarbons. Most alcohols, which contain the polar hydroxyl group, —OH, dissolve readily in water, as do amines, characterized by the amino group, $—NH_2$. Organic acids, containing the carboxyl group, —COOH, also tend to dissolve. When these polar groups are found in molecules with large hydrocarbon segments, then the solubility goes down because the bulky hydrocarbon part dominates the overall molecular polarity and the associated solubility. Thus ethyl alcohol, $CH_3—CH_2—OH$, is completely soluble in water in any desired proportion, a fact that permits alcoholic beverages of all different strengths to be prepared. In contrast, butyl alcohol, $CH_3—CH_2—CH_2—CH_2—OH$, with a hydrocarbon chain ($CH_3—CH_2—CH_2—CH_2—$) that is twice as long as the one ($CH_3—CH_2—$) in ethyl alcohol, is limited to a maximum solubility of about 8 g per 100 g of water at normal temperatures. This trend is almost universal—the larger the hydrocarbon segment in a molecule, the less the solubility, all other things being equal.

Natural surface water always contains a trace of dissolved organic material from decaying organisms, plant and animal. Even leaves blowing into a lake will have some organic compounds leached into the lake water; others will dissolve as the leaf decays. Ocean water contains dissolved organic materials at levels of about 0.5 ppm by weight.[4] Concentrations are usually higher in lakes and rivers, and they reach levels as high as 100 ppm in mangrove swamps and brine pools. A wide spectrum of organic compounds is represented in this dissolved material, including amino acids, sugars, alcohols, aldehydes, organic acids, amines, and slight amounts of hydrocarbons and other substances. (The basic chemical makeup of these various classes of compounds is defined in Chapter 4.)

Sediment Matter in Water. True solutions of dissolved substances require that molecules be thoroughly mixed. In addition to substances that dissolve and mix in water at the molecular level, some materials enter water as small, undissolved particles. They make

---

[4]In water-pollution studies, concentrations that are expressed in parts per million are invariably expressed as ppm by weight, in contrast to ppm by volume, used predominantly in air-pollution work. Throughout the rest of this chapter, "ppm" will mean ppm by weight. Another common concentration unit in water studies is milligrams (mg) per liter (ℓ). However, since 1 ℓ of water weighs very close to 1 kg ($= 10^6$ mg), then mg/ℓ units are equivalent to ppm by weight (see Chapter 1).

up the *sediment* (*silt*) in water. The sediment particles may settle out to become bottom sediments; they may be chemically degraded (altered) to a soluble form; or they may slowly dissolve in the water. Naturally occurring particles may be mineral or organic matter, eroded from the land or from decaying organisms.

The sediment load of streams varies from about 100 ppm to 100,000 ppm. Values in excess of 1900 ppm are considered high and are found mainly in arid regions where a thinner vegetation cover fails to hold the soil in place.

High sediment levels can destroy aquatic organisms by covering breeding habitat and food and by blocking sunlight and photosynthesis. Sediments also fill reservoirs and harbors; they cause turbine wear in hydroelectric power projects; and they generally make water so unesthetic that people no longer enjoy it.

Sediment is a natural part of surface waters, but it is enhanced greatly by man—most notably by his abuse of the land and destruction of protective vegetation.

Other Natural Contaminants. Any natural water contains literally hundreds of impurities in addition to those mentioned. There are gases dissolved from the atmosphere, two of which—oxygen (O_2) and carbon dioxide (CO_2)—have major importance to aquatic life. (The major role of O_2 will be discussed later in this chapter.) There are plant nutrients, including phosphates, nitrates, and trace metals. There are low concentrations of contaminants such as radioactivity and heavy metals.

There are also many inorganic minerals in water, and a number of acids and bases. These two types of impurity are so important that we shall examine them closely later on.

A PERSPECTIVE ON WATER POLLUTION

The sources of water pollution are unbelievably varied, involving almost every significant human activity. Most apparent is the direct dumping of wastes—sewage, industrial effluents, litter, and the like—into waterways. Somewhat less obvious are the wastes left on the ground that are eventually washed into water—wastes like animal manure, used crankcase oil, insecticides, and mine tailings. Similarly, disturbances of the soil mantle—by cultivation, road building, strip mining, and stream channelization—break the protective vegetation cover and encourage soil washout and the silting of water. Less obvious still are the air pollutants, many of which, like the oxides of sulfur and nitrogen, become acidic contaminants in the rain that falls on vital watersheds.

The chemical and physical forms of water pollutants are even more varied than are pollution sources. One sewer outfall can put thousands of chemical compounds into the water. The array of specific water pollutants is so wide that to study them rationally we must classify them into groups.

There are two broad types of water pollution. In one, man simply adds something to the water that was already there in moderate amounts. In another, he adds chemicals and materials strictly of his own invention that are new entrants into the environment. These two pollutant types are classified further in Table 8-3:

Table 8-3. **THE MAJOR TYPES OF WATER POLLUTION**

Materials of a Kind Already Present in Water	Materials New to the Environment
1. Microorganisms	10. Pesticides
2. Organic wastes	11. Many industrial chemicals
3. Plant nutrients	
4. Sediments or silt	
5. Inorganic minerals	
6. Acids and bases	
7. Heat	
8. Radioactivity	
9. Heavy metals	

The categories of the table are summarized as follows:

1. Microorganisms. In the past, disease microorganisms in water were the most serious form of water pollution—spreading typhoid, cholera, dysentery, and parasites across the land. This type of pollution has now been largely controlled by public health programs in the advanced nations.

2. Organic Wastes. These wastes presently constitute the most significant water pollution problem in the United States, most experts think. The next five sections of this chapter will deal with a wide spectrum of problems that ultimately are related to organic wastes.

3. Plant Nutrients. Phosphates, nitrates, and others contribute to the growth of unwanted algae, which in the end simply adds more organic waste to the water.

4. Sediments. These pollutants arise largely from the massive disturbances of land to which civilized countries seem addicted. There is an industrial contribution, also.

5. Inorganic Minerals. Already abundant in most of the world's waters, they are further concentrated by evaporation from reservoirs and by faulty agricultural practices.

6. Acids and Bases. Excessive quantities of acids and bases in the water come from industry and from the drainage waters from mines containing sulfur minerals.

7. Heat. Many waters are heated unduly by the waste heat from power-generating facilities. The problem is destined to grow rapidly because electric-power consumption is doubling every 10 years.

8. Radioactivity. Wastes from the mining and milling of uranium contain radioactive components that often wash into streams. Discarded radioactive tracers used in industry and research also contribute to water pollution, as does the nuclear-power industry. (The special problems of radioactive contamination in the environment will be treated in Chapter 11.)

9. Heavy Metals. Toxic heavy metals—such as mercury, lead, and cadmium—are transmitted to water through a variety of industrial, agricultural, and automotive sources. (Heavy metals in the environment will be treated in Chapter 9.)

10. Pesticides. These chemicals in environmental waters come almost entirely from agricultural applications. (Pesticides, too, are given separate treatment in this book; see Chapter 10.)

11. Industrial Chemicals. Many of man's synthetic chemicals are alien to our planet; living systems have not adjusted to their presence by evolutionary change. Some of these chemicals escape to the environment and are eventually washed into waterways, with unknown effects.

Finally, most of man's water projects—canal building, stream channelization, and the dredging of harbors and rivers—not only add pollutants (particularly silt) to water, but also constitute a form of physical pollution in themselves, as damaging to the ecology as chemical pollution. A channelized stream, for instance, is not compatible with the normal web of life that has strands tied to quiet pools, streamside vegetation, and well spaced riffles.

Any one of the above pollution types can significantly impair water quality. The greatest threat to water occurs in industrial regions, but agricultural lands, dense human populations, and Congressional pork-barrel projects are contributing factors. The Environmental Protection Agency (EPA) has prepared a map (Figure 8-6) showing the location of waters most seriously polluted. They are scattered throughout the country, but as expected, pollution on the eastern seaboard is most intense.

In the remainder of the chapter we shall discuss the chemical basis of some of the major classes of pollutants described in this section. We begin by exploring a nonpolluting substance, dissolved oxygen, for it has a crucial role in the chemical behavior of a great many of the serious pollutants of water.

DISSOLVED OXYGEN— A KEY TO CLEAN WATER

In setting out to discuss oxygen (O_2) dissolved in water (H_2O), we are examining together the two most important substances in the environment of man. These two substances work together to produce the priceless liquid we know as "clean water." Yet O_2 and H_2O

Figure 8-6. Principal areas of water pollution in the United States. Note that every area in the country has serious water-pollution problems which interfere with many desirable and beneficial water uses. Nearly all the nation's fresh-water resources are polluted to some degree. This map, prepared by the Water Quality Office of EPA, identifies the major trouble spots.

are quite unlike each other in most chemical and physical properties. Among other things, they are unlike in polarity. This important dissimilarity alone can be presumed to have controlled the formation of fossil fuels; the accumulation of O_2 in the atmosphere; the chemical properties of oceans, lakes, and rivers; and thus the evolution of life from the time oxygen became common on this planet nearly 2 billion years ago.

The oxygen molecule, O_2, is totally nonpolar because neither of its two oxygen atoms can pull charge away from the other. Water, in contrast, is highly polar. The expected consequence, based on the rule that like dissolves like, is that O_2 does not dissolve to an appreciable extent in H_2O. This deduction is borne out by the facts. Despite the enormous reservoir of O_2 in the atmosphere and its high relative concentration in air (21 percent), the O_2 that dissolves from air into water reaches only about 10 ppm at full saturation.[5] Yet on this thin chemical thread are suspended both the life of most aquatic creatures

---

[5] We note further that nonpolar O_2 can be expected to dissolve far more readily in nonpolar hydrocarbons than in polar water. Indeed, O_2 is 80 times more soluble in hexane (C_6H_{14}) than in H_2O. Like most equilibrium parameters, these solubilities vary with temperature, pressure, and other factors. In a later section we shall discuss how heat (thermal pollution) drives O_2 out of H_2O. Also, the reduced atmospheric pressures that occur with increasing elevation give proportional reduction in dissolved O_2. High mountain lakes, then, have less dissolved oxygen in their waters. Furthermore, salty water will absorb less oxygen than will fresh water, a factor important in oceans and other high-salt waters.

(including all fish) and the remarkable self-purifying characteristics of water that are ultimately responsible for water's main biological and human uses. The fragile O_2—H_2O thread sometimes breaks, even under natural stresses, and with man's increasing abuse of water, this life-support thread is now endangered in waters throughout the world.

We shall discuss the principal threat to dissolved oxygen (often abbreviated DO) in the next section. Here we shall explain more fully the biological and chemical roles of DO in water.

Biological Role of DO. Oxygen is needed by most plants and animals for respiration—a process in which "food" is oxidized so that the organism can obtain energy. However, there are some anaerobic microorganisms that derive their energy needs from other chemical reactions, not involving O_2.

It is clear that oxygen depletion will deprive many species of essential oxygen and will destroy those species. This destruction creates ecological havoc in the water. Serious imbalances begin when DO levels fall to about half of their saturation value. Cold-water fish, for instance, require about 5 ppm DO, while warm-water fish need at least 4 ppm O_2 in their water habitat.

Chemical Role of DO. Oxygen, whether in the atmosphere or in water, creates a strong oxidizing environment. Many substances that are found in water are profoundly changed by oxidation. Most organic contaminants, for instance, are oxidized to carbon dioxide (CO_2) and water. Compounds of sulfur are oxidized to the sulfate ion (SO_4^{2-}). Table 8-4 compares the chemical forms assumed by some important elements under oxidizing and reducing conditions. Reducing (anaerobic) conditions exist when O_2 is absent.

The lesson of Table 8-4 is that nearly every important element subject to oxidation and reduction in water exists in a less offensive form when fully oxidized. Carbon in its reduced form consists of various organic compounds, some toxic, that contribute to the bad taste and odor of contaminated water. Fully oxidized carbon, CO_2, is usually harmless, and it quickly escapes to the atmosphere. Sulfur is toxic and ill smelling in its reduced form, hydrogen sulfide (H_2S), a substance known for good reason as rotten-egg gas. The oxidized sulfate species (SO_4^{2-}, HSO_4^{-}) are, by contrast, neither harmful nor offensive at reasonable concentrations.

Similarly, reduced nitrogen exists as ammonia (NH_3) and various amines (R—NH_2), all foul smelling. Oxidized nitrogen becomes the relatively docile nitrate ion (NO_3^{-}). The partly reduced form of iron (Fe) is the iron(II) ion, Fe^{2+}. Compounds of iron(II) are fairly soluble in water, enough to give the water an "irony" taste. Iron(III) compounds, which occur in aerobic water, are not usually soluble enough to impair water quality.

While the importance of maintaining O_2 in waterways is abundantly clear, the means are largely beyond our control. Oxygen

Table 8-4. THE CHEMICAL FORM ASSUMED BY SOME IMPORTANT CHEMICAL ELEMENTS IN WATER, WITH AND WITHOUT DISSOLVED OXYGEN

Element	O_2 Absent (anaerobic water)		O_2 Present (aerobic water)	
	Oxidation (natural cleansing) $\longrightarrow$ (Most oxidized states)			
	(Most reduced states) $\longleftarrow$ Reduction			
	(Chemical species: toxic and offensive)		(Chemical species: inoffensive in moderate amounts)	
Carbon (C)	Organic (reduced) carbon: CH_4 Aldehydes, Organic acids, etc.		$[CO]^a$	CO_2 HCO_3^-
Nitrogen (N)	NH_3	N_2	NO_2^-	NO_3^-
Sulfur (S)	H_2S	$[S]$	SO_3^{2-}	SO_4^{2-}
Iron (Fe)	Fe^{2+}		Fe^{3+}	

[a] [ ] = not usually present in water.

enters water from the atmosphere, at a pace dictated by diffusion rates and water circulation. Its penetration is aided by the stirring action of waves and riffles. A secondary source of O_2 is photosynthesis, also largely out of our control. These natural sources must offset O_2 depletion if water is to remain viable. Oxygen depletion is caused mainly by organic matter in the water. Here man can take a hand by ending his excessive discharges of organic materials and the nutrients that cause unreasonable organic growth.

ORGANIC CONTAMINANTS AND BOD

The most significant element in organic material, as we have explained, is carbon. Most organic carbon in water is slowly oxidized to carbon dioxide:

$$C \quad + \quad O_2 \quad \rightarrow \quad CO_2$$

organic carbon	diatomic oxygen	carbon dioxide

This oxidation consumes oxygen, as the above reaction shows. Since O_2 is in such short supply in water, a small amount of organic contamination is enough to deplete the oxygen and make the water putrid.

The oxidation of organic material in water is caused by microorganisms. These tiny creatures actually are using the organic debris in water as food, oxidizing it with dissolved O_2 to obtain energy, much as we oxidize foodstuffs in our bodies, using inhaled O_2, for energy. The oxidation proceeds with the aid of enzymes. However, enzymes are not available to oxidize (degrade) some types of organic materials, particularly some synthetic materials that are new to the environment. These materials are called *nondegradable*, and they persist for a very long time in water. The materials that can be oxidized are called *degradable* or *biodegradable*. Our concern at the moment is with degradable organic materials, because it is these materials that combine with O_2 and cause its disappearance.[6]

Mass Analysis. Our mass analysis of carbon oxidation, $C + O_2 \rightarrow CO_2$, shows that the pertinent atomic and molecular weights are 12.0 for C, 32.0 for O_2, and 44 for CO_2. The ratio of O_2 mass to C mass is therefore 32/12, or 8/3, as we pointed out on several previous occasions (in Chapter 5, for instance). In the present context, this ratio simply means that each weight unit of degradable organic carbon will deplete 8/3, or 2.67, weight units of dissolved O_2. We are not far wrong if we replace the weight of organic carbon by the weight of dried organic matter, which is principally carbon anyhow. We can use, therefore, the approximate working ratio

$$\frac{8 \text{ parts of oxygen}}{3 \text{ parts of dried organic matter}}$$

for our estimates of environmental degradation. Other elements besides carbon in organic material (most notably hydrogen, oxygen, and nitrogen) will alter this ratio slightly, but 8/3 approximates reality well enough to show quite clearly the disaster we invite when we discard organic wastes into natural waterways.

Let us examine, for instance, the impact of adding degradable organic pollutants to water at the very small concentration of 3 ppm. The amount of O_2 ultimately consumed by this organic matter will be 3 ppm times the 8/3 ratio: 8 ppm of O_2 depletion.[7] Since water rarely contains over 10 ppm O_2 to start with, a loss of 8 ppm will virtually destroy the quality of the waterway. A bit more organic matter would wipe out the O_2 supply completely.

This simple calculation shows the power of trace levels of organic contaminants to defile the water. But many things can happen in water to modify these initial conclusions.

Organic matter in water is not oxidized instantaneously. Oxidation

---

[6] Our focus on degradable material is not intended to slight the importance of nondegradable organic matter. Indeed the latter often contains toxic substances that are environmentally damaging. And while nondegradable matter does not make an immediate demand on dissolved oxygen, some environmental price must eventually be paid.

[7] We ordinarily work problems of this type by setting down all units in order to check our work:

$$3 \text{ ppm C} \times \frac{8 \text{ parts } O_2}{3 \text{ parts C}} = 8 \text{ ppm } O_2$$

occurs by a complex set of chemical reactions, and all chemical reactions, even those aided by enzyme catalysts, take time to occur (Chapter 5). Therefore most degradable organic matter disappears only gradually, over a period of time from a few days to a few weeks. If oxygen in the water is methodically replaced, slight amounts of organic matter may not cause an oxygen crisis after all.

Lakes and Rivers. Lakes and rivers are quite opposite in the speed with which they assimilate replacement O_2. Lake water is quiet, without flow. Deep lakes tend to become thermally stratified (develop layers of different temperatures), particularly in the summer; there are no vertical currents to carry O_2 from the surface to the depths. Nor does photosynthesis add O_2—light fails to penetrate to the deep parts. The deep waters, if depleted of O_2, remain that way for months. Deep ocean water, whose fragile supply of O_2 is quickly used up by natural organic contamination and not replaced by fresh currents for centuries, is permanently devoid of oxygen. Many seas and deep lakes are also without oxygen in their deepest parts. Africa's mile-deep Lake Tanganyika has no O_2 below a depth of 80 ft. Now man's prolific wastes are reaching into new territory, significantly enlarging the anaerobic waters of this world. For instance, the Baltic Sea, a shallow sea 15 times larger than Lake Erie, has suffered steadily declining oxygen levels (Figure 8-7) through this century because of increasing sewage and industrial effluent from 20 million people.

Rivers, because of the churning currents caused by flow, have a

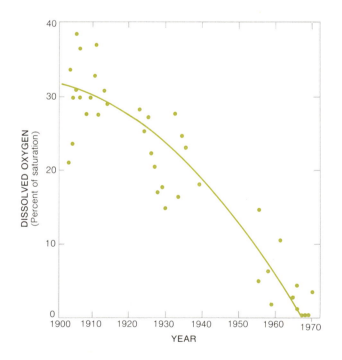

Figure 8-7. Oxygen levels have decreased steadily in this century near the bottom of the northern Baltic Sea.

much faster oxygen replenishment than do lakes. The rate of O_2 uptake is greatest for fast, shallow rivers and slowest for deep, sluggish rivers. The favorable O_2 replacement of rivers is offset, unfortunately, by the fact that rivers are the natural sewer lines of civilization. A constant chemical battle pits streams of organic wastes against replenishment O_2. The outcome depends on the size and characteristics of the river and on the magnitude of the waste streams. When the organic wastes win the battle, the river is a virtual cesspool for many miles. When O_2 dominates, the river water maintains most of its desirable characteristics, although it will still retain traces of offensive and sometimes toxic organic substances for some distance, until inexorable oxidation eats them up.

The O_2–organic-waste battle in rivers is often displayed by *oxygen sag curves* (Figure 8-8). Organic wastes from a city or industry enter the river at some outfall. The wastes mix with the river water and gradually deplete O_2 as the water moves downstream. The wastes are themselves slowly depleted until, at some point, O_2 replenishment gets the upper hand. At this point, the O_2 concentration in water starts increasing back to normal. The greatest damage to aquatic life occurs at the point of maximum decline, or sag, in oxygen content—many miles downstream of the outfall.

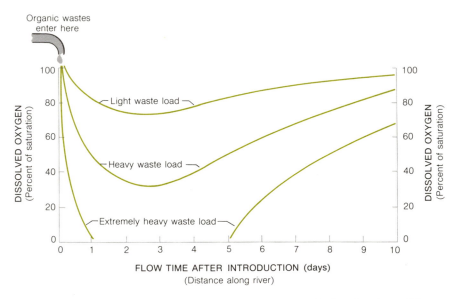

Figure 8-8. Oxygen sag curve for the Ohio River below Pittsburgh, where different loads of organic wastes were introduced.

BOD. The potential damage to water's dissolved O_2 is measured by the *biochemical oxygen demand (BOD)* of organic wastes. The BOD is the amount of O_2 required to degrade wastes.[8] Its name

---

[8] Since BOD is defined as an "amount of oxygen," it always carries the units of weight of oxygen, concentration of oxygen, etc. The "demand" comes from organic carbon, but it is a demand for O_2 and must be so expressed.

comes from the "demand" that organic carbon makes for O_2 through biochemical (enzyme) oxidation.

BOD—demanded oxygen—can be likened to a debt (demanded dollars) that must be paid. Dissolved oxygen (DO) can be thought of as the currency used to pay the debt. If DO (assets) exceeds BOD (debt), aerobic conditions (economic solvency) will prevail. If BOD exceeds DO, bankruptcy may ensue. Bankruptcy, in this case, is depletion of water's chief asset, O_2. Keep firmly in mind that *the debt represented by BOD is always measured in units of oxygen*, just as the national debt is expressed in dollars.

Our previous reasoning enables us to calculate approximate BOD values. Thus 3 g of degradable organic matter (dry weight) demands O_2 in an amount that is 8/3 times the 3 g of organic material—a total of 8 g of BOD. By the same logic, 3 ppm of organic carbon makes a demand for 8 ppm of O_2—the BOD is 8 ppm. The ratio 8/3 is our guide for all such considerations.[9]

Note that the 8/3 ratio used to calculate BOD is valid only for organic material subject to oxidation, $C + O_2 \rightarrow CO_2$. We do not count the nondegradable organic compounds because they are not oxidized and do not consume oxygen.

In a standard test for BOD—*the five-day test*—contaminated water is placed in closed bottles, and the amount of O_2 lost in 5 days of storage at 20°C is measured. (We must know the exact temperature because the rate of organic oxidation, like the rate of all reactions, is greatly accelerated by increased temperature—see Chapter 5.) The five-day test accounts for most of the ultimate oxygen demand but not all; a small fraction of slowly degrading organic material is not yet oxidized in 5 days. The 5 day BOD is generally about three quarters of the ultimate BOD.

Sources of BOD. Degradable organic matter comes from many sources. Domestic sewage, for instance, contains great quantities of organic wastes that use up oxygen. Untreated sewage generally has a BOD level ranging from 100 to 300 ppm. Treatment does not remove more than 80 percent of this BOD on the average, so even

---

[9]Two examples help show the significance of approximate BOD calculations. These examples concern a small lake, 0.5 mi across and 8 ft deep, that contains a total of 16 tons of O_2. Imagine, first, that 3 tons of degradable organic carbon is dumped into the lake. The BOD is

$$BOD = 3 \text{ tons C} \times \frac{8 \text{ parts } O_2}{3 \text{ parts C}} = 8 \text{ tons } O_2$$

Thus 8 tons of the total 16 tons of O_2 will be used up (barring rapid replenishment). The reduction of O_2 to half its normal level will make fish gasp, but most of them will survive. Ultimately, chemical and biological balance will be restored.

Consider, second, the consequences of discarding 60 tons of such wastes into the water.

$$BOD = 60 \text{ tons C} \times \frac{8 \text{ parts } O_2}{3 \text{ parts C}} = 160 \text{ tons } O_2$$

The demand is now 160 tons of O_2, but the supply is only 16 tons. This imbalance is environmental bankruptcy. The water will turn putrid and stay that way until O_2 replenishment slowly corrects the situation. Meanwhile fish will die and the lake's total biological community will suffer adverse changes. Recovery may take many years.

treated sewage water still carries 20 to 60 ppm BOD, well in excess of the fragile O_2 supply of about 10 ppm. Dilution reduces the BOD to manageable levels, but a nagging question remains—how many wastes can be diluted in the limited surface waters of earth before dilution becomes a hollow concept? Sewage—some raw, some treated—is now burdening the waters of this country with about 13.5 million lb of BOD every day; and sewage is not the only villain.

Both agriculture and industry add significantly to the organic outpouring. The manure from feedlots is a notorious source of BOD for nearby waterways. Other agricultural wastes contribute significantly. At the same time, United States industries generate over 60 million lb of BOD per day. Much of it is treated, but treatment is spotty and the BOD of some escaping waste streams is remarkably high (Table 8-5), enough so to deplete the oxygen from a stream of fresh water 1000 times as large as the waste stream.[10]

Table 8-5. THE BOD OF WASTES FROM CERTAIN INDUSTRIES HAVING HIGHLY ORGANIC DISCARDS[a]

Industry	BOD of Wastes
Molasses distilling	20,000–30,000
Pulp and paper (sulfite cooker)	16,000–25,000
Grain distilling	15,000–20,000
Meat packing	600–2,000
Tannery	500–5,000
Brewing	500–1,200
Beet sugar refining	450–2,000

[a]Data from P. H. McGauhey, *Engineering Management of Water Quality* (New York: McGraw-Hill, 1968).

Along with organic matter from the bowels of civilization, the limited O_2 dissolved in water must contend with natural plant debris. Here man adds still another burden. In his streams of wastewater are nutrients—phosphorus and nitrogen—that stimulate the growth of aquatic plants. When these plants die, they too must be degraded by dissolved oxygen. In the next few sections we shall look at the role of plant nutrients, starting with the one most talked about of all—the phosphorus contained in detergents.

DETERGENTS AND PHOSPHORUS

Detergents are cleansers of a special sort. Detergent action is based on the unique characteristics of certain long-chain *surfactant* molecules—chainlike molecules having one polar end and one nonpolar

---

[10]A BOD of 10,000 ppm, exceeded by several of the entries in Table 8-5, is 1000 times larger than the O_2 supply of clean water—10 ppm. It would therefore take 1000 streams of clean water to degrade the BOD in one such waste stream, all streams being the same size.

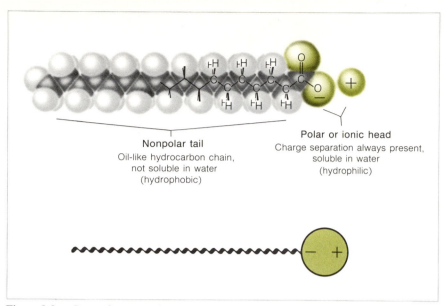

Figure 8-9. General nature of surfactant molecules.

end. Surfactant molecules have the form shown above in Figure 8-9. When grease-covered objects (hands, clothes, dishes) are immersed in water containing a detergent, the two-sided character of surfactant molecules become dramatically apparent. The nonpolar end of the surfactant molecules seeks to dissolve in the nonpolar grease, following the like-dissolves-like rule. The polar end has a much stronger attraction to the polar water. This conflict is resolved in a way that suits both parts of the molecule at once: detergent accumulates at the surface between the grease and the water, with surfactant molecules dipping their long, nonpolar tails into the grease layer but keeping their polar heads stuck out into the water. This surface accumulation and activity leads to the special name "surfactant."

As surface accumulation occurs, a remarkable chemical effect takes place. The surface of the grease, formerly nonpolar and water repellent, becomes polar and thus attractive to water. Water soaks in and around the grime, and small bits and droplets break loose. The droplets, now compatible with water because of their polar coating (Figure 8-10, next page), become waterborne, leaving clean clothes or clean dishes behind.

Detergents of two types are in common use; soaps and synthetic detergents, or "syndets." (In common practice the word "detergent" describes only the synthetic variety, but in a strict sense, soaps are detergents too.) The surfactant in soaps is derived from animal fats, while that in synthetic detergents is synthesized from petroleum chemicals.

Soap. The surfactant of soap is formed by the chemical breakdown of molecules of animal fat (fats are described in Chapter 4).

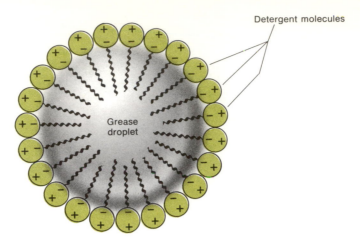

Grease
droplet

Figure 8-10. The formation of a grease droplet by the surfactant molecules of a detergent. The droplet has a polar coating and is water-compatible.

The fatty material is *saponified*—broken down with a strong base (alkali). The chemical reaction is

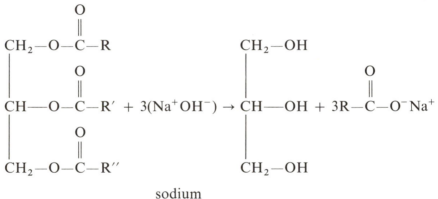

triglyceride:	sodium hydroxide:	glycerin	soap
a fatty molecule	a strong base		surfactant

The soap surfactant is composed, as expected, of a nonpolar end,—R, and a polar end, —COO$^-$Na$^+$. The hydrocarbon radical, —R, differs slightly from molecule to molecule, having generally from 13 to 17 linked carbon atoms.

Soaps have two drawbacks that interfere with their use. First, the surfactant molecules (shown above) are sodium salts of the fatty acid R—COOH, and in acidic waters the excess hydrogen ions (H$^+$) combine with the RCOO$^-$ ion to give this acid. The acid is not effective as a detergent.

Second and most important, all natural water is to a certain degree "hard," which means that it contains calcium ions (Ca^{2+}) and magnesium ions (Mg^{2+}). Water hardness will be discussed more fully later: the basic problem is that soap surfactant combines with hardness ions to form a gummy precipitate, (R—COO)$_2$Ca, most obvious as "bathtub ring." This scummy material is a considerable nuisance in clothes and washing machines, and further, formation of the precipitate depletes the soap, requiring that more be added. Synthetic detergents do not have these problems.

Synthetic Detergents.   The surfactant used in synthetic detergents is most commonly a linear alkyl benzene sulfonate (LAS) of the form

$$R—\langle\!\!\langle\rangle\!\!\rangle—SO_3^-\,Na^+$$

where R is a straight-chain (linear) hydrocarbon segment. Before 1965, the hydrocarbon segment was a branched (nonlinear) chain, resistant to oxidation by microorganisms. Because they were non-degradable, detergent surfactants (abbreviated ABS, the L for "linear" no longer appropriate) accumulated in waterways and water supplies, producing towering foam edifices on rivers and foamy heads on glasses of drinking water.

The present-day LAS surfactants can be chewed up by micro-organisms and dissolved oxygen, but, unfortunately, there is more to detergents than simple surfactants. Modern synthetic detergents contain a hodgepodge of chemicals, including the following:

1. *Surfactant:* the heart of detergent action, as explained.
2. *Builder:* a water softener and an aid to detergent action.
3. *Optical brightener:* absorbs ultraviolet light and converts it to visible light, making clothes appear brighter.
4. *Perfume:* to "sweeten" up the detergent and the clothes washed in it.
5. *Bleach:* to increase whiteness.
6. *Enzymes:* to remove protein-based soil and stains from clothes.

Of these components, builders are at the center of the environmental controversy that now surrounds detergents. The usual builder is sodium tripolyphosphate (STPP), which has the chemical composition $Na_5P_3O_{10}$. It ionizes in water to give five sodium ions (Na^+) and one tripolyphosphate (TPP) ion:

TPP ion

Unshared electron pairs on the oxygen atoms of this ion combine with other wash-water ions, causing the desired effects in the washing machine.

First, TPP ions tie up ions of calcium (Ca^{2+}) and magnesium (Mg^{2+}), both of which interfere with detergent action. Thus the primary function of TPP ions is simply to soften the water. In addition, the unshared electron pairs of TPP extract hydrogen ions (H^+) from water, leaving hydroxide ions (OH^-) and a basic solution behind. The hydrolysis reaction can be represented by

$$TPP + H_2O \rightarrow H^+TPP + OH^-$$

The basic, or alkaline, wash water produced by this reaction is necessary for good detergent action.

Phosphorus-containing builders are highly effective. Many substitutes have been proposed or tried, with varying levels of success.[11] Most of the substitute builders are quite alkaline—somewhat more so on average than the phosphate builders—and may be hazardous if swallowed. The most promising substitute builder is nitrilotriacetic acid, or NTA. The ion formed from NTA in alkaline washwater is predominantly

$$N \begin{cases} CH_2-COO^- \\ CH_2-COO^- \\ CH_2-COO^- \end{cases}$$

This ion is nearly as effective as the TPP ion, but it may interact with heavy metals in water to enhance their teratogenic (birth-defect) characteristics. Thus for the present, phosphorus remains a prominent part of the modern wash-day miracle—the one that the detergent industry spends a billion dollars a year telling us about.

Phosphorus in the Environment. From the narrow environment of the washing machine, phosphorus is carried by sewers and drains to the complex environment of earth. Few modern-day treatment plants can stop it: they are designed to handle silt and organic matter. But on the journey to the open waterways, chemical change takes place. The tripolyphosphate ion hydrolyzes to the phosphate-type ions shown here:[12]

$$P_3O_{10}{}^{5-} \quad + \quad 2H_2O \rightarrow 2HPO_4{}^{2-} \quad + \quad H_2PO_4{}^-$$

| tripolyphosphate ion | water | hydrogen phosphate ion | dihydrogen phosphate ion |

These ions, then, bring phosphorus to the environment.

Most detergents are loaded heavily with STPP—25 to 60 percent

---

[11]One of the most effective phosphate substitutes is washing soda or sal soda: sodium carbonate (Na_2CO_3). This compound can be used with ordinary soap. The carbonate ions ($CO_3{}^{2-}$) produced by washing soda soften the water by forming harmless, granular precipitates of calcium carbonate ($CaCO_3$) and magnesium carbonate ($MgCO_3$). The $CO_3{}^{2-}$ ion provides alkalinity by hydrolysis.

Another approach is to use a water softener so that no builder is needed at all—just soap.

[12]The phosphate ion, $PO_4{}^{3-}$, often symbolizes the phosphorus problem in water. However, it rarely occurs without associated hydrogen, as in the above species, $HPO_4{}^{2-}$. The ratio of $HPO_4{}^{2-}$ and $H_2PO_4{}^-$ is not always 2 to 1, as shown by the balanced reaction: it depends on

by weight. The average detergent in 1970 was nearly 40 percent STPP. Thus the average content of the element phosphorus in detergents was almost 10 percent—one fourth that of STPP.[13] With United States detergent sales exceeding 5 billion lb per year, the amount of phosphorus reaching our waterways is understandably large—close to 500 million lb annually. Most of it enters the surface waters; in 1967 it was estimated by federal authorities that 370 million lb of detergent phosphorus poured into our rivers and lakes.

Synthetic detergents are not the only source of phosphorus in waterways—but they are the largest single source. Of the 3.5 lb per capita of phosphorus in wastewater, it was estimated in 1970 that 2.3 lb came from synthetic detergents and the remainder—1.2 lb—from human excrement. A much larger quantity of phosphorus—70 percent of that produced, as opposed to only 13 percent for detergents—is used in farm fertilizers. However, phosphorus becomes chemically bound to soil and very little leaches out into the water under normal circumstances. Nonetheless, agricultural phosphorus dominates the waters of rural areas—this phosphorus comes from fertilizers applied inappropriately to frozen farmlands in the fall and from animal excrement.

An abundant supply of synthetic phosphate fertilizers is one of the crowning technological achievements of this age. Because phosphorus is essential to all life, but ordinarily in short supply, these fertilizers have stimulated crop growth around the world. Phosphorus stimulates plant growth in water, as well. However, here the effects are most often disastrous: among the benefiting plants are blue-green algae—obnoxious to man, deadly to other aquatic life, and responsible for the premature aging of lakes.

ALGAE, NUTRIENTS, AND EUTROPHICATION

Algae and other plants are complex living systems containing thousands of special molecules and ions to promote and regulate growth,

---

whether the water is acidic or basic. If the water is slightly acidic, the free H^+ ions combine with HPO_4^{2-} to give mainly the $H_2PO_4^-$ form:

$$H^+ + HPO_4^{2-} \rightarrow \underbrace{H_2PO_4^-}_{\text{acidic species}}$$

If it is slightly basic, the OH^- ion takes H^+ away from $H_2PO_4^-$ to form HPO_4^{2-} as the predominant species:

$$H_2PO_4^- + OH^- \rightarrow \underbrace{HPO_4^{2-} + H_2O}_{\text{basic species}}$$

[13] The ratio of STPP ($Na_5P_3O_{10}$) weight or weight percentage to phosphorus (P) weight or weight percentage is simply the ratio of the molecular weight of $Na_5P_3O_{10}$, which is 368, to the sum of atomic weights of the three phosphorus (P) atoms contained in $Na_5P_3O_{10}$, which is $3 \times 31.0 = 93.0$. This ratio, then, is $(368/93.0) = 3.96$, or approximately 4. The ratio of phosphorus to STPP is the reciprocal of this—about one fourth.

reproduction, photosynthesis, and respiration. These molecules together contain at least nineteen different elements.[14] These elements must be obtained from the medium—soil, water, and to a limited extent, air—in which the plants grow. These essential elements are termed plant *nutrients*.

Plants will stop growing if deprived of any one of their essential nutrients. And they can never grow faster than the supply of the nutrient that is scarcest relative to their needs. In general, a shortage of one or more nutrients is the limiting factor in the rate of growth of plants.

Most nutrients, especially certain metals, are required in such minute amounts that the environment nearly always provides enough. The best evidence suggests that the growth-limiting elements are usually among the major five elements of plant cellular material: carbon (C), hydrogen (H), oxygen (O), nitrogen (N), and phosphorus (P). These elements assemble in the atomic ratio expressed approximately by the chemical formula

$$C_{106}H_{181}O_{45}N_{16}P$$

(represents the composition of dry plant material)

This formula is a convenient representation of the composition of dry plant material, but it does not represent any actual molecule. Instead it portrays the average content of major elements in the thousands of living molecules in plants—it can be thought of as the hypothetical "average" molecule of dry plant tissue.

Of the five major elements, the three needed in the greatest quantity are provided abundantly by water: hydrogen and oxygen from H_2O and carbon from dissolved CO_2. The remaining two, nitrogen and phosphorus, are much scarcer and most often limit the growth of plants in the waterways of the world.[15] Growth in the oceans is limited mainly by the amount of nitrogen available, and growth in inland waters usually is controlled by phosphorus, but sometimes by nitrogen. The ratio of inland waters in which growth is controlled by one or the other of these two elements is in hot dispute, but as we shall see, this argument does not affect the practical courses of action open to man.

Eutrophication.   Lakes are born in a nutrient-poor or *oligotrophic* condition. Oligotrophic lakes support little plant life. As a result, their biological productivity is limited and their waters are clear. The lack of decaying vegetation keeps organic matter at a minimum, and the waters are saturated with dissolved oxygen.

---

[14] Among the essential elements for algal growth are B, C, Ca, Cl, Co, Cu, Fe, H, K, Mg, Mn, Mo, N, Na, O, P, S, V, and Zn.

[15] Nitrogen as N_2 is the most abundant gas in the atmosphere. Unfortunately, very few plants can use nitrogen in this chemical form. A few species can *fix* N_2—convert it to usable nitrates (NO_3^-) or other forms—but not in sufficient quantities to make it an abundant nutrient.

As the nutrient supply of a lake increases from the buildup of surrounding soils and vegetation, the lake passes into a stage that is called eutrophic.[16] A *eutrophic* lake is simply a nutrient-rich lake. The added nutrients that characterize a eutrophic lake cause fundamental changes in the lake's biology and chemistry and eventually in its very existence as a lake.

With the buildup of nutrients in a lake, the lid that holds plant growth in check is removed. Water weeds become abundant. Finally, thick mats of blue-green algae take over, crowding out other plant and animal life. Dying algal material sinks into the water, and its decay consumes the dissolved oxygen. The stench of anaerobic chemicals wells up from the water. The rain of dead algae, no longer oxidized, forms layers on the bottom; the lake becomes shallow; and thus is set in motion the process by which the lake fills—and dies.

Eutrophic lakes are highly undesirable to man. The mats of algal growth interfere with recreation and choke out game fish. The water has a bad appearance, a bad odor, and a bad taste. Also, water treatment filters become clogged with algae.

Man's outpouring of phosphorus and nitrogen has accelerated eutrophication in bodies of water around the world. Switzerland's beautiful Lake Zurich suffered its first eruption of blue-green algae in 1896. Man-made (cultural) eutrophication in America is more recent, and it has picked up substantially since the use of phosphate builders exploded in the 1950s. The phosphorus concentration in Lake Erie increased threefold from 1942 to 1968 (35 to 50 percent of the present phosphorus input to Lake Erie is from detergents). In this period, blue-green algae increased enormously and whitefish, blue pike, and walleye fish populations declined or disappeared. Oxygen has now disappeared from one third of the lake's water. It has been estimated that Lake Erie has aged as much in 50 years as it would have in 15,000 years without man's interference.

The estuary (mouth) of the Potomac River has suffered similar eutrophication. The stages of this process and the corresponding growth in phosphorus input are shown in Figure 8-11 (next page). Eighteen times more phosphorus is being added to this estuary now than was added in 1913. This change is due to a combination of factors; an eightfold increase in the population served by sewers and a doubling factor caused by synthetic-detergent use.

The Battle Against Eutrophication.   Clearly, detergent phosphorus alone is not responsible for the eutrophication of waterways. Population growth, synthetic fertilizers, and the addition of other nutrients play a substantial role. In casting about for a solution to this many-sided problem, most experts have concluded the following. Eutrophication can be halted only if we choke off a critical nutrient.

---

[16]An intermediate stage of the lake's existence is called *mesotrophic.* The word eutrophic derives from the Greek, and means "well nourished."

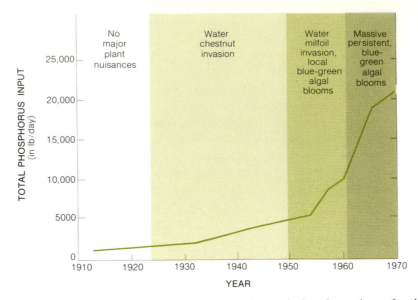

Figure 8-11. Stages in the growth of eutrophication and phosphorus input for the Potomac estuary from 1913 to 1970.

Phosphorus is the only candidate for reduction. There are, for instance, too many diffuse natural sources for nitrogen: it is moderately abundant in runoff water and seepage, and it can even be converted (fixed) into a biologically useful form by some aquatic plants that use the diatomic nitrogen (N_2) of air. In other words, the supply of nitrogen is so widespread that it can never be choked off very completely. Phosphorus, on the other hand, comes mainly from the drainpipes of civilization and a few farms.

Attempts to control phosphorus—while more promising than any other approach—are fraught with difficulty. First of all, control must be exceedingly tight. Algal blooms can get started in water containing concentrations of phosphorus as low as 0.01 ppm. Levels in many lakes have soared far beyond this critical level. The western basin of Lake Erie contained 0.04 ppm in 1968. Elimination of detergent phosphates would perhaps cut this figure in half—to 0.02 ppm—hardly a total solution to the problem. The only prospect for total success lies in a multiple approach, and this approach would logically start with phosphate builders, since they are the largest single source of phosphorus.

Detergent phosphorus can be attacked on several fronts. The search for alternate builders must be pressed. Manufactures could lower the phosphorus content selectively in soft-water areas—for phosphate builders are primarily water softeners. Public education might also help—perhaps Americans would settle for clothing a shade less than twinkling white in order to save their lakes.

There are other weapons that may aid in the difficult eutrophication battle. Dredging operations—which bring sediment nutrients to the surface—could be used more sparingly. Sewage water can be diverted around critical lakes (this has already been done to help preserve Lake Tahoe). Good farming practices should be encouraged. Proper zoning in critical watersheds would help, as would efforts to limit human population density. All these approaches may be needed, plus one more.

Not least of the tools available to fight eutrophication are up-graded plants for sewage treatment. Wastewater plants now capture very little phosphorus, but advanced methods, while expensive, are capable of retaining 90 percent or more of the phosphorus in sewage. In the section to follow, we shall describe the means and limitations of wastewater plants in removing organic carbon, phosphorus, and other undesirable contaminants of water.

WASTEWATER TREATMENT

A remarkably large quantity of raw sewage is let loose on the environment of earth. Many rivers are mere drain lines whose incredible load of filth has grown with the human population along their banks. Even in the United States—the most technically advanced nation on earth—the sewage from over 50 million people flows untreated into rivers, oceans, lakes, and groundwater.

Most of the treatment that does exist is crude: the image of waste-water treatment plants as mysterious places where technological magic is performed on sewage water is all wrong. Some "magic" is performed, but the show is directed by tiny microorganisms that existed on earth millions of years before man and his technology arrived. At present, then, sewage treatment is based overwhelmingly on natural processes, aided by crude mechanical devices. In large part, the wastewater plant is designed simply to encourage natural purification.

Wastewater treatment plants are arranged to have sewage flowing from one stage to another, each stage using some mechanical device to handle a special type of contamination. Prominent among the devices employed are the following:

1. *Mechanical screens:* through them wastewater is passed to remove grit and large particles.
2. *Sedimentation (settling) tanks:* giant holding tanks in which solid matter drops to the bottom (as it would in a lake) for removal; thus the wastewater is clarified.
3. *Skimming devices:* mechanical aids used to remove floating scum and grease from the water.
4. *Trickling filters:* giant beds of rocks, open to the atmosphere, through which the wastewater is trickled. Tiny organisms attached to the rocks and in the water gobble up organic matter with the aid of O_2 that circulates freely with air between

the rocks. The trickling filter is little more than an aeration device, providing intimate contact between the trickling water and the air's oxygen so that natural aerobic decay (oxidation) is speeded up.

5. *Activated sludge tanks*: tanks in which aeration is achieved simply by the bubbling of air (or pure oxygen) through tanks of wastewater.

6. *Anaerobic digestors*: large tanks sealed off from air, to which the bottom sludges of settling tanks are delivered. In these tanks, anaerobic microorganisms chew up the organic molecules without oxygen's help. The process is called anaerobic digestion, or anaerobic decay. The product of anaerobic digestion consists of reduced gases, most notably methane (CH_4), as opposed to the oxidized carbon dioxide (CO_2) produced aerobically. Anaerobic digestion is a natural process, occurring in the bottom sediment of waterways where oxygen is absent.

Trickling filters at the Salt Lake municipal plant. Eight such filters, each 173 ft across, remove most of the dissolved organic material from up to 32 million gal of waste water per day. (*Photo by Alexis Kelner.*)

Primary and Secondary Treatment. Various combinations of the above devices are used in most wastewater plants (Figure 8-12). *Primary treatment*—the sole treatment for wastes from about fifty million people in the United States—is a remarkably unsophisticated approach that employs only screening, settling, and skimming. The sludge, of course, is anaerobically digested or burned, and its residue is discarded or used as fertilizer.

Secondary treatment is added to the primary treatment of sewage from over 100 million people. *Secondary treatment* enlists the aid of microorganisms in trickling filters or activated sludge tanks to

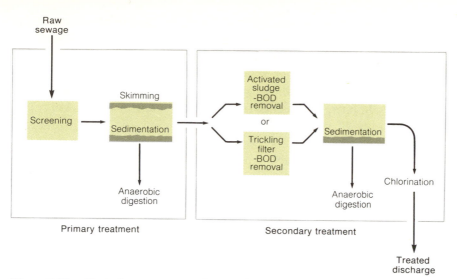

Figure 8-12. Typical arrangement of stages for primary and secondary wastewater treatment.

oxidize the dissolved organic matter not touched by primary treatment. Following this step, the wastewater undergoes further sedimentation, and it is then usually chlorinated (to kill microorganisms) and released.

Table 8-6 shows the relative efficiency of primary and secondary treatments in removing contaminants from water. The dismal performance of primary treatment is apparent. Primary processes remove most of the sediment and thus clarify the wastewater to a more attractive condition; but only one third of the organic matter is removed, and virtually none of the nitrogen and phosphorus nutrients.

Secondary treatment deals quite well with organic material. But it, too, fails to stop the steady nutrient flow to our waterways.

Advanced Wastewater Treatment. Additional treatment stages tacked on to secondary treatment are called *tertiary treatment*. When additional steps are integrated into the entire treatment process, the treatment as a whole is termed *advanced wastewater treatment*. Advanced or tertiary treatment can reach high levels of refinement, producing water fit to drink, or it can focus on the removal of certain

Table 8-6. THE AVERAGE PERCENT REMOVAL OF VARIOUS CLASSES OF WATER CONTAMINANTS BY SEWAGE TREATMENT PLANTS[a]

Contaminant	Removal Efficiency, Percent	
	Primary	Primary plus Secondary
Degradable organics (BOD)	35	90
Total organic matter	30	80
Sediment	60	90
Nitrogen	20	50
Phosphorus	10	30

[a]Data from *Cleaning Our Environment: The Chemical Basis for Action* (American Chemical Society, 1969).

troublesome substances not removed in secondary treatment, such as heavy metals or phosphorus. As more impurities are removed, the cost soars. Each new step in waste removal that is applied nationwide costs billion of dollars per year, as we shall now show.

About a hundred billion or 10^{11} gallons of water are used by industries, homes, and farms to carry away wastes every day. (The exact number of gallons depends on how dirty the water must get before treatment is considered desirable.) Primary treatment costs about 5 cents per 1000 gal, which would be $5 million per day for 10^{11} gal of wastewater, or nearly $2 billion per year.[17] Secondary treatment adds another 5 cents for each 1000 gal, costing the nation $2 billion more. It is estimated that the removal of phosphorus to the 90 to 95 percent level would add still another 5 cents and would bring the annual bill to nearly $6 billion. Here is what this money would buy.

Complete nationwide secondary treatment, upgraded to remove 90 percent of all organic matter, would reduce the daily BOD load from sewage in this country from 13.5 million lb to 2.7 million lb per day. This reduction would greatly alleviate the pressure on dissolved oxygen in our waterways—a valuable first step.

The removal of 90 to 95 percent of wastewater phosphorus would slash the phosphorus input to waterways in half—more than half in heavily urbanized regions. This step would be another excellent beginning.

The removal of phosphorus by advanced treatment steps is of great importance, as we have seen. Chemical precipitation is most promising for phosphorus removal. Added aluminum ions (provided by alum, $Al_2(SO_4)_3$) will precipitate phosphate by the reaction

$$Al^{3+} \quad + \quad PO_4^{3-} \quad \rightarrow \quad AlPO_4 \downarrow$$

| aluminum ion | phosphate ion | aluminum phosphate |

where the vertical arrow ($\downarrow$) indicates that an insoluble, collectable precipitate is formed. Soluble phosphate can also be precipitated by calcium ions (Ca^{2+}) and iron ions (Fe^{3+}). A biological technique in which microorganisms alternately absorb and yield excess phosphorus also shows promise.

Other advanced techniques include the adsorption of contaminants by charcoal beds; chemical oxidation of organic matter by powerful ozone (O_3); removal of inorganic ions by reverse osmosis;

---

[17] This figure is calculated by the sequence of conversions

$$\frac{1 \times 10^{11} \text{ gal}}{1 \text{ day}} \times \frac{\$.05}{1000 \text{ gal}} \times \frac{365 \text{ days}}{1 \text{ year}} = \frac{\$1.8 \times 10^9}{1 \text{ year}}$$

The figures above are approximate, intended only to show the basis for multibillion-dollar treatment costs. They include the amortization of capital, but initial capital costs would be much higher.

and many more. The trend is to use chemical and physical techniques, with less reliance on natural biological processes. Advanced treatment is now in a stage of rapid development and testing. The costly process of applying it nationwide has yet to be faced.

THERMAL POLLUTION

Heat added to water changes the characteristics of that water just as surely as do chemical contaminants. Aquatic species alter their behavior from one temperature range to another, just as they respond to fluctuating levels of sediment, minerals, and dissolved oxygen. And extremes of temperature, like extremes of chemical concentrations, will kill organisms outright.

The first threat of waste heat, then, is simply in disturbing the existing web of life and its natural cycles. Even small additions of heat—leading to a slight increase in temperature—may change the timing of a fish hatch, bringing young fish into the world before their natural food arrives. The reproductive cycle of fish—spawning and egg development—is generally hit first as temperature increases, as shown in Figure 8-13. Later, fish themselves can neither grow nor survive.

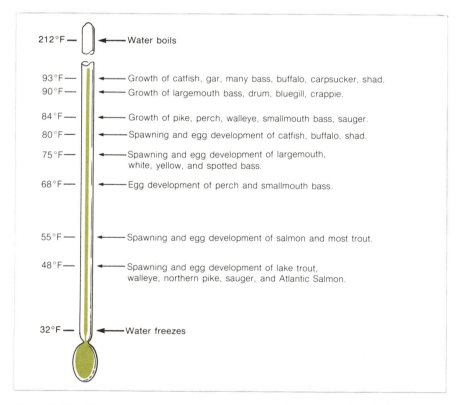

Figure 8-13. Maximum water temperatures suitable for the growth and development of various fish.

Hot Water and Dissolved Oxygen. The most fundamental problem with warm water is chemical—heated water holds less dissolved oxygen (DO) than does cold water. Figure 8-14 shows how the DO level plummets with gains in temperature. Heating drives the O_2 out of water, and by the time the temperature reaches $100°F$, fish begin to suffocate (among their other problems) for lack of essential O_2.

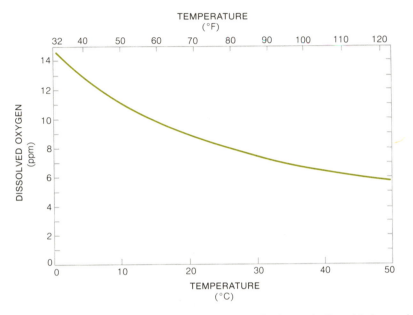

Figure 8-14. The concentration of dissolved O_2 falls dramatically with increasing temperature.

Heat ejects O_2 from water, and to make the situation worse, it sets in motion certain processes that make an increasing demand on water's dwindling supply of O_2. Increased temperature, as we saw in Chapter 5, speeds up chemical reactions, including most of those in living systems. Fish are more active when the water is warm, and they require more DO. Worse, the microorganisms of decay become more active, and they accelerate their use of O_2 to oxidize organic matter. Many a stream is suitably balanced between slow organic decay and O_2 replenishment from air, but when a heated waste stream mixes with its waters, its decreased O_2 supply is rapidly gobbled up by enlivened microorganisms. The result is foul, anaerobic water below the heat source.

And additional O_2 disturbance occurs when heated water is discharged to lakes. The hot water encourages the formation of stratified layers and reduces the periodic mixing that renews oxygen in deep waters.

Proposals are widespread to harness hot water discharges to extend the growing season or otherwise enhance agricultural and aquatic production. Catfish and oyster farming, for instance, can be done with warm water. Indeed, if we are willing to allow broad changes in water ecology (changes in aquatic species), heat will often increase total biological productivity (harvestable algae, for instance, will grow faster in heated water). Those who have an eye solely on productivity give the name *thermal enrichment* to heated discharges. Whether added heat is "pollution" or "enrichment" is sometimes a matter of point of view. But when thermal additions go so far that they make our natural waterways anaerobic, almost everyone agrees that the proper term for added heat is *thermal pollution*.

Sources and Solutions.   Nearly all the waste heat entering our waterways comes from the furnaces of industry. Eighty percent of the water used to quench this heat is employed for a single activity—electric-power generation. As we pointed out in Chapter 1, for every unit of energy that flows along electric-power lines, about two units enter the environment as heat. Water is the most effective absorber of this heat, as we have already explained. But heat absorption causes the temperature to rise and leads to the many side effects we have mentioned.

A single 1000 megawatt nuclear generating plant needs a million gallons of water per minute for cooling—a moderate sized river in its pipes. The emerging water is hotter by 20°F than the intake water.

Electric-power production is doubling every 10 years. This rate means that by present procedures the amount of waste heat entering the environment will also double every decade. The environment cannot long take abuse that grows *exponentially*—doubles in a fixed time—because very soon one has twice as much abuse, then 4 times as much abuse after second doubling period, and then 8, 16, 32, 64, 128, 256 . . . times as much abuse with the passage of each successive doubling time. Such rapid growth may make it necessary to use one sixth of the total fresh-water flow in the United States for cooling as early as 1980. The cooling-water requirements of 1990, 2000, and later hardly need stating.

Various approaches are under study to limit the pouring of heat into our waterways. If the heat is used—diverted to agriculture—it will dissipate finally to the atmosphere. Giant cooling ponds will serve the same function, but they require 4 mi^2 of land area for one large nuclear plant (2 million kilowatts). Evaporation losses are significant in either case.

Heat also can be diverted to the atmosphere through cooling towers. Wet towers dissipate heat by evaporation. They lose a great deal of vaporized water—up to 14,000 gal per min, enough possibly to increase local cloud formation, fogginess, ice formation, and precipitation. Dry towers, by contrast, save water by transferring heat directly to air; but they are unbearably expensive.

The oceans, like the atmosphere, provide a much greater sink for waste heat than do surface waters. Many nuclear plants—which are more heat wasteful than are fossil-fuel plants—are going up along our coastlines. California, by the turn of the century, is expected to have a nuclear plant along each 10 mi of its Pacific coastline so that its waste heat can be discarded to the ocean currents.

Dissipation of heat to the atmosphere and oceans will lessen the impact of heat on our lakes and rivers. But in the end, there is no part of our environment, no matter how large, that can cope with endless exponential growth. Soon man must come to grips with the limits of growth. He must seek a steady-state, no-growth world of limited population and human activity or face the threat of having his fragile life-support system crack up under expanding human pressures.

INORGANIC MINERALS IN WATER

The high polarity of the water molecule allows water to dissolve variable amounts of ionic mineral substances—the bulwark of our earth's crust. We shall use common table salt, sodium chloride (NaCl), as an example to show how this process occurs.

In Chapter 3 we learned that NaCl is a crystal in which sodium ions (Na^+) and chloride ions (Cl^-) are held together by ionic bonds— the electrostatic bonding attraction of positive charge for negative charge. However, since the water molecule has charged ends, it, too, can attract ions. The negative oxygen (O) atom of H_2O attracts Na^+ with a significant force, one that is strong enough to pull Na^+ away from its crystalline Cl^- neighbor. Once pulled loose, the Na^+ ion is surrounded by other H_2O's, each tending to bind the Na^+ through its negative oxygen atom. In the same way, Cl^- is bound by the positive part of several H_2O molecules. The binding of ions by the charged ends of the H_2O molecule is called *hydration* (Figure 8-15),

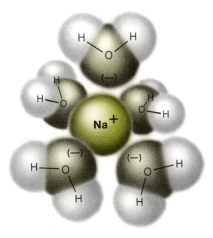

Figure 8-15. The hydration of ions (here the sodium ion, Na^+) brings into play the attraction of opposite charges, helping pull ions into water.

and the cluster of H_2O molecules with an ion in the center is called a *hydrated ion.*[18]

Sodium chloride dissolves, then, because H_2O binds Na^+ and Cl^- tightly enough to pull them out of the NaCl crystal. Any crystalline substance placed in water has a similar tug-of-war going on: H_2O is acting to pull the ions loose and the ionic bonds of the crystal are acting to keep the ions in place. Sometimes one force is predominant and sometimes the other, depending on the exact nature of the crystalline substance. Hence some crystalline minerals are highly soluble, and others are relatively insoluble. In general, the further left we find the positive ion in the periodic table, the greater the solubility of the mineral. In accordance with this rule, compounds of sodium are highly soluble. This fact, combined with the abundance of NaCl in nature, explains why the oceans are salty and why NaCl is a frequent contaminant of inland waters as well.

Dissolved minerals greatly affect the uses of water and the type of aquatic life it supports. Next we shall describe two important categories of water mineralization.

Saline Water.   *Salinity* is a measure of the total of all dissolved ionic substances (salts) in water. Waters are classified according to their content of dissolved ionic solids, as shown in Table 8-7. Natural

Table 8-7.   THE CLASSIFICATION OF WATER ACCORDING TO THE LEVEL OF DISSOLVED IONIC SOLIDS, OR SALTS

Name Given Water	Dissolved Ionic Solids (ppm)
Fresh	0–1000
Brackish	1000–10,000
Salty	10,000–100,000
Brine	Above 100,000

waters occur in all four categories. Table 8-8 (p. 316) shows examples.

Waters of high salinity do not support most forms of life; the natural brine lakes have a very simple ecological web with fewer species than are present in fresh-water lakes. Even brackish waters cannot be used by humans or for agricultural irrigation.

Salinity, a measure of total dissolved solids, does not specify which ions are in the water. It is found that natural waters of both very high and very low salinities are dominated by the sodium cation, Na^+ (the dominant action is Cl^-). Waters of low-intermediate salinity—100 to 500 ppm—are usually dominated by the calcium cation, Ca^{2+}. This cation is the main cause of hardness in water.

Water Hardness.   *Hard* water is water with a high concentration of cations with multiple charge—mostly the cations of calcium

---

[18] The H_2O on some hydrated ions is attached to the ion by coordinate covalent bonds (as well as electrostatic forces), in which unshared electrons on the water's oxygen atom are shared with the outer shell of the metal atom.

Table 8-8.   THE SALINITY OF SOME NATURAL WATERS

Source	Dissolved Solids (ppm)	Water Classification
Rio Negro	10	
Lake Tahoe	70	
Lake Michigan	170	Fresh water
Yukon River	280	
Missouri River	360	
Colorado River	700	
Pecos River	3000	Brackish water
Baltic Sea	7000	
Black Sea	20,000	Salty water
Oceans	35,000	
Dead Sea	250,000	Brine
Great Salt Lake	266,000	

(Ca^{2+}) and magnesium (Mg^{2+}). Hard water is so named because it is *hard* to form suds or lather in such water if you are using soap. Hardness ions such as Ca^{2+} and Mg^{2+} combine chemically with soap to form a scummy precipitate, as we noted before. This action depletes the soap and thus prevents sudsing.

The hardness of water depends on the rock and soil over which the water runs. If an abundance of limestone ($CaCO_3$) is present, enough $CaCO_3$ ordinarily dissolves to make the water hard (dolomite rock supplies both Ca^{2+} and Mg^{2+}). Thus hardness depends on the local geology, and it varies significantly from place to place. Figure 8-16 shows the variations of water hardness in municipal water supplies in the United States and explains the classification system for water hardness. Note that hard water contains only 120 to 250 ppm $CaCO_3$; this figure is low compared to the total salt content of some waters as measured by their salinity—over 10,000 ppm. The hardness ions fail to reach such high concentrations because their compounds are generally less soluble than, for instance, sodium chloride (NaCl). This difference is expected, because these ions appear in a region further to the right (group IIA) in the periodic table, where solubilities are generally reduced.

The hardness ions in water are usually considered undesirable because they form scale in pipes and boilers; they waste soap and form a gummy precipitate with it; and they have therefore encouraged the headlong rush into synthetic detergents, which have contributed to the ecological devastation of many waterways.

Man's Influence on Water's Minerals.   Mineral ions in water are concentrated by evaporation or by leaching from the soil. Thus reservoir and crop evaporation, along with agricultural drainage, contribute to high mineral levels in water. The Colorado River, as it reaches Mexico, is so salty as a result of these multiple abuses

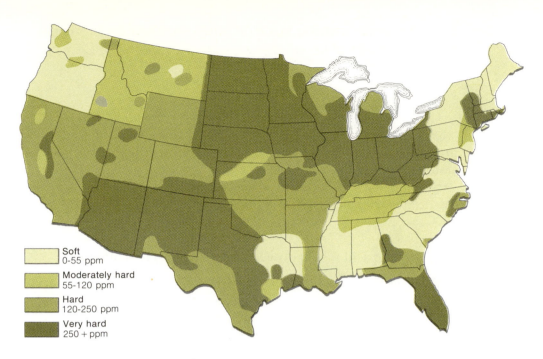

Figure 8-16. The distribution of water hardness in municipal water supplies in the United States.

Soft
0-55 ppm

Moderately hard
55-120 ppm

Hard
120-250 ppm

Very hard
250 + ppm

that it damages Mexican agriculture and strains United States–Mexican relations.

In addition, the diversion of water from some rivers is enough to lower their flow and permit seawater to penetrate upward—often to the intake for municipal water-supply systems. In a similar way, the excessive depletion of fresh groundwater often encourages the migration of salt water to fill the depleted aquifers.

Another disturbance is caused by the 6 million tons of salt used to de-ice this nation's roads every year—up 1800 percent in 30 years. Salty runoff has endangered roadside trees, vegetation, water supplies, and natural waterways. De-icing salt used in Rochester, New York, has increased the salt level of Irondequoit Bay five times; the increased density of the salt water has caused water stratification and has thus retarded the mixing that helps renew dissolved O_2.

ACIDIC AND BASIC WATER: THE pH SCALE

Many substances that occur naturally in water are acidic or basic; they release hydrogen ion (H^+) or hydroxide ions (OH^-). For instance, a number of organic acids are released from soils and decaying vegetation, including formic acid, acetic acid, and the humic acids. These acids release H^+ by *dissociation* (breakup into ions), a reaction represented by

$$\overset{O}{\underset{\|}{R-C}}-OH \rightleftarrows \overset{O}{\underset{\|}{R-C}}-O^- + H^+$$

where, recall, R stands for a general organic group. The double arrow shows that the reaction proceeds both ways and that dissociation is therefore not complete: the acid is a *weak* acid. By contrast, very small amounts of sulfuric acid (H_2SO_4) in natural waterways dissociate completely:

$$H_2SO_4 \quad \rightarrow \quad H^+ \quad + \quad HSO_4^-$$

sulfuric acid   hydrogen ion   hydrogen sulfate ion

Sulfuric acid is termed a *strong* acid because of its complete dissociation. Fortunately, H_2SO_4 from natural sources is usually very dilute, so that it does not appreciably upset the acid balance of waterways.

The most important acid affecting natural waters is carbonic acid, a weak acid produced in moderate abundance by dissolved (CO_2):

$$CO_2 \quad + \quad H_2O \quad \rightarrow \quad H_2CO_3$$

carbon dioxide   water   carbonic acid
dissolved in water

This acid dissociates by the reaction

$$H_2CO_3 \quad \rightleftarrows \quad H^+ \quad + \quad HCO_3^-$$

carbonic acid   hydrogen ion   bicarbonate ion

Occasionally the bicarbonate (or hydrogen carbonate) ion will undergo an additional dissociation:

$$HCO_3^- \quad \rightleftarrows \quad H^+ \quad + \quad CO_3^{2-}$$

bicarbonate ion   hydrogen ion   carbonate ion

However, this dissociation is very slight, because the carbonate ion (CO_3^{2-}) has a voracious appetite for H^+, and it immediately recombines with H^+, as shown by the longer of the two arrows.

Natural waters are occasionally basic, or *alkaline*, because substances that release OH^- overwhelm the acid components. For instance, when slight amounts of limestone—a mineral composed of calcium carbonate ($CaCO_3$)—dissolved in water, carbonate ions (CO_3^{2-}) are freed into the water. Carbonate ions, as we noted, have an appetite for H^+, and they will steal H^+ from water, leaving OH^- behind and making the water alkaline:

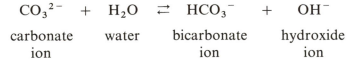

$$CO_3^{2-} \quad + \quad H_2O \quad \rightleftarrows \quad HCO_3^- \quad + \quad OH^-$$

carbonate   water   bicarbonate   hydroxide
ion   ion   ion

This same hydrolysis reaction gives household washing soda (Na_2CO_3) and other carbonates their alkalinity, making them good substitute detergent builders.

Another natural base is ammonia (NH_3), given off by decomposing animal and plant material. Ammonia in water forms ammonium hydroxide (NH_4OH) by the reaction

$$NH_3 \quad + \quad H_2O \quad \rightarrow \quad NH_4OH$$

ammonia   water   ammonium
hydroxide

The NH_4OH partially dissociates to provide OH^- ions, which creates a basic condition in the water:

$$NH_4OH \quad \rightarrow \quad NH_4^+ \quad + \quad OH^-$$

ammonium   ammonium   hydroxide
hydroxide   ion   ion

The pH: A Measure of Acidity and Alkalinity. We said in Chapter 4 that acids and bases cannot exist together because H^+ and OH^- combine almost completely to form water: $H^+ + OH^- \rightleftarrows H_2O$. Only one of the two can be dominant at a time. In acidic water, for instance, where the H^+ concentration is high, the OH^- concentration is extremely small.

In pure water (neither acidic or basic), there are 10^{-7} g of hydrogen ions (H^+) in a liter of water. They are generated by the very slight dissociation of water:

$$H_2O \rightleftarrows \text{---} H^+ + OH^-$$

This dissociation produces H^+ ions and OH^- ions in equal but small numbers.

The concentration level of H^+ and OH^- in any water is most conveniently represented by the pH value. The *pH* is defined as the negative exponent of the hydrogen-ion concentration in water, measured in moles (grams) per liter.[19] Thus

$$\text{concentration } H^+ \text{ (g/}\ell\text{)} = 10^{-pH}$$

The pH of pure water is 7, which reflects the exponent, minus 7, in the 10^{-7} g of H^+ in a liter of pure water. Likewise, an acid with an H^+ concentration of 10^{-2} g/ℓ has a pH of 2.

In alkaline water, OH^- is in such high concentrations that it reacts with nearly every H^+ ion available. Thus the H^+ concentration plummets below its value in pure water, already a low 10^{-7} g/ℓ. In

---

[19] Recall (Chapter 5) that 1 mole of a substance has in it exactly as many grams as the atomic weight of that substance. The H^+ ion has an atomic weight of 1.01, which we can approximate as 1. Thus 1 mole of H^+ contains about 1 g: that is, for H^+, moles and grams are roughly equal and can be interchanged. While pH values are defined in terms of moles, we use here the more familiar notation of grams.

strong bases, the concentration of H^+ may fall as low as 10^{-12} g/ℓ to 10^{-14} g/ℓ, corresponding to pH's from 12 to 14.[20]

The pH scale, then, is one for which a value of 7 represents neutral water—neither acidic nor basic—while scale values below 7 represent acidic water and values above 7 represent alkaline waters. Bear in mind that the *pH is an exponent*, and therefore *each change of one unit in pH represents a drastic tenfold change in acidity or alkalinity*.

Figure 8-17 shows the pH range of most natural waters, and for comparison, the pH of some familiar solutions.

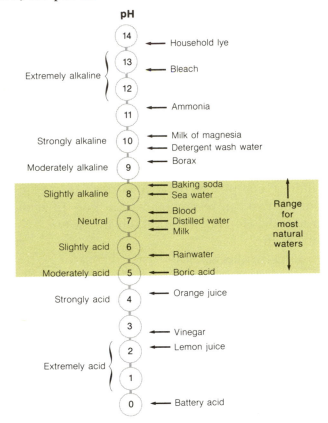

Figure 8-17. The pH scale and pH values for natural waters and various common solutions.

The internal pH of most organisms is a critical, finely tuned parameter. Substantial variations will lead to death. These organisms cannot stand great variations in the water they drink or live in. For example, a pH of 4 will destroy most aquatic life. Acidic water also causes considerable corrosion of stone, metals, and so on.

pH Pollution. The acid-base balance of liquids not only is critical to life, but is a prime factor in fixing chemical properties and

---

[20]The concentrations of H^+ and OH^- are fixed by the equilibrium point of the water dissociation, $H_2O \rightarrow H^+ + OH^-$. The equilibrium relationship takes the simple form (concentration of H^+) × (concentration of OH^-) = 10^{-14}, where concentrations are in moles/ℓ. The concentration of H^+ is 10^{-12} moles/ℓ at pH = 12—by the definition of pH—and the OH^- concentration is therefore 10^{-2} moles/ℓ. This relation shows how OH^- concentrations can be deduced from pH.

the rates of chemical change. Many industries use concentrated acids and bases because of their strong chemical activity. The wastes from such industries are frequently acidic or basic, enough to cause fish kills and other severe aquatic disturbances.

In recent decades, acid rain has exerted a telling influence on water pH throughout wide regions, threatening waterway ecology. The source is air pollution—oxides of sulfur and nitrogen that turn to acid upon contact with water. This is but one dramatic example of the closely knit relationship of air, water, and living organisms. (Acid rain is further discussed in Chapter 7.)

A major source of acid pollution is acid mine drainage. Mining exposes long-covered sulfide compounds to air. The sulfide mineral iron pyrite (FeS_2) is common in both coal deposits and underground metal ores. Such sulfides are oxidized to sulfuric acid (H_2SO_4) by the air's O_2 when moisture (H_2O) is present:

$$2\,FeS_2 + 7\,O_2 + 4\,H_2O \rightarrow 4\,H_2SO_4 + 2\,FeO$$

Drainage from the mine then carries the powerful acid H_2SO_4 to the nearest waterway.

An estimated 8 million tons of sulfuric acid are discharged yearly by coal mines alone. The worst offenders are abandoned coal mines, in which acid formation can continue for decades. Some 10,000 miles of streams in Appalachia have been degraded by this source. Most plans to stem acid mine drainage call for flooding or sealing abandoned mines to prevent O_2 circulation. Once the acid is formed, the only recourse is to neutralize it with artificial barriers of alkaline materials, such as limestone ($CaCO_3$) or lime (CaO), spanning drainages and streams. Thousands of miles of scattered acid-producing tunnels are a formidable challenge to such abatement plans. It is estimated by federal authorities that the cleanup of acid mine drainage would cost between $1.7 billion and $6.6 billion.

WATER TO DRINK

We turn now to the small fraction—about 1 part in 4000—of man's water that is used for drinking. Needless to say, man is fussier about the water he drinks than about water he uses for any other purpose. However, the water that sprinkles his lawn and washes his clothes comes in through the same pipe as the water he drinks, so drinking-water standards actually are applied to a substantial quantity of water.

Both biological and chemical hazards can exist in drinking water. Many great epidemics of the past were caused by disease-contaminated drinking water. All too commonly in the past, there was no clear distinction between sewer drainage and drinking supplies. Deadly microorganisms were passed from person to person and city to city with ease. Even today, many rivers receive raw or poorly treated sewage at one point and provide drinking water at a point below. The main difference from the past is that water is now widely treated before use to kill microorganisms.

The four steps commonly employed for municipal water treatment are disinfection (chlorination), sand filtration, sedimentation, and chemical coagulation. The water emerges clear, and it is free of most biological contamination. However, the degree to which certain viruses, including the infectious hepatitis virus, survive conventional water treatment is not yet certain.

Special water-treatment steps are often added to the basic water-treatment program for esthetic and practical reasons. Odor and taste control, water softening, corrosion control through pH adjustments, and water fluoridation are widely employed.

As the biological hazards of the past subside, the chemical hazards of drinking water emerge to prominence. Chemical pollution is a threat that is in some ways more insidious than biological pollution. There is no single technique that will remove all or even most of the toxic chemicals in water (as chlorination or sterilization will destroy nearly all microbes). Each chemical pollutant needs separate consideration. Treatment is often difficult, and in practice, chemical removal is very limited. Meanwhile, the ongoing wheels of progress provide our waters with an increasing variety of chemical contaminants.

At present the most serious chemical pollutants in drinking water are herbicides and insecticides (of which there are hundreds), nitrates, fluorides, selenium, arsenic, mercury, and lead. Nitrates are perhaps most serious with a potential for causing methemoglobinemia and death in infants.

Chemical standards have been developed by the United States Public Health Service (USPHS) to make drinking water safe. Two levels have been designated: a recommended limit which should not be exceeded if any suitable alternative can be found, and a mandatory limit that should not be exceeded under any circumstances. These limits are shown in Table 8-9 for several toxic contaminants. In addition, standards exist for radioactivity, fluoride additions, and other materials in water. These federal standards have been adopted by most states. Nonetheless it is estimated that one half of this

Table 8-9. **DRINKING-WATER QUALITY STANDARD OF THE U.S. PUBLIC HEALTH SERVICE, 1962**

Contaminant	Recommended Limit (ppm)	Mandatory Limit (ppm)
Arsenic (As)	0.01	0.05
Cadmium (Cd)	–	0.01
Chromium (Cr^{6+})	–	0.05
Cyanide (CN^-)	0.01	0.2
Lead (Pb)	–	0.05
Nitrate (NO_3^-)	45	–

nation's people drink water that does not meet federal standards. We still have some distance to go in improving water treatment, seeking new water sources, and, in general, getting the filth out of the nation's waterways.

THE OCEANS

We end this chapter, most appropriately, with the enigmatic oceans. These vast reservoirs contain 97 percent of the world's water, but man cannot drink it. Distributed evenly, the oceans would submerge the earth in 9000 ft of water, yet our planet is spotted with deserts; and while man scratches the continents for scarce water, far away ocean currents regulate his success by influencing world climate and precipitation.

Man's domain of land is one third of the earth's surface. His endless hunger for expansion and conquest has drawn his attention to the oceans, beneath which two thirds of the earth lies. He would like to subdue and farm the oceans, as he did the land 10,000 years before. But before he harvests great riches of food, he must overcome a desert—a special chemical desert—that is as vast as thirty Saharas.

It is ironic that the surface waters of earth are threatened with ruin by excess nutrients—nutrients that promote choking tangles of plants—because the oceans are literally starved for nutrients. Nutrient deserts are plant deserts, and plant deserts are fish deserts. The open oceans—90 percent of the total ocean surface—produce less than one tenth of 1 percent of man's seafood. The coastal zone—10 percent of the area—provides the remainder. Small coastal regions in the coastal zone, comprising a mere 0.1 percent of the ocean surface, produce 99 percent of the seafood. These coastal areas contain the world's *estuaries*—regions where rivers pour into and mix waters with the sea—and coastal upwellings. Their productivity stems from their nutrient-rich waters: estuary nutrients are brought in by the rivers, and coastal upwellings bring nutrients to the surface from the depths of the oceans.

Estuaries are among the richest biological regions on earth.[21] This wealth is shown in Figure 8-18, which compares the rates of production of green plants over different regions of land and water. The estuaries support abundant fish and mollusk populations in their rich waters, and they are fertile breeding grounds for many species.

The estuaries and coastal regions—the breadbaskets of the ocean —receive the brunt of man's abuse. First of all, they are crisscrossed with shipping lanes, where collisions, groundings, and intentional

---

[21] Excess richness (nutrient content) is biologically devastating, as we have learned, but a vast majority of the earth's land and water is so nutrient poor that one need not be concerned that they will suffer from nutrient overbundance. In such places, nutrients are the key to supporting a substantial biological community.

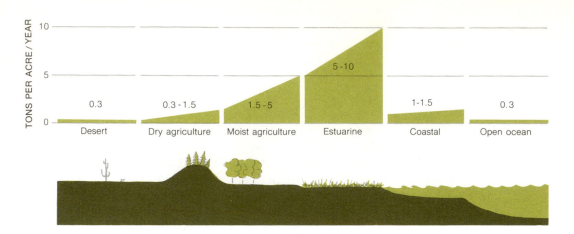

Figure 8-18. A comparison of the rate of production of living material in various land and water environments. The estuaries are most productive of all.

discharges release noxious cargoes, including oil. Increasing sea traffic does not bode well for the future. Soon, million-ton oil tankers will be navigating coastal waters; their inevitable breakup will cause oil spills dwarfing any of the past.

Second, coastal areas are preferred dumping grounds for any waste too abundant or too toxic to dump anywhere else. In 1968, 48 million tons of wastes were dumped at sea. These wastes ranged from dredging spoils (38 million tons) and sewage sludge (4.5 million tons) to outdated explosives (15,000 tons).

Third, nearly all wastes, no matter where they are dumped, are finally washed to sea by the leaching action of rain and the steady flow of rivers. All of man's worst debris is channeled down the world's rivers—right into the productive estuaries.

These many sources of chemical pollution are aided and abetted by excessive physical damage to coastal areas. Overfishing, dredging, and the reclaiming (destruction) of coastal swamps and marshes by landfill are examples of man's shortsighted physical abuse. With so much chemical and physical destruction focused on the narrow life-producing belts of the sea, it is easy to understand Jacques Cousteau's observation that 30 to 50 percent of life in the ocean has been damaged in the last 20 years.

In his nobler moments, man may aspire to manage the enigmatic oceans and feed the world's hungry people, but by his everyday activities, he is making the oceans less productive, less pleasant, and less beautiful. Man, it seems, is the ultimate enigma.

Exercises

1. How many tons of steel can be made with the aid of 1 million gal of water, without recycling? With total recycling?

2. The manufacture of 1 gal of gasoline uses about 25 gal of water. How much water is used to produce gasoline for a 3000 mi vacation drive to Yellowstone, at 15 mi per gal of gasoline? Are you more concerned about the water use or the fuel use? Why?

3. Of the 4.5×10^9 acre-ft of groundwater estimated to lie beneath Arizona, 7×10^8 acre-ft are potentially useful to man (J. W. Harshburger et al., Geological Survey Paper No. 1648). This water is being withdrawn at a net rate of 4×10^6 acre-ft/year (5×10^6 acre-ft gross withdrawal and 1×10^6 acre-ft natural replenishment). At present net withdrawal rates, how long will Arizona's water last? What then?

4. The Mississippi River discharges about 2 billion tons of water a day into the Gulf of Mexico. The water has an average silt content of 1000 ppm. How many tons of silt are carried to the Gulf each day?

5. Which compound of the following pairs would you expect to dissolve most abundantly in water?

   a. methyl amine (CH_3-NH_2) or propyl amine ($CH_3-CH_2-CH_2-NH_2$)
   b. nitrogen (N_2) or hydrogen fluoride (HF)
   c. hexane (C_6H_{14}) or decane ($C_{10}H_{22}$)

6. A 1 ℓ water sample (weighing 1 kg) is found to contain 0.006 g of O_2. How many milligrams of O_2 does it contain? What is the O_2 concentration expressed in parts per million?

7. The small lake referred to in footnote 9 has 16 tons of O_2. If O_2 is present in the lake water at 10 ppm, how many tons of water does the lake contain?

8. One gal of water (8.3 lb or nearly 3.8 kg) at a normal 10 ppm O_2 contains only 38 milligrams of oxygen. About how much organic matter is needed to deplete this oxygen? How does this amount compare to the mass of organic material in a drop of oil, about 30 mg?

9. The average daily weight of oxygen taken up by the sewage discharged per person in the United States is 0.13 lb. How many tons of water, at 10 ppm O_2, are depleted of oxygen by the discharge of raw sewage from a community of 5000 population?

10. A lumber company in Alaska was cited by the Environmental Protection Agency (EPA) in 1971 for discharging 240,000 lb of BOD into Silver Bay, causing two fish kills. How many tons of water containing 10 ppm O_2 would be made anaerobic by such discharges?

11. Using the modern kraft pulping method, paper mills produce 80 lb of BOD for each ton of paper (older plants create more BOD). This BOD results from the rejection of some sugarlike organic components unsuitable for paper. At 10 ppm O_2, how many tons of water would be needed to supply the BOD produced with a ton of paper?

12. Serious proposals have been made to eliminate sulfur dioxide (SO_2), a serious air pollutant, by diverting stack gases into the sea. The O_2 dissolved in seawater would oxidize SO_2 to unobtrusive sulfate ($SO_4{}^{2-}$) by means of the reaction

$$2SO_2 + O_2 + 4OH^- \rightarrow 2SO_4{}^{2-} + 2H_2O$$

How many tons of dissolved oxygen would be depleted in oxidizing 480 tons of SO_2, a typical pollution load produced by a large power plant in just one day? How many tons of seawater containing 8 ppm

O_2 could be made anaerobic in one year by the effluent from this single power plant?

13. In 1963, wastes from all manufacturers in the U.S. produced 22 billion lb of BOD (much of it was treated before release). How many pounds of water at 10 ppm O_2 would be needed to consume this much untreated waste? How many tons?

14. (Refer to Problem 13.) Total stream flow in the U.S. is about 1250 billion gal/day, equivalent to 450 trillion gal or 3800 trillion lb in a year. What is the oxygen content of the water flowing yearly, assuming 10 ppm O_2? How does this figure compare (larger or smaller) with the total 1963 BOD reported in Problem 13?

15. (This question is based on figures given in Problems 13 and 14.) If all 1963 industrial wastes, untreated, had been dumped evenly into U.S. streams, to what would the ppm oxygen level of the stream water be reduced, assuming no O_2 replenishment from the air? (Actually there is considerable O_2 renewal in rivers because of their churning, but offsetting this effect is the fact that waste dumping is localized in character, putting a bigger O_2 demand than that calculated on some rivers.)

16. The city of Chicago has banned the sale of laundry detergents containing more than 8.7 percent of the element phosphorus since February 1971. What maximum percentage of STPP are Chicago detergents allowed to contain?

17. Trisodium phosphate (Na_3PO_4) is occasionally used as the phosphate builder in detergents. What weight percent of the element phosphorus is in a detergent containing 53 percent Na_3PO_4?

18. In highly alkaline water, sodium tripolyphosphate (STPP) hydrolyzes to the phosphate ion (PO_4^{3-})

$$Na_5^+ (P_3O_{10})^{5-} + 4OH^- \rightarrow 3PO_4^{3-} + 2H_2O + 5Na^+$$

Sometimes detergent phosphorus is reported as phosphate rather than as STPP. Use this chemical equation to calculate the hypothetical phosphate (PO_4^{3-}) content of a detergent that contains 40 percent STPP.

19. A small lake (about 1 mi across, 12 ft average depth) contains 10 million tons of water. The water contains the "critical" level of 0.01 ppm phosphorus (1 part of phosphorus by weight per 100 million parts water). What weight of phosphorus does the lake contain?

20. Assume that dry algal material can be represented by the chemical formula $C_{106}H_{181}O_{45}N_{16}P$. Would nitrogen (N) or phosphorus (P) be more likely to limit growth in a pond containing 5×10^{27} P atoms and 1×10^{29} N atoms? Which would control growth if there were 1×10^{28} P atoms and 1×10^{29} N atoms?

21. Each year in the United States, 2.5 billion lb of phosphorus (P) goes into detergents. How many tons, dry weight, of algal growth would this phosphorus stimulate if all of it were to become incorporated in algae? (Use $C_{106}H_{181}O_{45}N_{16}P$ for the approximate composition of dry algae.)

22. Phosphorus in treated sewage wastewater amounts to about 3 lb per person per year. If the wastewater enters a lake and the phosphorus

in it enters algae with other elements in the ratio $C_{106}H_{181}O_{45}N_{16}P$ (exclusive of water), what dry weight of algae will be produced from the per capita 3 lb of phosphorus? What wet weight, assuming the living algae to be 80 percent water?

23. (Refer to Problem 22.) If a city of 200,000, by eliminating phosphate detergents, is able to reduce phosphorus to 1 lb per person per year in waste water, what reduction could be hoped for in the total wet mass of lake algae? (The calculated value will be a maximum, because the mass depends on the limiting role of phosphorus to algal growth, complete phosphorus uptake, and so on.)

24. If power-plant cooling requires one sixth of the fresh-water flow of the United States in 1980, and if cooling needs continue to double every 10 years, in which decade (1980's, 1990's, 2000's, 2010's . . .) would we suddenly find that our entire fresh-water flow was needed for cooling?

25. How many tons of iron pyrites (FeS_2) are oxidized each year to produce the 8 million tons of coal-mine acid (H_2SO_4) that enters U.S. streams?

26. Lime (CaO) neutralizes sulfuric acid (H_2SO_4) by the reaction

$$H_2SO_4 + CaO \rightarrow H_2O + CaSO_4$$

How many tons of lime would be needed to exactly neutralize the yearly flow of H_2SO_4 from coal mines? (In practice, a considerable excess of lime is needed to allow for waste and ensure total neutralization, so the figure calculated will underestimate actual needs.)

27. Calculate the pH of 30 million ℓ of stream water (the daily flow of a small stream of 10 ft^3/sec) made acidic by 30 metric tons of sulfuric acid from coal-mine drainage. Assume that all H^+ is produced by the single dissociation

$$H_2SO_4 \rightarrow H^+ + HSO_4^-$$

For simplicity, use 100 for the molecular weight of H_2SO_4 and 1 for H^+.

Glossary

Activated sludge tank   tank in which wastewater is aerated.
Advanced wastewater treatment   treatment process into which refined techniques have been integrated to remove certain substances not removed by secondary treatment.
Alkaline   basic; containing substances that release OH^-.
Alkyl benzene sulfonate (ABS)   nondegradable surfactant used in synthetic detergents before 1965.
Anaerobic degestors   large tanks sealed off from air, in which anaerobic microorganisms break down organic molecules without oxygen.

Biochemical oxygen demand (BOD)   the amount of O_2 required to degrade wastes.

Biodegradable   capable of being broken down or oxidized by biological organisms; degradable.

Degradable   capable of being broken down.

Detergent   a cleanser whose action depends on surfactant molecules.

Dissociation   breakup of an acid or base into ions.

Dissolved oxygen (DO) O_2 present in water; necessary for breakdown of wastes.

Estuary   regions where rivers pour into and mix with water from the sea.

Eutrophic   rich in nutrients, leading to excess algal growth and lake death.

Exponential growth   growth in which doubling occurs in successive time periods.

Five day test   standard test for BOD, in which loss of O_2 during five days by a contaminated water sample in a closed bottle is measured.

Hard water   water containing a high concentration of cations of multiple charge, mainly Ca^{2+} and Mg^{2+}.

Hydration   the binding of ions by the charged ends of water molecules.

Hydrological cycle   the cycle, powered by the sun's energy, in which water is moved from air to land, to the oceans, and back to air.

Hydrophilic   attracted to water: soluble in water.

Hydrophobic   repelled by water: water insoluble.

Linear alkyl benzene sulfonate (LAS)   degradable surfactant molecule used in present-day synthetic detergents.

Mesotrophic   intermediate stage between oligotrophic (nutrient poor) and eutrophic (nutrient rich).

Nitrilotriacetic acid (NTA)   a compound that may be an effective substitute for phosphorous-containing detergent builders.

Nondegradable   not capable of being oxidized or degraded.

Nutrients   elements essential for growth of plants.

Oligotrophic   describes a body of water that is nutrient poor.

Oxygen sag curve   a graph that shows how O_2 is depleted and then replenished downstream of an outfall.

pH   a measure of acidity in water, defined as the negative exponent of the hydrogen-ion concentration.

Primary treatment   wastewater treatment involving screening, settling, and skimming.

Salinity   a measure of the total of all dissolved ionic substances in water.

Saponification   breakdown of fatty molecules by strong base to produce soap.

Secondary treatment   wastewater treatment stage added to primary treatment, including trickling filters or activated sludge tanks and sedimentation.

Sediment (silt)   small, undissolved particles in water.

Sedimentation tank   giant tank in which solid matter is allowed to settle out of wastewater.

Sodium tripolyphosphate (STPP)   the most commonly used builder in synthetic detergents.

Strong acid   an acid that is highly dissociated in water.

Surfactant molecule   a molecule that has one polar end (water soluble) and one nonpolar end (grease soluble); the active ingredient of detergents.

Teratogenic   causing birth defects.

Tertiary treatment   additional treatment stages following secondary treatment of wastewater; may remove any of several substances.

Thermal enrichment   heat discharged into water, from the point of view of increased biological productivity.

Thermal pollution discharge of excess heat into waterways, making them anaerobic and causing other undesirable changes.

Trickling filter an aeration device in which microorganisms break down organic matter in wastewater as it trickles over beds of rocks.

Weak acid an acid whose dissociation in water is incomplete.

Weather modification attempts by man to control rainfall for his own benefit.

Additional Reading

Starred selections are most suitable for the nontechnical reader.

*Bardach, John, *Harvest of the Sea.* New York: Harper & Row, 1968.

*Carr, Donald E., *Death of the Sweet Waters.* New York: Norton, 1966.

Ciaccio, Leonard L., ed., *Water and Water Pollution Handbook.* New York: Dekker, 1971.

Cleaning Our Environment: The Chemical Basis for Action. Report by the Subcommittee on Environmental Improvement, Committee on Chemistry and Public Affairs. Washington, D.C.: American Chemical Society, 1969.

*Giddings, J. Calvin, and Manus B. Monroe, eds., *Our Chemical Environment.* San Francisco: Canfield, 1972.

Hynes, H. B. N., *The Ecology of Running Waters.* Toronto: University of Toronto Press, 1970.

*Leopold, Luna B., and Walter B. Langbein, *A Primer on Water.* Washington, D.C.: U.S. Department of the Interior, Geological Survey, 1960.

**Phosphates in Detergents and the Eutrophication of America's Waters.* 23rd Report by the Committee on Government Operations. Washington, D.C.: 1970.

*Renn, Charles E., *Our Environment Battles Water Pollution.* Chestertown, Md.: LaMotte Chemical Products Company, 1969.

*U.S., Department of the Interior, *Clean Water for the 1970's.* June, 1970.

Velz, Clarence J., *Applied Stream Sanitation.* New York: Wiley-Interscience, 1970.

Warren, Charles E., *Biology and Water Pollution Control.* Philadelphia: Saunders, 1971.

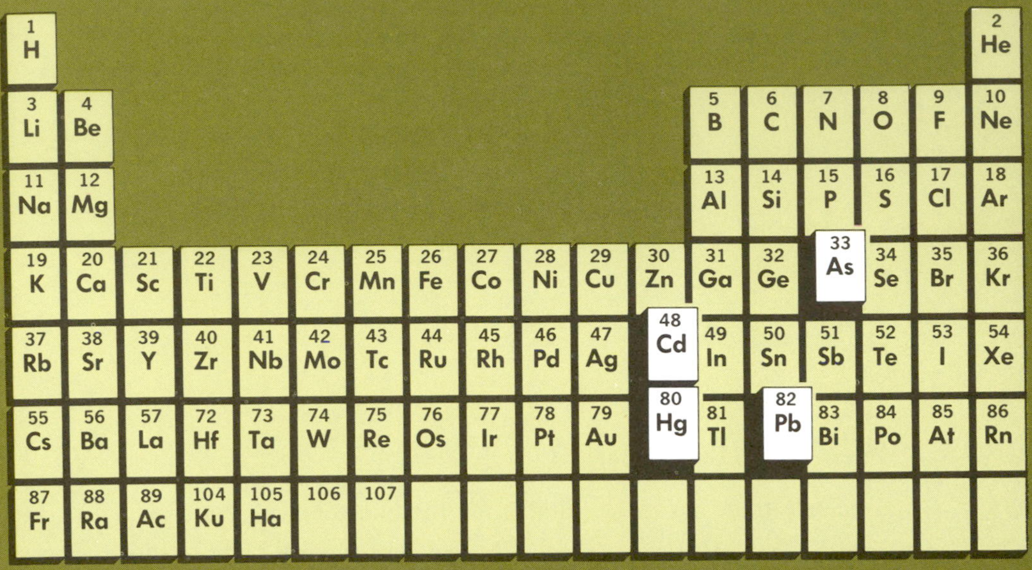

ARSENIC, CADMIUM, MERCURY, LEAD The four toxic elements displayed above have had unique importance in the history of heavy-metal pollution.

9 ENVIRONMENTAL CONTAMINATION BY HEAVY METALS

OUR METAL-RICH ENVIRONMENT

Our earth is a chemical oasis in space. We have an unusual abundance of water (H_2O) and diatomic oxygen (O_2), two compounds of overwhelming importance to life. We also have vast reserves of metals, a rare thing in a universe so dominated by the nonmetals hydrogen (H) and helium (He). The core of earth, for example, is metallic—mostly iron (Fe) and nickel (Ni). The crust of earth is nearly one quarter metal. The abundance and variety of metals on earth make our environment chemically rich because metals dominate the periodic table. A total of 83 elements are metals: thus 79 percent of the atomic building blocks of matter are metallic.

Metals in the crust of the earth—and therefore in the living environment—vary greatly in abundance. A few metals, such as aluminum (Al) and iron (Fe), are found at abundance levels of about 5 percent. They can be extracted in vast quantities from the earth and used for the structural framework of civilization. We shall never run out of these metals.[1] Altogether there are seven metals among

---

[1] An environmental price must be paid as we use the abundant metals, even though the resource is not threatened with depletion. The mining, extraction, and milling of metal ores leaves the land defaced, the air polluted, and the water murky. Enormous energy is consumed, particularly in the preparation of aluminum metal. Scrap-metal wastes adorn the landscape. As more metal is used, we must seek lower-grade ores: to do so, we disturb more of the earth's surface and use more energy for each ton of metal produced. Recycling would help, although energy dissipation and pollution will eventually limit even the amount of recycled metal in circulation.

the ten most abundant elements, as shown in Table 9-1. These seven include sodium and calcium, both important in establishing water quality, as we saw in the last chapter.

Table 9-1.   SEVEN METALS ARE FOUND AMONG THE TEN MOST ABUNDANT ELEMENTS IN THE CRUST OF THE EARTH

Rank	Element	Weight Percent	Rank	Element	Weight Percent
1	oxygen (O)	49.5	6	sodium (Na)	2.6
2	silicon (Si)	25.7	7	potassium (K)	2.4
3	aluminum (Al)	7.5	8	magnesium (Mg)	1.9
4	iron (Fe)	4.7	9	hydrogen (H)	0.9
5	calcium (Ca)	3.4	10	titanium (Ti)	0.6

At the other end of the spectrum are the *trace metals*, which are present at only a few parts per million or less in the earth's crust. Though lacking abundance, these metals do not lack significance. Uranium (U) and thorium (Th) may someday quench the energy thirst of man through their energy-rich atomic nuclei (Chapter 11). Mercury (Hg) and lead (Pb) are widely used servants of technological man, but they are so toxic that minute quantities destroy life. Soon we shall study the important environmental role of some of the trace metals.

CHEMICAL BONDING OF METAL ATOMS

The significance of metals in the environment hinges invariably on the way they chemically bond. Metals are versatile bond formers—ionic bonds, metallic bonds, covalent bonds, and coordinate covalent bonds all play a prominent role in the environmental chemistry of metals.

Metals Held by Ionic Bonds.   Most metal atoms in the crust of the earth are combined with nonmetals as ionically bonded compounds. Water is able to dissolve some of them, breaking them into their constituent ions, as we saw in the last chapter. Water especially dissolves sodium chloride, or "table salt" ($NaCl$). The $NaCl$ level in natural water influences the kinds of life that water will support. Salt-water species and fresh-water species are quite distinct from one another, mostly because of their specific adaptations to the salt content of the water they live in.

Other metal ions, notably those of calcium (Ca^{2+}) and magnesium (Mg^{2+}), cause the "hardness" of water. In this way they establish a demand for hardness-combating detergent phosphates, and by this chain of circumstances they are involved in the deterioration of lakes.

Metallic bonds. Metals rarely exist uncombined in nature. They are usually found in chemical combination with nonmetals such as oxygen and sulfur, although the occasional discovery of gold (Au) nuggets and uncombined copper (Cu) metal shows that exceptions exist. After they are extracted from their ores, metals in pure and mixed forms are used abundantly by man. The metal atoms are held together by the so called *metallic bond.* Here the outer electrons are stripped from metal atoms and added to a common sea of electrons, in which the cores of the metal atoms are imbedded. The sea of electrons forms the metallic bonds which hold these atoms together. The mobile electron sea is also responsible for the characteristic heat conductance, electrical conductance, and luster of metals— properties we discussed in Chapter 2. These features are unique to metals because metals alone lose electrons easily enough to form common pools of electrons.

Covalent Bonding of Metals. Covalent bonds are also important for some metals, particularly those near the center of the periodic table. Compared with the electropositive elements to the far left of the periodic table, the central metals are more electronegative: they lose electrons with greater reluctance. They are therefore some- what less inclined to form ions and ionic bonds than are the electro- positive metals. The chemical bonding of these metals with non- metals is often covalent or a resonance mixture of covalent and ionic bonding. Many toxic metals from this central region bond covalently. For instance, the prominent mercury compound dimethyl mercury has the following covalent bonding structure.

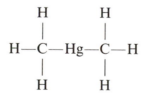

dimethyl mercury

One of the most toxic of all chemical substances is methyl mercury, existing all too commonly in water as the polyatomic methyl mercury ion, bonded covalently as follows:

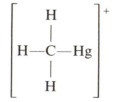

methyl mercury ion

Most mercury in natural water ends up in such a highly lethal form.

The gasoline additive responsible for most lead (Pb) pollution, tetraethyl lead, is also bonded covalently:

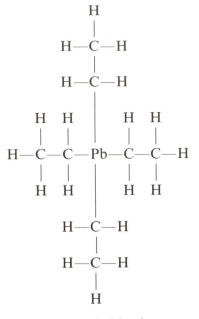

tetraethyl lead

Coordinate Covalence. Coordinate covalent bonds between metals and nonmetals also play an important environmental role. The electronegative metals at the center of the periodic table undergo coordinate covalent bonding readily. Ions of these metals attract electrons strongly and thus form bonds with the unshared electrons that are on nitrogen and oxygen atoms in many compounds. Some molecules can literally wrap around a metal ion, forming coordinate covalent bonds with the metal from several directions. These multiple-bond formers are called *chelating agents* (pronounced key'-lating), after the Greek word *chele*, meaning "claw." Chelating agents can hold metals in a powerful chemical vise. Many enzymes hold crucial metal ions with these bonds. The detergent builders sodium tripolyphosphate (STPP or $Na_5P_3O_{10}$) and nitrilotriacetic acid [NTA or $N(COOH)_3$] hold the hardness ions Ca^{2+} and Mg^{2+} with coordinate covalent bonds.

Hemoglobin, the O_2-carrying protein of the red cell, contains four iron (Fe) atoms. Each Fe atom is held by coordinate covalent bonds, and Fe in turn holds O_2 by a reversible coordinate covalent bond. The chelating agent in hemoglobin is known as heme, shown in Figure 9-1. Two coordinate covalent bonds with Fe are shown by arrows in this figure, and two more are formed in the complete protein, one of them with molecular oxygen, O_2.

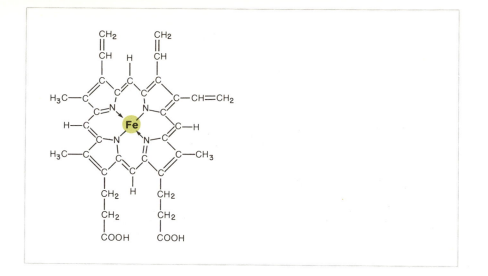

Figure 9-1. Heme, the iron-containing group of hemoglobin, responsible for O_2 transport. Coordinate covalent bonds are shown as arrows.

HEAVY METALS AND SULFUR

In Chapter 2 we explained that atoms are smaller as we move to the right across rows of the periodic table. With each step to the right, a new proton is added to the nucleus. This proton tugs on the electrons, drawing them closer and holding them tighter. Elements thus become more electronegative with each step right. Also the atoms become smaller with their increasingly compressed electron shells. The smaller size of the atoms, combined with their increasingly heavier nuclei, makes the successive elements denser. Therefore metals that have a high density also have a high affinity for electrons: they are the most electronegative of all the metals. (They are not, of course, as electronegative as the nonmetals, located further to the right in the periodic table.) The electronegative metals with a density greater than 5 g/cm³ are termed the *heavy metals*. Figure 9-2 (p. 336) shows the many metals in this category. These metals are generally less abundant than are the remaining metals, the *light metals*: of the seven most abundant metals in Table 9-1, six are light metals and iron (Fe) alone is a heavy metal.

Arsenic (As), element 33, occupies a unique position because it is not metallic, yet it is next to the heavy metals in the periodic table and its chemical-ecological effects resemble those of the heavy metals. In environmental discussions, therefore, As is usually classified as a heavy metal.

The heavy metals are of great importance despite their lack of abundance. One of their chemical characteristics is their relative inertness. Because they hold electrons rather tightly, these metals

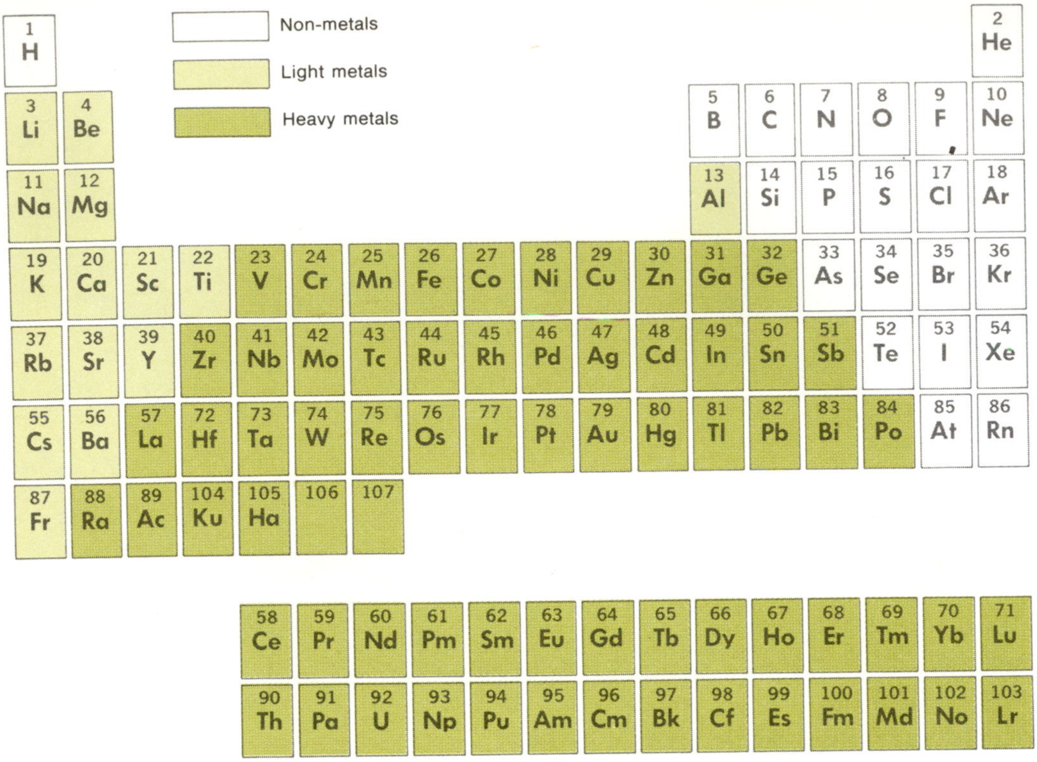

Non-metals

Light metals

Heavy metals

Figure 9-2. The heavy metals, defined as those having a density greater than 5 g/cm^3. The heavy metals are the most electronegative metals of the periodic table.

do not form ions with ease, and they do not readily yield their electrons to electron-hungry elements like oxygen. Therefore heavy metals tend to resist oxidation. The *noble* or *coinage metals*—gold (Au), platinum (Pt), and silver (Ag)—are heavy metals that keep their metallic luster for ages because oxidation occurs only slowly or not at all.

Some heavy metals—most notably copper (Cu), zinc (Zn), cobalt (Co), and chromium (Cr)—are essential to life in trace (small) quantities. They are called the *essential trace metals*. They are vital in the molecular architecture of various proteins, enzymes, and vitamins.

Affinity of Heavy Metals for Sulfur. Perhaps the most significant chemical activity shared by most heavy metals is their strong bonding to sulfur (S). The origin of this tenacious sulfur bonding is best seen if we look first at the way light metals bond with S. Electropositive light metals, such as magnesium (Mg), lose their electrons to sulfur to form ions and ionic bonds

$$Mg \colon\ +\ \colon\!S\colon\ \to\ Mg^{2+}\ +\ \colon\!\ddot{S}\colon^{2-}\ \to\ \underbrace{Mg^{2+}S^{2-}}$$

ionic bond

Compare this action of Mg^{2+} with that of the typical heavy metal ion mercury(II), Hg^{2+}. The Hg^{2+} ion is relatively small because of its position to the right of most metals in the periodic table. The sulfide ion (S^{2-}) is relatively large, so much so that the outer-shell electrons are not held tightly. The small, intensely charged Hg^{2+} ion is therefore able to pull electrons back to its own outer shell occasionally, as it would in covalent bonding. The bonding structure of the HgS crystal may therefore be represented as a resonance hybrid of ionic and covalent bonding:

$$Hg—S \longleftrightarrow Hg^{2+} S^{2-}$$

The resonance gives the bond unusual strength and stability. This fact has considerable environmental significance, as we shall now see.

Heavy Metals and "Living" Sulfur. The extraordinary chemical affinity of most heavy metals for sulfur (S) encourages them to combine with S at every opportunity. Our foremost concern with this sulfur affinity is the growing abundance of heavy-metal pollution and its threat to the sulfur atoms that are part of living systems. We found in Chapter 4 that some proteins carry one or more molecular units of the sulfur-containing amino acid cysteine. Enzymes are particularly rich in cysteine and thus rich in sulfur. The sulfur atoms of these sulfur-rich proteins are bonded as follows:

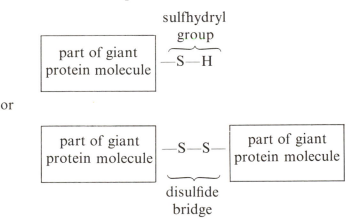

or

The disulfide bridge is of particular importance in shaping proteins, as we noted in Chapter 4. Most heavy metals will bond with and alter the critical functioning of the sulfur atoms in the disulfide bridge or in the sulfhydryl group.[2] For instance, the methyl mercury ion ($CH_3—Hg^+$) will combine with proteins to form new molecules of the form

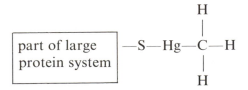

_____

[2]The affinity of mercury and sulfhydryl groups has long been recognized. Sulfhydryl compounds are called mercaptans (Chapter 4), a word derived from "mercury capture."

Any organism in which enough of these proteins are altered will soon die. This chemical process is the chief (but unfortunately not the sole) threat of heavy-metal pollution.

Heavy Metals and Sulfide Ores.   The second reason heavy-metal-sulfur compounds concern us is that when man seeks the heavy metals for his various uses, he almost invariably finds them combined with sulfur. These natural compounds—the *sulfide minerals*—are sent to smelters, where the pure metals are extracted. In the process the chemically attached sulfur is released to the atmosphere, usually as sulfur dioxide (SO_2). Heavy-metal smelters are therefore a major source of SO_2 pollution (see Chapters 5 and 7). Copper smelters are the worst offenders, but the smelting of lead and zinc and to a lesser extent cadmium, mercury, silver, and molybdenum may also release locally heavy amounts of sulfur dioxide. Table 9-2 lists some of the important sulfide minerals and the metals extracted from them.

Table 9-2.   SOME IMPORTANT SULFIDE MINERALS THAT LEAD TO SULFUR DIOXIDE (SO_2) POLLUTION UPON SMELTING

Mineral (common name)	Formula	Metal Extracted
Chalcocite	Cu_2S	Copper (Cu)
Chalcopyrite	$CuFeS_2$	
Zinc blende (sphalerite)	ZnS	Zinc (Zn)
Galena	PbS	Lead (Pb)
Cinnabar	HgS	Mercury (Hg)
Argentite	Ag_2S	Silver (Ag)
Greenockite	CdS	Cadmium (Cd)
Molybdenite	MoS_2	Molybdenum (Mo)

HEAVY-METAL POLLUTION— NATURE AND BIOLOGICAL EFFECT

Heavy metals circulate in natural cycles through the environment—from rock to soil to living organisms to water to sediments to rock again. Their concentrations at each stage are usually a few ppm or less. Living organisms are exposed to slight concentrations of the heavy metals in several of the stages. Some natural human exposure, for instance, occurs in the eating of food. A slight level of natural exposure stems from heavy metals in drinking water.

Man has greatly amplified the concentration of heavy metals in his environment. His careless use patterns and throw-away philosophy lie behind his widespread scattering of heavy metals in water, soil, and living organisms. He has pumped untold tons of heavy metals into the air, where, with the exception of mercury, they usually are very rare. Eventually the natural cycles will take these excess metals to sea and bury them in rock again. But steps in the cycle for metals require from decades to millions of years. Thus heavy-metal pollution on the short time scale of man can often be regarded as permanent, in sharp contrast to most man-made pollutants. The reasons are easy to understand.

The threat of heavy-metal pollution is the heavy-metal atom itself. This atom is not destroyed despite the many ways it chemically bonds with other atoms. Most pollutants—insecticides, air pollutants, organic contaminants in water—are a threat because of the unique bonding arrangement *among* atoms in molecules. When the molecules are altered or degraded, the pollutant is destroyed. When molecules containing heavy metals are altered, the problem-causing metal atoms still abound in the new chemical form.

Disturbance of Worldwide Heavy-Metal Cycles.   The degree to which man has amplified some heavy-metal concentrations in the world environment is shown in Table 9-3. This table compares the rates at which man brings metals to the earth's surface by mining with the rate at which they are uncovered naturally by erosion. The factor by which man has multiplied heavy-metal concentrations is given in the right-hand column. The table shows, for instance, that thirteen times more lead (Pb) is brought to the world environment by mining than by erosion.

Table 9-3 shows that man has grossly disturbed the environmental concentrations of many heavy metals. The situation is more serious than this table indicates, because (1) the disturbance is much greater for populated regions than for the earth as a whole, and (2) mining activities are growing at 5 percent each year, a trend that will see the man-made exposure double every 14 years.

Effect of Heavy Metals on Life.   Most heavy metals are toxic in small amounts, but the exact toxicity varies considerably. Most mercury and lead compounds, for instance, are much more toxic than are iron compounds. Toxicity also varies with chemical form:

Table 9-3.   NATURAL VERSUS MAN-MADE CIRCULATION OF SOME HEAVY METALS THROUGH THE WORLD ENVIRONMENT[a]

Metal	Amount Uncovered per Year (thousands of metric tons)		Amplification Factor (Mining Rate ÷ Erosion Rate)
	Erosion	Mining	
Lead (Pb)	180	2330	13
Mercury (Hg)	3	7	2.3
Copper (Cu)	375	4660	12
Zinc (Zn)	370	3930	11
Nickel (Ni)	300	358	1.2
Silver (Ag)	5	7	1.4
Manganese (Mn)	440	1600	3.6
Molybdenum (Mo)	13	57	4.4
Tin (Sn)	2	166	83

[a]Data from *Man's Impact on the Global Environment—Report of the Study of Critical Environmental Problems (SCEP)* (Cambridge, Mass.: MIT Press, 1971).

mercury metal (Hg) can be swallowed in amounts up to a pound without serious effect, while a few milligrams of methyl mercury chloride (CH_3—Hg^+Cl^-) cause ill effects. The point of exposure makes a difference too: the pound of metallic Hg that can be safely swallowed is deadly if a mere trace of its vapor is inhaled into the lungs.

The nutritional role of the essential trace metals adds to the complications. These metals are needed in low concentrations but cause death at high concentrations. The balance between *not enough* and *too much* of these metals is precarious.

The chemical-biological damage caused by heavy metals has various origins. The chemical attack on enzyme sulfur atoms, mentioned earlier, is probably most important. Heavy metals also attack the free amino (—NH_2) and organic acid (—$COOH$) groups found in proteins. A few heavy metals combine with and precipitate some of the body's vital phosphate compounds; other heavy metals catalyze the decomposition of phosphates. Several of the most toxic heavy metals (Hg, Pb, Cd, Cu) combine with cell membranes and interfere with the transport of chemicals in and out of the cell "microfactory."

The combined result of this assault by heavy metals is a variety of diseases, ranging from cancer to heart disease. Table 9-4 summarizes the health effects and also pinpoints the major man-made sources of prominent heavy metals.

Table 9-4. SOURCES AND HEALTH EFFECTS OF SOME PROMINENT HEAVY METALS[a]

Element	Sources	Health Effects
Mercury	Coal, electrical batteries, other industrial	Kidney damage, nerve damage, and death
Lead	Auto exhaust, paints	Brain, liver, and kidney damage; convulsions, behavioral disorders, death
Cadmium	Coal, zinc mining, water mains and pipes, tobacco smoke, burning plastics	High blood pressure and cardiovascular disease, interferes with zinc and copper metabolism
Nickel	Diesel oil, residual oil, coal, tobacco smoke, chemicals and catalysts, steel and nonferrous alloys	Lung cancer
Arsenic	Coal, petroleum, detergents, pesticides, mine tailings	Hazard disputed, may cause cancer
Germanium	Coal	Little innate toxicity
Vanadium	Petroleum (Venezuela, Iran), chemicals and catalysts, steel and nonferrous alloys	Probably no hazard at current levels
Antimony	Industry	Shortened life span in rats

[a]Data from *Chemical & Engineering News 49* (July 19, 1971), 29–33, and other sources.

HEAVY METALS IN
AIR, EARTH, AND WATER

Table 9-4 establishes a remarkable fact of heavy-metal pollution: coal and petroleum products are a major source of nearly every heavy metal. Coal is laced with chemical elements, as explained in Chapter 7, and these elements are released upon combustion. Most heavy metals form ash; a substantial fraction of these metals enter the atmosphere as particles of fly ash. Mercury is more volatile and enters the air as vapor.

Airborne Heavy Metals. Once airborne, heavy metals are frequently inhaled. Airborne metal contamination is serious because the lungs lack some of the defenses that other parts of the body have developed.

The hodgepodge of heavy metals in the air—each present at a different average concentration and each with its own special health effects—makes it difficult to compare the threat of individual airborne metals. A study by the Midwest Research Institute has shed light on this tangle. The authors have compared metal concentrations in urban air with industrial *threshold limit values* (TLV's)—maximum values allowed for healthy industrial workers in an 8 hour day. Table 9-5 shows their conclusions. This table is their ranking of the hazard of metal particles in average urban air. Over some cities, the amounts of individual metals soar far above their average levels: Hg, for instance, was measured at 28 percent of its TLV value on one occasion.

Metal Fallout. The hazard in the air is often multiplied when the metal particles fall to earth. In the air, metals are spread out and

Table 9-5. HAZARD FACTOR RANKING AND TLV VALUES FOR METALS IN THE AIR[a]

Rank	Metal	TLV[b] ($\mu g/m^3$)	(Average Metal Concentration) ÷ (Metal TLV) × 100
1	Lead (Pb)	100–200	0.4
2	Mercury (Hg)	10–50	0.2
3	Iron (Fe)	1000	0.031
4	Beryllium (Be)	2	0.025 or less
5	Vanadium (V)	500	0.017
6	Cadmium (Cd)	100	0.013
7	Zinc (Zn)	1000	0.013
8	Copper (Cu)	1000	0.009
9	Arsenic (As)	200–500	0.004
10	Nickel (Ni)	7–1000	0.003

[a] Note that all hazardous metals but beryllium are classified as heavy metals.
[b] TLV values for some metals depend on the chemical form of the metal. In this case the range of TLV values is shown.

diluted over vertical miles of sky, but with fallout, they concentrate on thin surface layers of soil and concrete (Figure 9-3). They accumulate at the earth's surface as fallout continues, exposing children and entering green plants and thus foods. We can see how grave the situation is from a recent finding that urban street dirt contains about 1 percent lead—creating a toxic blanket over the habitat of man and the playground of his children.

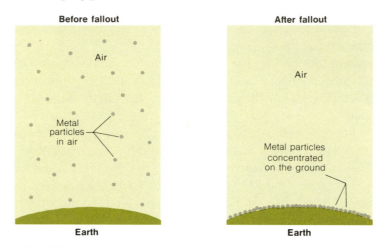

Figure 9-3. With fallout, metal particles become concentrated at the surface of the earth. This diagram shows how metal particles get packed much closer together—become more concentrated—with fallout. The effect is worse than indicated because fallout occurs again and again, building up the surface layer of concentrated metal.

Biological Concentration. Metals that fall into waterways or find their way to water through industrial wastes experience another increase in concentration that greatly expands their hazard level. They undergo *biological concentration.* By this means, aquatic organisms collect nearly all elements to levels higher than those in surrounding water. Heavy metals are concentrated most of all, largely through chelation (multiple coordinate covalent bonding) by the organic material of living systems. Through this intensive concentration, living organisms become islands rich in heavy metals.

A predator who is high in the food chain, as man is, gets particularly high metal doses because predators concentrate heavy metals from prey organisms (plants or animals) that are already metal rich. The increasing concentration along links in the food chain is *biological* or *food-chain magnification.* Biological concentration followed by food-chain magnification makes even low-level heavy-metal pollution dangerous because of excessive accumulation in living systems. Table 9-6 shows the dramatic concentrating effect of living organisms on zinc (Zn), a typical heavy metal in seawater. The last column of this table shows that Zn concentrations are multiplied up to 150,000 times by these unique accumulations in living systems. Next we shall show how a similar multiplication of mercury levels has been disastrous for man.

Table 9-6. THE BIOLOGICAL CONCENTRATION AND MAGNIFICATION OF ZINC (Zn) FROM SEAWATER[a]

Site	Zn Concentration (ppm)	Magnification Factor (Zn concentration relative to seawater)
Seawater	0.01	1
Marine plants	150	1500
Marine animals	6 to 1500	600 to 150,000

[a]Data from *Water Quality Criteria*, Federal Water Pollution Control Administration, 1968.

MERCURY

Minamata is a tiny fishing village in Japan that has a kind of fame no one wants. Starting in 1953, many residents of Minamata were struck by a mysterious disease that crippled, deranged, deformed, and killed. Named *Minamata disease*, this frightening affliction was eventually traced to mercury (Hg) wastes discharged into Minamata Bay from a giant chemical plant nearby.

The plant on Minamata Bay used an abundance of mercury to catalyze essential chemical reactions in the production of plastics. Some of the mercury became converted into deadly methyl mercury (CH_3Hg^+) before discharge into the bay. The methyl mercury was diluted so much by the enormous volume of the bay that concentrations of Hg did not exceed 2 to 4 parts per billion (ppb)—a level so low that the water could be drunk. All seemed well, but beneath the surface living organisms were incorporating atom after atom of Hg in organic tissue—concentrating the mercury to levels thousands of times its waterborne value. Fish and shellfish pulled from the bay by Minamata fishermen contained 5000 to 20,000 ppb (5 to 20 ppm) Hg. Thus through biological concentration, the harmless Hg of Minamata Bay was multiplied about four thousand times—to a level that devastated this quiet village. The Minamata death toll is now 52. Nearly a hundred people suffered crippling neurological damage from which they will never recover. Many are helpless and blind. Altogether 900 people in the area of the bay have shown some symptoms of Hg poisoning.

Americans slept through the distant Minamata disaster and a smaller episode at Niigata, Japan. No one became concerned when Swedish birds and fish turned up with high mercury levels, which forced a 1967 ban on fishing in 40 Swedish rivers and lakes. After all, most mercury wastes in the United States were inorganic—containing Hg in the metallic form or in the form of simple inorganic salts like mercury(I) chloride (Hg_2Cl_2) or mercury(II) chloride ($HgCl_2$). These substances were expected to chemically combine with or be absorbed on the bottom sediments of waterways and stay there—away from the biological activity in the overlying water.

A wave of concern swept the nation in March 1970 when graduate student Norvald Fimreite at the University of Western Ontario revealed to Canadian authorities his discovery of Hg levels up to 7 ppm in the fish of Lake St. Clair on the Canadian-U.S. border northeast of Detroit. This Hg level is as high as that found in some of the seafood from Minamata Bay and is 10 times higher than the Food and Drug Administration (FDA) limit of 0.5 ppm Hg in seafood. Commercial fishing was promptly banned in Lake St. Clair and other Great Lakes waters. Further study revealed elevated Hg levels in the waterways of thirty-three states; fishing restrictions were quickly imposed in eighteen states. No human damage has yet been documented from Hg in Lake St. Clair or other U.S. waterways, but as usual, the simmering, long-term impact of toxic chemicals on health and longevity are uncertain.

The surprise of Lake St. Clair makes us wonder how many other major environmental surprises are ahead of us. Clearly Mother Nature and her tangled cycles are still in charge here on earth: it might be to our advantage to learn how to live less abrasively with the temperamental old lady.

Methyl Mercury. Where did the mercury in Lake St. Clair fish come from? Bottom sediments, as we have mentioned, hold inorganic mercury: a variety of chemical, ionic, and intermolecular forces bind Hg to the clays and organic materials of bottom muds. However, as early as 1968 a University of Illinois biochemist, Dr. John Wood, discovered that the microorganisms in natural waterways have their own chemical influence on mercury. These organisms, it seems, have a strong inclination to *methylate*—attach methyl ($-CH_3$) groups—to atoms of mercury. Some of the chemical steps of methylation are shown in Figure 9-4. By methylation, dimethyl mercury ($CH_3-Hg-CH_3$) and methyl mercury (CH_3-Hg^+) are formed. These organic forms of mercury escape from the sediments into the water above and eventually enter fish and other aquatic life.

Mercury in all chemical forms is toxic if absorbed by the system, but methyl mercury (CH_3-Hg^+) is toxic in the extreme. This substance is retained for months in the body after it is ingested. It passes with ease across the protective blood-brain barrier to affect man's most unique organ—the brain. The crippling neurological (nervous system) disorders common in victims of methyl mercury poisoning are therefore explained. Methyl mercury also bridges the placental barrier to infest unborn infants. Even healthy mothers with subclinical (subtoxic) levels of CH_3-Hg^+ in their blood can give birth to deformed young.

Methyl mercury is the most mutagenic (mutation-causing) chemical known—1000 times more potent than its closest rival, colchicine. Fruit flies given food containing only 0.25 ppm methyl mercury have offspring that carry one extra chromosome.

Now that we understand the mercury crisis, why not clean up

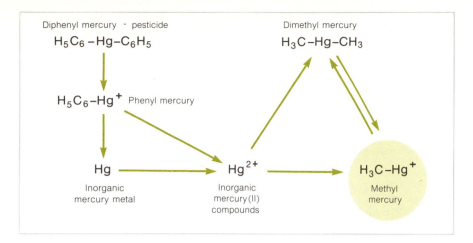

Diphenyl mercury - pesticide

$H_5C_6 - Hg - C_6H_5$

$H_5C_6 - Hg^+$ Phenyl mercury

Dimethyl mercury

$H_3C - Hg - CH_3$

Hg

Inorganic
mercury metal

Hg^{2+}

Inorganic
mercury (II)
compounds

$H_3C - Hg^+$

Methyl
mercury

Figure 9-4. Some of the chemical pathways which in water lead to deadly methyl mercury. (These steps merely show some of the chemical forms of Hg in its transformation to methyl mercury: balanced chemical reactions are not shown.)

the mess? Unfortunately, man's environmental mistakes are not always easy to erase. Inorganic Hg wastes have accumulated in sediments from our centuries-old habit of treating waterways as cesspools. The slow methylation of mercury in these sediments will infest the water for a long time to come. The mercury now in San Francisco Bay, for instance, is believed to be a memento of gold rush days, when Hg was used to extract gold from its ores. Lake St. Clair, by some estimates, will remain contaminated for a thousand years because of its mercury-rich sediments.

The Sources of Environmental Mercury. Mercury, like its heavy metal cousins, is found at low levels in all soil, water, and air. Ocean water contains about 0.1 ppb Hg—a concentration roughly thirty-fold below the disastrous level of Hg in Minamata Bay. The ocean's concentration of Hg increases with depth. The mercury content of rivers in America varies from a value under 0.1 ppb to 6 ppb (the Maumee River), but some of the Hg may have human sources.

Mercury vapor in air exists naturally at levels slightly under 10^{-9} g/m^3. Over Palo Alto, California, Hg reached concentrations up to 50×10^{-9} g/m^3—still not much. However, the air may be a major route for the worldwide movement of Hg.

In soil and rock, mercury generally varies in concentration from 3 to 250 ppb, but occasionally it reaches 4600 ppb (4.6 ppm).

Populated regions have suffered gross increases in environmental Hg. The reasons are generally two: (1) Hg is a useful metal employed in quantity, and (2) sloppy and unthinking habits have led to frequent environmental release.

Mercury, although scarce and expensive, is a valuable machine-age aid. As the only metal that is a liquid at room temperature, Hg is important in electrical switches, control devices, and electrodes. Because Hg creates havoc with the molecules of living systems, its compounds are powerful pesticides. These compounds are used, for instance, to combat fungus diseases in seed grains and eliminate slime in papermaking. Mercury, in addition, is valuable as a catalyst; as an industrial chemical; as a component of dental fillings; as a drug; and as the working liquid of thermometers, barometers, and other precision instruments. Table 9-7 shows how mercury consumption is divided among its various tasks.

Table 9-7. ANNUAL MERCURY CONSUMPTION IN THE UNITED STATES[a]

Use	Tons Consumed 1969	Tons Consumed 1974–1975 (est.)
Electrical:		
Electrodes for chlorine production	788	869
Electrical control equipment	710	863
Pesticidal:		
Agricultural	102	101
In paint	370	407
Paper and pulp mills	21	9
Catalyst	112	89
Amalgamation	7	9
Dental materials	116	144
Drugs	27	25
Industrial controls	265	351
Laboratory use	78	79
Other	410	226
	3006	3172

[a] From *Hazards of Mercury*, Special Report to the Secretary's Pesticide Advisory Committee, Department of Health, Education, and Welfare, November 1970.

A good deal of Hg escapes in use. The high Hg level in the Lake St. Clair fish, for instance, is commonly blamed on discharges from plants making chlorine (Cl_2) and caustic soda (NaOH) with the aid of Hg electrodes. Most spillage is easy to control: chlorine plants were able to reduce their Hg discharges in a few months following the Lake St. Clair incident. Unfortunately, great quantities of mercury from decades of careless operation now cling to the bottom sediments, waiting their turn to undergo biological methylation and enter the web of life. No one knows for certain how long this legacy of carelessness will be with us.

While many kinds of mercury spillage can be controlled, other sources of environmental mercury are difficult or impossible to shut off. The sources that spew Hg into the air are particularly vexing. The mercury in coal and oil, for instance, is released to the air in a very dilute form with combustion. Annual worldwide releases from this source total about 3000 tons.[3] The roasting of sulfide ores may

[3] Estimates of the average Hg content in fossil fuels vary from about 0.2 ppm to 3.3 ppm. Assuming an average content of 0.5 ppm Hg and an annual fossil-fuel consumption of 6×10^9 metric tons, the mercury release is

$$6 \times 10^9 \text{ tons C} \times \frac{0.5 \text{ parts Hg}}{10^6 \text{ parts C}} = 3000 \text{ tons Hg}$$

Fresh earth exposed in thousands of earth-moving projects, including open pit mining as shown here, may produce most of the mercury released into the worldwide atmosphere. (*Photo by Paolo Koch/Rapho Guillumette.*)

release a similar quantity of mercury. According to Edward D. Goldberg and fellow workers at the Scripps Institute of Oceanography, the biggest man-made source of Hg may simply be man's extensive disturbance of soil and rock. Earth that is freshly exposed in man's growing attempts to remodel the face of the planet loses much of its volatile mercury to the atmosphere. Amounts as great as 100,000 tons may be released this way each year. This fact alone can account for the two- or threefold worldwide increase in atmospheric Hg that has been recorded in the precipitated layers of Greenland ice.

Compared to many other contaminants, the human release of mercury is not very large. But a combination of factors—natural methylation of mercury; the extreme toxicity of the end product, methyl mercury; and the intense biological concentration of Hg—combine to make mercury pollution one of the most serious environmental problems faced by man.

Tuna, Swordfish, and Seawater. The shock of finding mercury in inland fish was followed by a deeper concern at the discovery of up to several ppm Hg in ocean-dwelling tuna and swordfish. It is one thing to fill a few lakes with toxic mercury, but it is an altogether different matter to defile the vast oceans. Has man's pollution really been that abundant?

Seawater contains Hg at a minute concentration—about 0.1 ppb, or 100 parts per trillion (100 parts Hg/10^{12} parts water) near the surface, and somewhat more below. Despite the low Hg concentration, if all the water in the oceans—a staggering 1.5×10^{18} metric tons—contained this level of Hg, the total mercury content of the world's oceans would be 150 million metric tons.[4] There may be more in the oceans because of greater Hg concentrations in the depths, but the estimate of 150 million metric tons will do to illustrate an important point. The 150×10^6 metric tons of ocean Hg totally dwarfs the quantities of mercury used by man. About 3000 tons of Hg are used annually in the United States, as was shown in Table 9-6, and the worldwide usage is about 10,000 tons. It would therefore take about 15,000 years, dumping 10,000 tons of Hg per year in the oceans, to duplicate the present level of ocean mercury. Even the 100,000 tons of mercury that is thought to be vaporized from disturbed soil and later precipitated into the oceans will not make a significant impact for several centuries—unless the release accelerates rapidly because of the expansion of earth-moving projects. But for now, at least, the oceans seem safe from man's mercury. And similar reasoning would show that the oceans are *presently* immune to man's other metal and nonmetal contaminants.

But what about the mercury-infested tuna and swordfish? A study of museum tuna pulled from the ocean almost a century ago shows Hg levels close to those in fresh tuna. Elevated Hg concentrations were also found in the bony remains of prehistoric fish from sites in the United States and Peru. The best evidence indicates, then, that high mercury levels in tuna and swordfish occur naturally. Both fish are active ocean predators, at the top of the food chain, and they would be expected to magnify the concentration of Hg found in their food—other fish of the ocean. For some reason, swordfish have concentrated Hg most intensely—90 percent of swordfish exceed the FDA limit of 0.5 ppm, and some range up to 2 ppm. Only 5 to 10 percent of tuna exceed the FDA specifications.

Finally, our conclusion that man has not yet polluted the deep oceans must be put in perspective. We noted in Chapter 8 that small coastal regions, comprising only 0.1 percent of the ocean surface, produce 99 percent of the seafood. These regions contain the estuaries —the coastal regions where the rivers enter the sea. Here the oceans get the brunt of man's pollution—directly from his garbage-filled rivers. So while man is not yet capable of contaminating the vast oceans, he can and he is polluting its most important and productive parts—the coastal breadbaskets. This pollution could have unfortunate consequences: exploding populations will need every scrap of food the world can produce.

---

[4]This quantity is obtained as follows:

$$1.5 \times 10^{18} \text{ metric tons } H_2O \times \frac{100 \text{ parts Hg}}{10^{12} \text{ parts } H_2O} = 150 \times 10^6 \text{ metric tons Hg}$$

LEAD

Lead (Pb) is another heavy metal dominated by two clashing characteristics—extreme usefulness and extreme toxicity. Since man first began digging Pb out of the earth 5000 years ago, it has been accepted enthusiastically, used widely, and discarded carelessly. Lead poisoning has been recognized for centuries, and yet lead has been used copiously in the environment from which man gathers food, water, and air. Lead has long been employed in the manufacture of cooking vessels, pottery, metal plumbing, household paints, and pesticides. It has been speculated that the ruling class of ancient Rome was brought to disintegration by the use of leaden devices for handling food and drink.

In recent decades, considerable progress has been made in getting lead out of the household. Lead plumbing is now rare; lead glazes on pottery are being discouraged; and lead is now being phased out of household paints. But only in this decade have we thought seriously about getting waste Pb out of the general environment of earth.

The environmental cleanup of lead is past due, for with the growth in population and technology, discarded Pb has grown to staggering proportions. Lead is mined from the earth in amounts exceeding 2 million tons a year—most of it destined ultimately for the environment. As we saw in Table 9-3, the quantity of mined lead entering the worldwide environment is 13 times greater than the Pb uncovered by erosion.

Leaden Skies. The atmosphere has borne the brunt of man's lead assault. Much of man's lead—from automobiles, smelters, and burning coal—goes directly into the air. Airborne Pb particles— some circulating worldwide—pepper the earth with fallout. Lead fallout in the ice strata of Arctic Greenland shows how great man's global influence has been. Figure 9-5 (p. 350) documents the spectacular growth of atmospheric lead deposited in snow over the past two centuries and more.

The concentration of lead in Greenland air is dwarfed by that over urban centers. In city air the content of lead often exceeds 1 microgram (μg) per cubic meter (m^3) and sometimes reaches 40 μg/m^3 near traffic arteries. Air over remote Pacific sites contains only 0.001 μg Pb/m^3.

Lead in Man. Most lead-containing particles in the air are small—0.2 to 0.4 microns (μ)—and thus penetrate to deep parts of the lung (Chapter 7). It is estimated that 40 percent of inhaled Pb is captured and absorbed by the body. Less than 10 percent of the lead in food and water is absorbed.

A typical urban dweller inhaling 20 m^3 of air per day polluted with 2 μg Pb/m^3 of air will take in 40 μg of lead and retain 40 percent

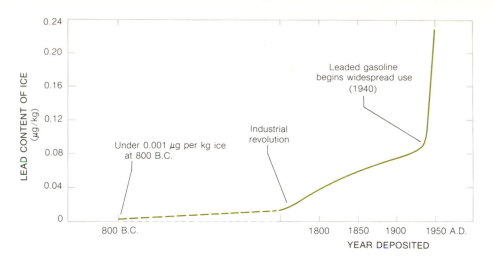

Figure 9-5. Man's dramatic influence on worldwide atmospheric lead is recorded in the strata of Greenland ice. [Data from M. Murozumi, T. J. Chow, and C. C. Patterson, *Geochim. Cosmochim. Acta 33*, 1247 (1969).]

of that—16 μg.[5] This amount is over one-third of the present daily Pb intake of most city residents. The remainder is absorbed from food and drink.

The absorption of airborne Pb is not unduly serious by itself, but when added to the Pb intake from food and drink, it creates a level of human exposure that concerns many scientists and medical authorities. Most big-city residents have Pb in their blood at levels only slightly below those that are known to bring on clinical symptoms of Pb poisoning. Figure 9-6 shows the concentration of lead in the blood of different Philadelphia residents. Many downtown residents have Pb levels that reach the proposed thresholds for toxic effects. The average lead level in urban blood, about 0.2 ppm, exceeds by a factor of almost 100 the estimated level of 0.0025 ppm in the blood of primitive man. It is believed by noted British chemist D. Bryce-Smith that compared to lead, "no other toxic chemical has accumulated in man to average levels so close to the threshold for overt clinical poisoning."

The symptoms of low-level lead poisoning are mild anemia accompanied by vague muscle discomfort, fatigue, irritability, and headache. Who among us can say for sure that our occasional "bad days" do not stem in part from environmental excesses of lead? Or that our life will not be shortened a few months by our burden of lead? Indeed, the death of mice and rats is hastened by Pb levels equivalent to ours. Still the questions above, like most questions related to long-term, low-level toxic effects, are impossible to answer

---

[5] This amount is calculated as follows:

$$20 \text{ m}^3 \text{ air} \times \frac{2 \text{ μg Pb}}{1 \text{ m}^3 \text{ air}} \times \frac{40 \text{ parts Pb absorbed}}{100 \text{ parts Pb (total)}} = 16 \text{ μg Pb absorbed}$$

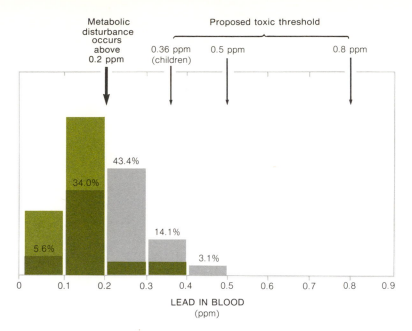

Figure 9-6. Percentage of downtown Philadelphia residents having the specified levels of lead in their blood. The dashed lines, for suburban dwellers, show a lower average blood lead level. [Data from J. H. Ludwig *et al., Am. Ind. Hygiene Assn. Journal 26,* 270 (1965) and D. Bryce-Smith, *Chemistry in Britain,* February 2, 1971, pp. 54–56.]

at this time. Until we know the answers, it would be prudent to reduce environmental Pb in any way we can.

A few individuals encounter lead in such large quantities that sickness or death ensues. Lead damages the brain and central nervous system along with the kidney, liver, and reproductive system. The most unfortunate source of such excesses is the Pb undercoating of old painted surfaces in slum areas, applied to building interiors in the 1940s and before. Malnourished children nibble at strange things— a condition called *pica*—and their diet may include flakes of leaded paint. Their high lead diet is enriched even more by the high (near 1 percent) Pb content of street dirt mentioned earlier. It is estimated that 6000 children acquire severe lead poisoning and permanent brain damage each year, and about 200 of them die of the malady. This dismal condition will not be cleared up until old buildings are totally repainted or torn down, and the thick fallout of lead on our streets is erased.

Excessive lead exposure comes from other sources as well. Not least is 50 million gal of moonshine whiskey distilled every year in automobile radiators and other lead-containing devices. Painters, welders, heavy smokers, and gasoline sniffers are also susceptible.

Lead in Gasoline. Nearly all gasoline contains small quantities of lead. The average content in 1971 was 2.2 g Pb per gal of gasoline in the United States. The Pb is added in organic form—tetraethyl lead, $Pb(C_2H_5)_4$, is the most common additive. About 250,000 tons of Pb in this form are added to this nation's gasoline each year. Lead raises the octane rating of gasoline and thus increases engine efficiency, as explained in Chapter 7.

Upon combustion, the organic Pb is oxidized first to lead(II) oxide (PbO).

$$2\,Pb(C_2H_5)_4 \;+\; 27\,O_2 \;\rightarrow\; 2\,PbO \;+\; 16\,CO_2 \;+\; 20\,H_2O$$

| tetraethyl lead | oxygen | lead(II) oxide | carbon dioxide | water |

The lead then reacts with halogen scavengers—coadditives containing bromine (Br) and chlorine (Cl)—forming particles of lead halides ($PbBr_2$, $PbBrCl$, $PbCl_2$) that escape readily through the exhaust pipe. By this means, about 80 percent of the lead in gasoline—almost 200,000 tons annually—is poured into the air. This amount is 98 percent of all the lead released to the atmosphere over the United States. The nonemitted lead accumulates as engine deposits.

The Pb added to gasoline accounts for less than one fourth of that used in the United States. Nearly half of the Pb is used in making lead storage batteries, and significant quantities also go into metal products, solder, and paint pigments. Most of this Pb ends up in the environment. But the Pb in gasoline is the only Pb channeled directly

This nation's fleet of automobiles, burning leaded gasoline, is responsible for almost 200,000 tons of lead released annually to the atmosphere. (*Photo by Jim Stuart/Photofind.*)

ENVIRONMENTAL CONTAMINATION BY HEAVY METALS

to air—and subsequently to city streets, sidewalks, and lawns by fallout. Hence the controversy over lead is focused on leaded gasoline.

Those concerned mainly about the health effects of lead in gasoline have a critical ally, for emitted lead destroys the catalytic activity of catalytic reactors (Chapter 7) designed to clean up carbon monoxide (CO) and hydrocarbon (HC) pollution from the exhaust of automobiles. Thus the removal of Pb from gasoline is the first essential step to broader air-pollution control. Despite the fact that conversion to unleaded gasolines will cost this nation over $2 billion in new refining facilities, the environmental protection agency (EPA) is working to gradually phase lead out of gasoline through the mid-1970s. This action may help us escape the slow drift toward nationwide lead poisoning that is aided so much by our expanding car population. If we can now moderate our frenzied use of more and more metals for increasingly useless gadgets, lead pollution from all sources might drop and the large burden of this toxic metal in our bodies might fall to a healthier level. Perhaps, just knowing this, we would have fewer bad days.

OTHER HEAVY METALS

The story of mercury and lead pollution is repeated over and over again with other heavy metals—but the effects occur on a reduced scale and with many variations. The uses of heavy metals vary and the human sources of metal pollution vary accordingly. The symptoms of poisoning vary slightly from metal to metal, even though the underlying molecular basis of toxic effects is about the same for all heavy metals, as we have noted. Few metals match environmental Hg and Pb in the number of people that they sicken and disable, but the same kinds of uncertainties arise when we try to decide when low-level toxic effects occur and the number of people are affected.

A considerable number of heavy metals in the environment have drawn recent attention. Chromium (Cr), cobalt (Co), zinc (Zn), cadmium (Cd), and arsenic (As) have been of greatest concern after Hg and Pb. Of these substances, Cd and As stand out as the newest and the oldest of recognized environmental threats, respectively.

Cadmium (Cd)—the newest metal threat—is a rare element in the crust of the earth. It is about as abundant as Hg but much less so than Pb. As with Hg, the environmental threat of Cd was first pinpointed in Japan. Since World War II, farmers in the Jintsu River region of Japan suffered a severe skeletal disorder known as osteomalacia. In 1970 this disorder was traced to cadmium in mine wastewater that was used for irrigating rice. At about the same time, Cd was found to exceed the Public Health Service (PHS) standard—0.1 ppm Cd—in many supplies of drinking water in the United States. Because adsorbed Cd is retained in the body for decades and is extremely toxic, this metal has gained its rightful place among the most serious contaminants of man's environment.

Arsenic (As), on the other hand, has been recognized—and used—as a human poison since ancient times. It is discharged to air by many metal smelters and it later contaminates soil and vegetation by fallout. The most prominent use of arsenic is in pesticides—where clear advantage is taken of its toxic nature. Arsenic-containing pesticides have been used for centuries and are still employed in significant amounts. Arsenic—like other heavy-metal contaminants—does not degrade; some orchards have accumulated so much As from continual spraying that the soil has been rendered permanently toxic and sterile.

Now arsenic pesticides (and pesticides made of other heavy metals) have been relegated to second rank importance by a modern arsenal of synthetic organic chemicals. These organic pesticides and the way they have changed life on earth will be discussed in the following chapter.

Exercises

1. A drug known as dimercaprol is injected at levels of 0.5–1 g/day into victims of poisoning by some heavy metals—mercury, arsenic, antimony, and others. The drug binds chemically with the heavy metals, allowing the organism to excrete them. Dimercaprol has the chemical structure

What chemical explanation can you give for the effectiveness of dimercaprol in tying up heavy metals in the body?

2. Thirty-eight power plants in Illinois consumed 30 million tons of coal in 1969. The coal has an average mercury (Hg) content of 0.2 ppm. How much Hg was emitted into the air over Illinois (assuming a 100 percent release of Hg) in 1969?

3. If mercury is present naturally in swordfish at levels exceeding the 0.5 ppm FDA limit, do you think that the FDA limit should be raised? (See Problem 4 for added perspective on this question.)

ENVIRONMENTAL CONTAMINATION BY HEAVY METALS

4. In 1971, a dieting New York woman was identified as the first un-mistakable victim of mercury poisoning from store-bought food. She acquired tremors, difficulty in writing and talking, numbness, and loss of equilibrium. Her problems were traced to a long-standing diet containing 12.5 ounces (350 g) of swordfish per day. Assuming that her swordfish contained the average swordfish concentration of 1 ppm Hg, how much Hg did this woman ingest per day? How does this amount compare with the observed lowest toxic intake for humans of 300 μg Hg/day?

5. Some fish from polluted Canadian lakes contain 5 ppm Hg (mostly in the form of methyl mercury). How much Hg would be ingested per day by a person eating 200 g (just under half a pound) of such fish daily? How does this amount compare with the lowest toxic intake of 300 μg Hg/day?

6. The average Hg concentration found in numerous samples of salmon, herring, pilchard, sardine, and mackerel was 0.04 ppm. How many lb of such sea fish would you need to eat each day to reach the lowest toxic human intake of 300 μg Hg/day?

7. Variable concentrations of Hg ranging up to 0.018 ppm have been found in the drinking water of 24 Chicago area communities. The average person drinks 2 ℓ (2000 g) of water a day. Does this amount pose a threat in view of the 300 μg Hg/day intake that causes the onset of toxic effects in man? (Presumably much of this Hg is inorganic, in contrast to the methyl mercury predominant in seafood. Only a fraction of inorganic Hg is absorbed into the bloodstream from the digestive tract. An offsetting fact is that Hg in food and air are con-tributing toward a larger total intake.)

8. Air in dental offices has been found to contain up to 180 micrograms (μg) of Hg per cubic meter (m^3) of air. (It comes from the Hg used in filling materials.) How much Hg would be inhaled in an 8-hour day by dental personnel breathing $1/3 \times 20{,}000 \; \ell = 6700 \; \ell$ of such air? How does this amount compare to the toxic threshold of 300 μg Hg/day? (Hg vapor entering the lung is almost totally absorbed into the blood-stream. At least one dental surgery assistant has died from Hg poison-ing.)

9. A health limit of 10 micrograms (μg) of mercury (Hg) per cubic meter (m^3) of air has been widely proposed for the general public. How many μg of Hg would be inhaled per day by the typical adult breathing 20,000 ℓ per day of air containing the maximum Hg level?

10. Ocean-dwelling fish like tuna and swordfish live in the top 100 m of ocean water, a layer containing 3×10^{16} metric tons of water world-wide. How much Hg is contained in this layer at the natural Hg con-centration of 0.1 ppb in seawater? Would man-made sources of Hg increase this concentration significantly in a 5 year period (the time it takes for surface water to mix with deeper ocean water)?

11. Mercury-containing fungicides at concentrations as low as 0.1 ppb are reported to reduce the photosynthesis of some phytoplankton [R. C. Harris *et al.*, *Science*, *170*, 736 (1970)]. Because phytoplankton are at the base of the food chain, their reduced photosynthesis would reduce the biological productivity of water. What weight of organo-mercurial fungicides would hinder the productivity of a small lake (0.5 mi wide, 8 ft deep) containing 1.6×10^6 tons of H$_2$O?

12. The average lead (Pb) content of gasoline was about 2.2 g Pb/gal (about 3 kg) of gasoline in 1971. What was the Pb concentration of this gasoline in parts per million by weight?

13. How many grams of tetraethyl lead, $Pb(C_2H_5)_4$, must be added to 1000 gal of gasoline to bring the Pb level up to 2 g Pb/gal of gasoline?

14. How many kg of lead are consumed by a typical car driven 100,000 mi in a 10 year lifetime on gasoline containing 2.4 g Pb/gal gasoline? The average mileage figure is 16 mi/gal gasoline.

15. A commuter drives 20 mi to work and 20 mi back on 250 working days in a year. His car runs 20 mi on a gallon of gasoline, and the gasoline contains 2 g Pb/gal. How much lead does this commuter use annually in commuting? Name three steps that society could take to reduce this source of environmental lead.

16. A child 2 years of age swallows 0.1 oz (3 g) of street dirt containing 0.5 percent lead. How many micrograms (μg) of lead are swallowed? How many μg are absorbed, assuming that 10 percent of the swallowed dose is absorbed by the child's body? (The normal absorption of Pb by a child this age is about 10 μg/day.)

17. How much lead would be inhaled per day by a traffic policeman who spends 8 hours at a busy intersection with an average lead content of 10 μg/m^3 in the air? How much lead would be absorbed into his body?

Glossary

Biological concentration process by which organisms concentrate certain substances to levels above those in their environment.
Biological (food-chain) magnification increasing biological concentration along the links of a food chain.
Chelating agent a molecule that can wrap around an ion, forming coordinate covalent bonds with it from several directions.
Essential trace metals those heavy metals that in small (trace) quantities are essential to life: they include Cu, Zn, Co, and Cr.
Heavy metals the electronegative metals with density greater than 5 g/cm^3.
Light metals metal elements having density less than 5 g/cm^3.
Metallic bond arrangement in which the cores of metal atoms are held together by a "sea" of common electrons.
Methylate to attach methyl groups (—CH$_3$) to an atom or molecule.
Methyl mercury CH$_3$—Hg$^+$, a lethal substance created when certain microorganisms methylate inorganic mercury.

Minamata disease name given to symptoms of mercury poisoning that appeared after residents of Minamata, Japan ingested mercury-contaminated seafood from Minamata Bay.

Mutagenic causing mutations.

Noble (coinage) metals Au, Pt, and Ag, so named because they are relatively inert and therefore keep their metallic luster.

Pica name given to a conditions in which undernourished children tend to nibble at foreign materials, thus ingesting lead and other toxic substances.

Threshold limit value (TLV) maximum amounts of toxins allowed for healthy industrial workers during an 8 hour day.

Trace metals metals present at a few ppm or less in the earth's crust.

Additional Reading

Starred selections are most suitable for the nontechnical reader.

Friberg, Lars, Magnus Piscator, and Gunnar Nordberg, *Cadmium in the Environment*. Cleveland, Ohio: CRC Press, 1971.

Friberg, Lars, and Jaroslav Vostal, eds., *Mercury in the Environment*. Cleveland, Ohio: CRC Press, 1972.

*Giddings, J. Calvin, and Manus B. Monroe, eds., *Our Chemical Environment*. San Francisco: Canfield, 1972.

Hazards of Mercury, Special Report to the Secretary's Pesticide Advisory Committee, Department of Health, Education, and Welfare, November 1970. New York: Academic Press, 1971.

Lee, Douglas H. K., ed., *Metallic Contaminants and Human Health*. New York: Academic Press, 1972.

Man's Impact on the Global Environment, Assessment and Recommendations for Action, Report of the Study of Critical Environmental Problems (SCEP). Cambridge: MIT Press, 1970.

*Montague, Katherine, and Peter Montague, *Mercury*. San Francisco: Sierra Club, 1971.

*Novick, Sheldon, and Dorothy Cottrell, eds., *Our World in Peril: An Environment Review*. Greenwich, Conn.: Fawcett, 1971.

*Schroeder, Henry A., *Metals in the Air. Environment 13*, 18 (1971).

HYDROGEN, CARBON, NITROGEN, OXYGEN, PHOSPHORUS, SULFUR, CHLORINE The seven elements highlighted above are the principal building blocks of modern organic pesticides.

10 CHEMICAL WARFARE AGAINST PESTS

PEST CONTROL AND ENVIRONMENTAL REALITY

Nothing illustrates the great power of technology in this century better than the story of chemical pesticides. And nothing shows quite so clearly how necessary it is to apply earth-shaking technology cautiously and with great concern and forethought for side effects.

For three decades chemical pesticides have been rained down on the earth's surface in a massive attempt to subdue part of our natural world. This warfare was initiated with meager knowledge regarding the ultimate effect on ecological systems and human health. It was a vast experiment with spaceship earth which, out of good fortune, was not totally disastrous.

Yet who would have done differently? The beginning was brilliant. In Italy a massive epidemic of typhus, a disease carried by lice, was halted in 1943 by DDT. Malaria was brought under control in many parts of the world through DDT control of the carrier mosquitoes. Crops were saved from insects and starvation was deferred. The extermination of all major insect pests was optimistically predicted.

Now, several decades later, we have contaminated the earth in even its most remote places with pesticides. The air we breathe, the food we eat, and the water we drink all contain DDT and related pesticides. Each of us has a variety of organic pesticides in his tissues, with no clear knowledge of their ultimate effect. Birds, including the bald eagle, are threatened with extinction, and the productivity and balance of the ecosystem have been seriously disturbed.

As the damage accumulates, the benefit declines. Insect resistance to insecticides has increased so much that chemical control is often difficult and occasionally impossible. We have created an accelerating chemical treadmill in which more and more insecticide exerts a decreasing control over fewer species. Meanwhile, the side effects become more severe.

The damage to wildlife, apart from its ecological implications, should concern us all. Man is not so different in chemical functioning from other animals to preclude a similar damage. Someone has pointed out how the miners at one time took canaries into mines to warn of bad air. Our canaries are dying, but now there is no uncontaminated place of escape.

PESTICIDE POLLUTION

Many pesticides escape from the local environment that they are meant to protect and circulate throughout the world environment. The result is unwanted accumulations in man and wildlife. The escaped pesticide is a pollutant as much as is any other toxic material released to the general environment. Pesticides are unique pollutants, however, in that they are not waste products of some human enterprise that are carelessly cast off to the environment. Instead they are expensive, synthetic materials applied intentionally and abundantly to a large portion of the earth's surface.

HISTORY OF CHEMICAL PEST CONTROL

Insect-killing chemicals were made by plants long before man appeared on earth. These natural organic substances were apparently the earliest insecticides used by man. It is thought that one of them, the still effective pyrethrum, was brought to Europe by Marco Polo from the Far East. In South America the sabadilla plant has been used for centuries to control lice. Nicotine, turpentine, and petroleum products were later added to the arsenal of natural organic insect killers.

Inorganic insecticides have also played a role in pest control, some for more than a century. Many of them are highly toxic compounds of arsenic, such as Paris Green, which contains $Cu_3(AsO_3)_2$, lead arsenate ($Pb_3(AsO_4)_2$), and sodium arsenite (Na_3AsO_3). Other compounds containing the heavy metals lead (Pb), mercury (Hg), copper (Cu), and zinc (Zn) have been used.

Synthetic organic insecticides are more recent. A few were developed around the turn of the century, but the true revolution in insect control was marked by the discovery in 1939 that DDT, a chemical substance known since 1874, was a potent insecticide. Paul Mueller, a Swiss chemist, was awarded a Nobel Prize for this discovery in 1948. DDT was first tested in the United States in 1942, was immediately pressed into military service, and then entered commercial arteries in 1945.

As early as 1946, houseflies showed resistance to DDT, and by 1948, the year the Nobel Prize was awarded, a dozen insects had developed resistance. The rapid increase of insect resistance to DDT prompted the development of many new pesticides. There are now about nine hundred such substances, a hundred of them used widely. However, over two hundred insects have developed a resistance to at least one insecticide, and several damaging insects can no longer be controlled chemically.

The advent of synthetic organic insecticides was greeted with great optimism, as we have noted. Now the optimism is muted and the concerns are multiplied. The transition in attitudes may be traced to Rachel Carson's book *Silent Spring*, published in 1962. The validity of this book's central theme—that pesticides pose a grave threat to man and other animals in the ecological web—is still widely debated, but no one can deny Rachel Carson a place in history for catalyzing a new way of thinking.

THE STRUCTURE OF INSECTICIDE MOLECULES

Pesticides are chemicals used to destroy pests. Pests include insects, plants, fungi, and rodents and are treated, respectively, by pesticides logically classed as *insecticides*, *herbicides*, *fungicides*, and *rodenticides*. Of these classes, insecticides have caused the most apparent ecological changes and have been at the focal point of people's concern over both human health and ecology. Therefore our emphasis will be on insecticides.

The three major groups of insecticides. DDT was the first in a group of *organochlorine* or *chlorinated hydrocarbon* insecticides. Most of these insecticides are classed as *persistent*, retaining their activities for over 2 years, and sometimes for 20 years or more. There are two major classes of *nonpersistent* insecticides, which generally remain active 1 to 12 weeks. They are the *organophosphorus* compounds and the *carbamates*. Many inorganic insecticides, particularly the still popular compounds of arsenic, are classed as *permanent*, and they remain in the soil indefinitely. The different levels of persistence are summarized in Figure 10-1, next page.

The classification of the organic insecticides as organochlorine, organophosphorus, or carbamate depends on the structure and composition of the insecticide molecule. The molecules of some

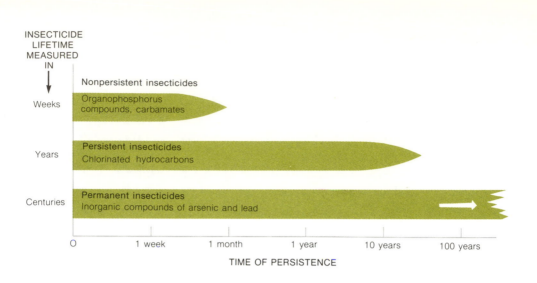

Figure 10-1. The persistence of various classes of insecticides under normal conditions.

important insecticides from these classes are shown in Figure 10-2. We shall be discussing mainly the insecticides shown in this figure, although a few other insecticides are equally widespread.

The names given to these insecticide classes are related to molecular structure in a logical way. Organochlorine insecticides contain chlorine and carbon, as we would expect from the name. The chlorine atoms, usually several in number, bond directly to carbon atoms as shown:

essential organochlorine component

where the free ends are bonded in various ways to other molecular fragments to yield the whole molecule.

Likewise, organophosphorus compounds contain phosphorus and carbon. The phosphorus (P) in such compounds has three covalent bonds and one coordinate covalent bond (Chapter 3), shown as (↑), in which the P atom contributes both of the bond's electrons. It is generally linked as follows, with slight variations in sulfur and oxygen positions from compound to compound:

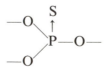

typical organophosphorus core

This arrangement is the central atomic core of such insecticides. Again, the rest of the molecule is formed by various attachments to the free bonds (Figure 10-2).

362   CHEMICAL WARFARE AGAINST PESTS

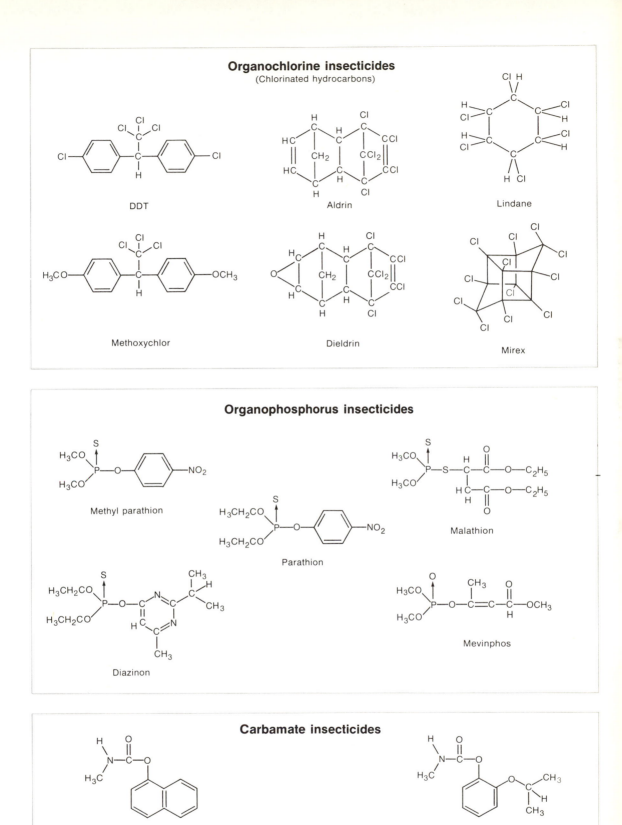

Figure 10-2. The structure of some important insecticides in the three major insecticide groups.

The carbamate insecticides, of which carbaryl or sevin is the outstanding member, all have the basic atomic core shown:

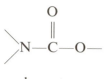

carbamate core

The name "carbamate" comes from carbamic acid, NH_2—$\overset{\overset{\textstyle O}{\|}}{C}$—OH, the simplest compound with the carbamate core.

A few organic insecticides overlap two groups. For instance DDVP or dichlorvos, the active component of Shell's popular No-Pest Strip, fits in both organochlorine and organophosphorus categories.

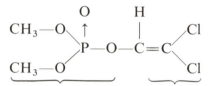

organophosphorus   organochlorine

Molecular Structure and Environmental Properties. The fore-going structural units, and to a lesser extent the structure of the remainder of the molecule, determine the toxicity, solubility, persistence, selectivity, and other relevant properties of the insecticide. All these functions are tied up in the molecular architecture and content. However, these molecules are rather complex, and we do not yet completely understand how they work. When they interact with the even more complex molecules of living systems, many questions are unanswered. What damage can insecticides do to the crucial molecules of human life? A few partial answers are known and will be discussed, but decades may pass before we unravel the chemical intricacies behind it all.

INSECTICIDE USES

The United States produces about 1 billion lb of pesticide yearly, half of the world's supply. Insecticides account for about half of the U.S. production, with herbicides and fungicides together making up the other half. Exports are considerable, about 20 percent of the total.

Agricultural Use. The major domestic use of insecticides is insect control on farm crops. Two thirds of these agricultural insecticides are applied to only three crops: cotton, corn, and apples.

Cotton by itself receives over one half of all agricultural insecticides—which tilts the geographical distribution of insecticide used heavily toward the south. The principal insect targets are the boll weevil (cotton), the bollworm (cotton), and the codling moth (apples), together recipients of almost half of the insecticides used on U.S. farms. Insecticides are generally applied a few pounds to the acre, most often by aerial spraying (photo).

For 20 years the major crop insecticide was DDT. The use of DDT in the United States has been largely curtailed by government restrictions, but widespread DDT spraying continues unabated in much of the world. Other crop insecticides are endrin (a close molecular relative of dieldrin) and methyl parathion (much like parathion), both used principally on cotton, and aldrin and heptachlor (related to aldrin), used especially on corn.

Area-wide Programs. Insecticides have been used abundantly in efforts to rid entire geographical regions of pests. Notable campaigns in the 1950's involved efforts to eradicate the gypsy moth in the East by spraying 3.5 million acres with DDT; the elm bark beatle, the carrier of Dutch elm disease, throughout the midwest and New England by DDT; the Mediterranean fruit fly by spraying a million acres in Florida with malathion; the spruce budworm over 15 million Canadian acres and many in the western U.S. by DDT; and the fire ant in the southeast using dieldrin and heptachlor on more than a million acres. For the most part these campaigns were failures, killing fish and wildlife and leading to pest resurgence rather than

A commercial spray plane releases an aerial swath of insecticide over cotton in Mississippi. (*Photo by Bern Keating/Black Star.*)

final eradication. As a result there are fewer such programs now. However, the fire ant, which easily survived the earlier attack, will be the target of a nine-state, 120-million-acre campaign involving mirex, if plans by the U.S. Department of Agriculture and state farm agencies materialize. The fire ant is a nuisance, but not a serious agricultural pest.

Domestic and Household Uses. Domestic insecticide uses, although they do not consume as much insecticide, lead to high exposure and may therefore have a disproportionate impact on human health. Various devices are used to discharge vapors directly into household air, including aerosol bombs, spray cans, automatic dispensers, impregnated furnace filters, and insecticide-laden hanging strips and shelf paper. Lindane and DDVP are popular components. Insecticides are used in the summer in some dry cleaning establishments to "moth proof" all clothes, giving the wearer direct contact exposure. New carpets and clothing often have been impregnated with long-lasting dieldrin. Aldrin, dieldrin, and other persistent insecticides are used in building foundations to control termites. Pest-control companies do a thriving business in ridding households of many pests by chemical means. We can see that only in exceptional cases will a person escape rather intimate contact with insecticides in everyday living, food sources not considered.

International. Foreign uses of insecticide are divided between agricultural and public health programs. Control of the malarial mosquito is the major public health application. For this purpose, AID and UNICEF export large amounts of insecticide, including a major fraction of the U.S. production of DDT. The success of malaria control hinges on the persistence and inexpensiveness of DDT. However, an increasing resistance of mosquitoes to DDT and other insecticides is making control more difficult.

World agriculture consumed insecticides in the yearly amounts shown in Table 10-1, according to a 1966 Shell survey.

Table 10-1. MAJOR WORLD CROPS AND THE AMOUNT OF INSECTICIDE USED IN THEIR PROTECTION

Crop	Organochlorines (lb)	Total Insecticide (lb)
Cotton	46,000,000	120,000,000
Rice	14,000,000	24,000,000
Other cereals	13,000,000	15,000,000
Vegetables	6,300,000	14,000,000
Potatoes	3,400,000	5,600,000
Sugar beets	2,700,000	4,800,000
Sugarcane	3,100,000	4,200,000
Tobacco	2,700,000	4,000,000
Oilseeds	2,900,000	3,800,000
Coffee	1,300,000	1,600,000

Factors in the Choice of Insecticide. Many factors determine the types of pesticides used for various purposes. Not least is the influence of commercial salesmen, unlicensed in this country, whose primary loyalty is to the manufacturer and to sales objectives and not to human health and ecology. Insect resistance is a big factor, and it was responsible in large part for the slowly declining use of DDT even before government restrictions. Toxicities to man, fish, wildlife, pollinating insects, and insect predators are considered, but not frequently enough. Persistence, effectiveness, ease of application, and cost are major factors. DDT's world-wide popularity is partly due to its low cost of about 20¢ per lb.

CHEMICAL BREAKDOWN

We may reasonably estimate that over 20 billion lb of insecticide have been applied to the earth in the last 25 years, at least 3 billion lb of it DDT. None of the insecticide has disappeared magically. Each and every molecule either exists in its original form somewhere on earth or has suffered a chemical breakdown to other molecules. Here and in the next section we trace the chemical and geographical fate of several insecticides.

The persistent pesticides—predominantly the chlorinated hydro-carbons—break down very slowly. Their long lifetime allows them to drift over the surface of the earth. DDT has been found in several species of distant antarctic wildlife. We shall discuss in the next section how the persistent insecticides are able to get so far from target.

The nonpersistent insecticides suffer chemical breakdown in a matter of weeks, as we noted earlier. While they can contaminate local environments, their early breakdown prevents worldwide circulation.

Chemical breakdown, like most aspects of insecticide chemistry, is not completely understood. However, some important features are known. The organophosphorus compounds and carbamates are perhaps simplest. They break down mainly by hydrolysis (reaction with water). Many large molecules, including the organophosphorus compounds and carbamates, are literally split apart by hydrolysis. Figure 10-3 (p. 368) shows the hydrolysis of parathion, in which a reaction occurs with three separate water (HOH) molecules. This hydrolysis reaction is summarized by the chemical equation

$$C_{10}H_{14}O_5PSN + 3 H_2O \rightarrow 2 C_2H_5OH + H_3PO_3S + C_6H_5O_3N$$

Four fragment molecules are produced, as shown, none toxic like parathion. The reaction proceeds fairly rapidly, which accounts for the nonpersistence of this and related insecticides. It proceeds most

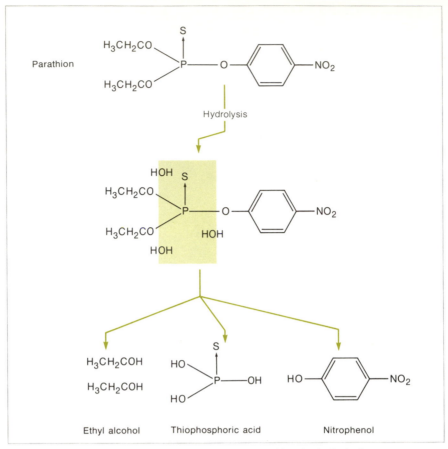

Figure 10-3. The breakdown of parathion by hydrolysis.

rapidly under moist conditions because H_2O is an essential ingredient of the reaction.

Parathion often undergoes a reduction (opposite of oxidation) reaction caused by soil yeasts. In this reaction the nitro group ($-NO_2$) is converted with the help of the yeast enzymes to an amino group ($-NH_2$). This different chemical path shows how complex insecticide breakdown is: one of several chemical routes may be taken depending on soil microorganisms, moisture, temperature, acidity, and so on.

The main breakdown product of aldrin is dieldrin (Figure 10-4). This reaction, too, is promoted by the enzymes in microorganisms. The conversion is clearly an oxidation reaction, which can be represented by the chemical equation

$$2 \text{ aldrin} + O_2 \longrightarrow 2 \text{ dieldrin}$$

Unfortunately, dieldrin is nearly as toxic as aldrin, more so with fish, so that biological damage can continue to occur after conversion. Dieldrin has caused one of the most serious insecticide residue problems, disappearing very slowly from the environment by additional breakdown steps.

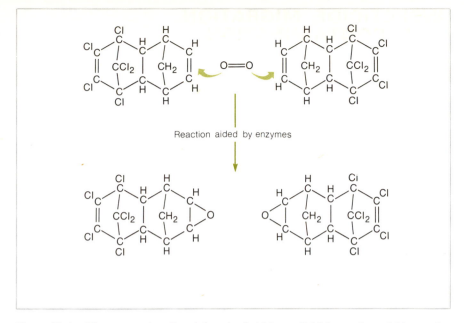

Figure 10-4. The conversion (breakdown) of aldrin to dieldrin, an insecticide nearly as toxic.

The most common first step in the breakdown of DDT yields primarily a substance called DDE:

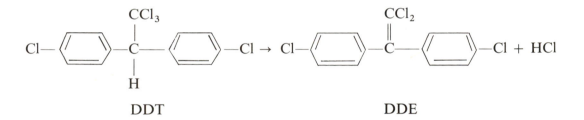

DDT   DDE

A similar substance, DDD, is formed in lesser amounts. DDE and DDD are themselves persistent insecticides. They remain in the environment, with some of the parent DDT, a long time before further breakdown. Reported levels of DDT often include the associated DDE and DDD.

Many reactions that would not otherwise occur are caused by the energy in sunlight. These are the *photochemical reactions.* Such reactions create the notorious photochemical smog, containing ozone (O_3), nitrogen dioxide (NO_2), and numerous organic molecules, which plagues cities like Los Angeles (Chapter 7). Photochemical reactions can also break down insecticides exposed to sunlight. This sometimes occurs with DDT, methoxychlor, aldrin, and perhaps others. The aldrin and dieldrin conversion just discussed can probably occur photochemically as well as through microorganism and plant enzymes.

INSECTICIDE MIGRATION
AND CONTAMINATION

Most insecticides are applied to green plants and soils. The target insects are poisoned as they eat or touch the toxic substances. However, during and after application the pesticide may contaminate air, water, foodstuffs, and wildlife, and it may migrate over great distances. The greater the persistence of the insecticide, the larger is its chance for such contamination and migration. Insecticide persistence, as we shall explain, invariably leads to unwanted accumulations in living creatures.

Insecticides may be carried through the atmosphere by air currents, by moving water, or by a chain of living organisms. We shall discuss these paths one at a time.

Air Contamination and Transport. Most insecticides have low volatility, or vapor pressure, which means they cannot vaporize to great concentrations in air. Large molecules such as insecticides, as we explained in Chapter 5, vaporize less readily than do small molecules such as water (H_2O) and sulfur dioxide (SO_2). Their large size causes them to cling tightly to solids and liquids. Insecticides themselves vary in volatility. Lindane, which Figure 10-2 shows to be a relatively small molecule, has a higher vapor pressure than do most insecticides. So does the DDVP used in No-Pest Strips. Enough vapor is present so that these insecticides can be relied on to kill insects in the surrounding air, without direct contact.

Despite the low volatility of most insecticides, the atmosphere is so enormous that it can absorb and transport large quantities in total. Aerial spraying encourages this transport. It is thought that most of the applied organochlorines finally enter the atmosphere, which at any one time may contain nearly 0.5 billion lb of insecticide worldwide.

The direct vaporization of insecticides is augmented by insecticides that cling to air-borne particles. Soil on which insecticides have been deposited, if whipped up by winds, can cause a significant fallout of insecticides on downwind locations (Figure 10-5).

The atmosphere is believed to constitute the main route by which remote antarctic seals, fish, and penguins have accumulated DDT. It is clear that insecticide contamination is not strictly a local problem.

Insecticides in Water. Just as insecticides do not reach high concentration in air, neither do they dissolve appreciably in water. Like most large, oil-like organic molecules, their solubility in polar water is very low. Here, too, there are individual variations, with lindane a hundred times more soluble than DDT. The organophosphorus compounds are more soluble than are the organochlorines, but they hydrolyze and thus disappear more rapidly.

Most water-borne insecticides are carried on small silt particles in rivers and lakes. The heavy silt loads washed into rivers by rain

Figure 10-5. Dust kicked up in a mammoth dust storm over Texas on January 25, 1965, left a deposit of pesticide fallout over a broad swath of the United States. [From J. M. Cohen and C. Pinkerton, *Organic Pesticides in the Environment* (Washington, D.C.: American Chemical Society, 1966), Chap. 13.]

water are mainly responsible for the massive fish kills which have occurred in this country.

Although insecticides are very dilute in unsilted water, rarely more than 2 ppb, living creatures in and near the water may contain in their tissues concentrations a million times higher. It is this *biological magnification* that makes pesticides a threat to the ecosystem and to human health.

Insecticides in Living Organisms. Insecticides have a considerable affinity for organic material, including living tissue, and they have a great affinity for the fatty tissues of animals. These oil-like insecticide molecules simply dissolve with great ease into the oil-like molecules of fats.

To begin with, then, plants absorb many insecticides from soil, water, or insecticide spray and transport them to leaves, stems, and fruits. Here they may be ingested by countless animals and man. The most dramatic accumulation of insecticides in humans and in predatory birds and animals results from the biological magnification of persistent insecticides in food chains.

In a typical *food chain* a plant is eaten by a small insect, which in turn is devoured by a predatory insect. The latter may then be eaten by a rodent, which subsequently is fed upon by an eagle. At each step in this chain, plant → insect → insect → rodent → eagle, the insecticide becomes more highly concentrated. This occurs as each creature collects the insecticide from food equal to many times its own body weight (Figure 10-6). Insecticides are thus particularly dangerous to meat eaters at the end of long food chains.

Examples of biological magnification abound. Oysters by themselves can concentrate insecticide from water 70,000 times. In a Lake Michigan food chain consisting of aquatic plants → insects → fish → herring gulls, DDT levels were increased 400,000 times

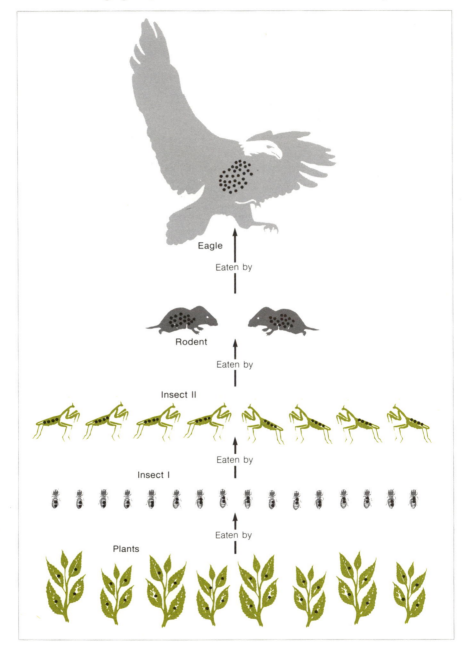

Figure 10-6. The increasing concentration of insecticide (shown by dots) with steps in the food chain. Predators such as eagles have suffered most because they have the highest insecticide concentration—much higher than in most insects themselves.

above those in bottom sediments, and much more relative to the surrounding water. The gulls ended with over 3000 ppm in their fat. This amount is nearly one part in three hundred, which is ex-extremely high for a biologically active substance such as DDT. Coho salmon from Lake Michigan, which form another food chain in which man occasionally is the last link, have been found to contain 10 to 20 ppm DDT.

It is believed that man gets his insecticides mainly through his foodstuffs, although home use and dust-laden air have been implicated too. Foods of animal origin contribute over half the total insecticide intake from food, as is reasonable because of the longer food chain. Fruits and vegetables, however, contribute significantly. At the beginning of this decade DDT, dieldrin, lindane, malathion, and heptachlor, in that order, were most abundant in foods. Along with DDT, dieldrin has caused considerable concern because it is both toxic and relatively abundant. The average person consumed about 0.005 milligrams (mg) per day of dieldrin, compared with 0.1 mg of DDT. With government action now in effect against DDT, the human consumption of DDT will drop. The drop will not be sudden because of the long persistence of millions of pounds of DDT already in the environment and because DDT is still being applied around the world.

With a steady supply from food, air, and water, man magnifies by one more step the levels of most insecticides. In the U.S., body fat has accumulated 5 to 20 ppm DDT and somewhat under 1 ppm dieldrin. Other countries show comparable amounts; even the remote Eskimo has 3 ppm DDT. We can estimate that human tissue, worldwide, contains about 500,000 lb of insecticide.

THE -CIDE OF INSECTICIDES

The suffix -cide (Latin of -cida) means "to kill." Insecticides are designed to kill insects, as pointed out earlier. Unfortunately, the molecular basis of life is so similar among bad insects, good insects (the honeybee and others), birds, and mammals that most insecticides have some negative influence throughout the community of life, including man.

Nerve Impulses. Most insecticides act as nerve poisons. By inducing continuous nerve action, different amounts can cause tremors, convulsions, and eventually death in insects, birds, or mammals.

The biochemical story is a fascinating one. A substance known as *acetylcholine* transmits nerve impulses from one nerve cell to another. After serving its function, acetylcholine is rapidly destroyed by the enzyme acetylcholinesterase (ACHE). Enzymes are highly specific biological catalysts (Chapters 4 and 5) that cause biochemical reactions to proceed at the proper rate, thus controlling the pace of life. They are giant protein molecules, each kind having its own chemical reaction to regulate. ACHE is the enzyme which controls the destruction of acetylcholine. This destruction occurs by the hydrolysis reaction:

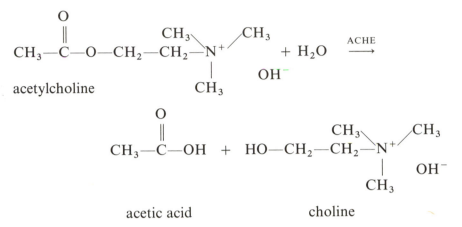

acetylcholine

acetic acid   choline

The giant ACHE molecule carries out its catalytic role by cradling the acetylcholine molecule in a close-fitting molecular cavity. The position of the cradled acetylcholine is such that the chemical bond to be broken in hydrolysis is close to a key hydroxyl group (—OH) on the enzyme (Figure 10-7). The electronegative oxygen atom of

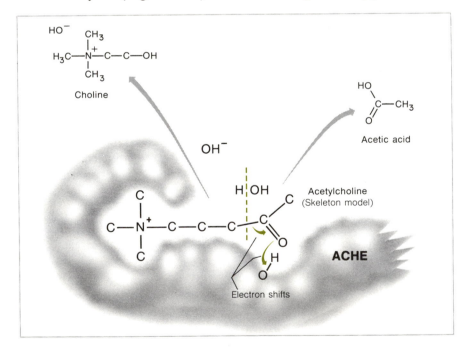

Figure 10-7. The giant enzyme, ACHE, cradles acetylcholine in a fixed position and uses a key hydroxyl group (constituting the so called "active site") to weaken the bond to be hydrolyzed in acetylcholine. This figure is merely a schematic one of a complex three-dimensional arrangement, but it illustrates why enzyme activity is so specific and is sometimes compared to the working of a lock and a key.

this hydroxyl group tugs at the electrons of the crucial chemical bond, weakening it. The weakened bond is hydrolyzed speedily; the broken fragments of acetylcholine depart; and ACHE is ready to go to work on fresh acetylcholine from the next nerve impulse.

Insecticides and Enzymes. Organophosphorus and carbamate molecules acquire their main toxic activity by attacking and inactivating ACHE.

$$\left.\begin{array}{c} \text{organophosphorus} \\ \text{or} \\ \text{carbamate} \end{array}\right\} + \text{ACHE} \longrightarrow \text{inactivated ACHE}$$

This inactivation occurs as the insecticide molecule bonds with the key hydroxyl group on ACHE, as shown in Figure 10-8. With the ACHE ineffective, acetylcholine accumulates and activates a stream of unwanted nerve impulses, leading to death.

Organochlorine compounds also work on the nervous system, but the mechanism is not entirely clear. It is thought that they dissolve in the oil-like membrane surrounding the nerve cell, there to interfere with the vital transport of sodium in and out of the cell.

The interplay of insecticides and enzymes does not end with nerve impulses. The liver produces, by complicated chemical re-

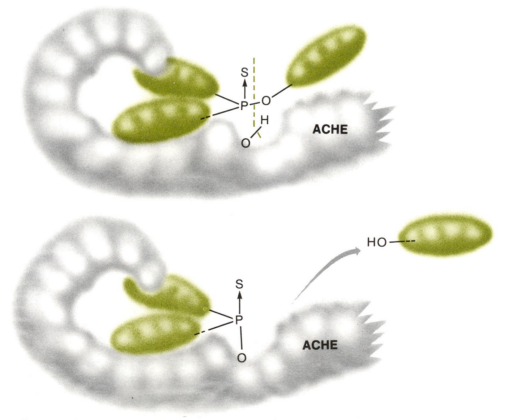

Figure 10-8. Organophosphorus insecticide molecules bind chemically with the key hydroxyl group on ACHE. With this, ACHE cannot cradle and hydrolyze acetylcholine molecules. This failure leads to death.

actions, many different enzymes which metabolize (break down) various molecules in the body. Insecticides somehow influence the production of these enzymes. The organochlorines stimulate enzyme production and organophosphorus molecules inhibit it. These effects upset normal metabolic activity. For instance, hormone levels can change because of abnormal hormone metabolism, thus influencing sexual and reproductive characteristics. This effect is behind the thinning of eggshells that has decimated bird populations around the world.

The liver enzymes have another role, which begins to show the intricacy of any chemistry involving living systems. They are the catalysts which break down toxic substances such as insecticides themselves. Malathion, for instance, is relatively nontoxic precisely because an enzyme destroys it rapidly before it can act. Other insecticides are destroyed at greater or lesser rates by these enzymes. Thus we see that one insecticide (call it insecticide A), by affecting the liver's production of enzymes, can affect the rate of breakdown of another insecticide (insecticide B). The resulting *synergism* of A and B can make the effect of two insecticides together totally unlike that of the insecticides acting individually. For instance, the organophosphorus compounds malathion and EPN together are 50 times more toxic than they are individually.

The effects of insecticides on one another (synergism) and on hormone levels are illustrated in Figure 10-9.

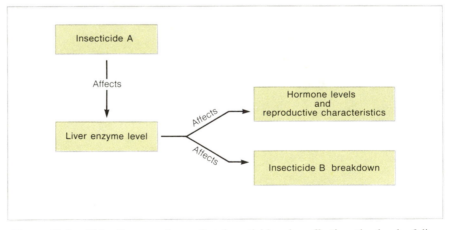

Figure 10-9. This diagram shows that insecticides, by affecting the level of liver enzymes, can alter normal hormone levels and can also affect the chemical breakdown of a second insecticide.

Enzymes are also largely behind the variable susceptibility of different species exposed to various insecticides. If a species has an enzyme which will rapidly destroy a given insecticide, that species will be less susceptible than will one whose enzyme arsenal does not include the appropriate enzymatic weapon (Figure 10-10). Insect resistance to insecticides may be explained on the same basis. The

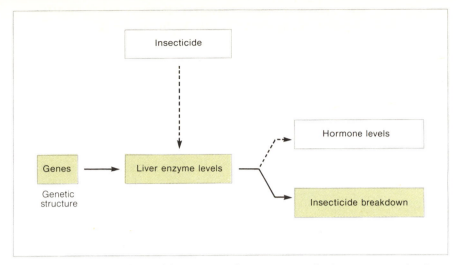

Figure 10-10. Genes have ultimate control over an organism's enzyme production. Genetically different organisms, then, have different abilities to break down and thus resist insecticides. Widespread insect resistance has been caused by the evolution of genetic structures that yield more insecticide-destroying liver enzymes.

individual insects within a species which survive a spraying are those with sufficient enzyme of the right type to combat the insecticide. With repeated spraying, the progeny of these genetically resistant few survive and multiply, providing eventually a resistant population of that species.

INSECTICIDES IN MAN AND ECOSYSTEM

Because insecticides have invaded earth, air, and water universally, their effect on the living world is considerable. However, long-term consequences are still unclear. Most insecticides have been around for less than a quarter century, and the majority of studies have covered only a small fraction of that time. It is now apparent that there are wide ecological disturbances, but the long-term effects on human health are uncertain. Here we shall try to understand the major consequences of insecticides in the living world.

Insecticides in Man. A human population is difficult to study because it cannot be controlled and it cannot be methodically poisoned. Our meager knowledge about the total effects of pesticides comes from very limited human experimentation, animal experiments, and the erratic but universal exposure of humans to insecticides in the environment.

DDT is most common and has been most studied. The enthusiasm of its initial acceptance stemmed in part from its relatively low direct toxicity to man and other mammals. In Table 10-2 we compare the acute (short-term) toxicity of some common insecticides to

Table 10-2. THE LD_{50} VALUES OF SOME COMMON INSECTICIDES FOR MAMMALS (mg INSECTICIDE/kg BODY WEIGHT)

Organochlorine		Organophosphorus		Carbamate	
DDT	200	Parathion	8	Carbaryl	500
Methoxychlor	6000	Methyl parathion	20		
Aldrin	40	Malathion	1000		
Dieldrin	50	Diazinon	80		
Lindane	100	Mevinphos	6		
		DDVP	60		

mammals. Toxicity is measured by an LD_{50} index, the dose quickly lethal to 50 percent of a test population. Methoxychlor and malathion are particularly low in toxicity (as indicated by their high LD_{50} values) because liver enzymes rapidly destroy them.

Although humans have similar enzyme systems, the extrapolation of the LD_{50} values for mammals to human populations is uncertain, but comprises the best information we have on acute toxicity. Some inadvertent human experiments confirm the chart's major features. Parathion is indeed very poisonous, and it has killed hundreds of people accidentally. Parathion-contaminated flour killed 80 people in an accident in Colombia in 1967. DDT is much less toxic, comparable to aspirin for a single dose. Workers in DDT plants have accumulated up to 1000 ppm in their fatty tissues with no apparent ill effect. Short-term human experiments with high DDT intake have shown similar results. However, the long-term effects of constant exposure cannot be predicted by short-term experiments. Some recent results are disquieting, and they show how little has really been known through these many years of relentless insecticide spraying.

First, many insecticides, including DDT, do alter the output of liver enzymes in man. The effect is generally small at normal exposures, but some individuals may be unusually susceptible to such disturbances.

Second, organochlorine concentrations are unusually high in victims of liver disease, brain disease, and cancer. This fact does not, of course, prove a cause-and-effect relationship, but only suggests that one may exist.

Third, it was recently discovered that high DDT doses cause cancer in mice. Other insecticides found to be carcinogenic (cancer producing) are aldrin, dieldrin, mirex, and heptachlor. Whether they cause cancer in man at normal exposures is not known.

Finally, valid questions have been raised about irreversible damage to the gene pool of mankind, a misfortune that would affect generations to come.

In sum, long-term effects are not known. People worldwide are receiving lifelong exposure with uncertain consequences. Increasing moderation in insecticide usage and enlarged research to show insecticide effects should be immediate goals for human health.

Insecticides in the Ecosystem. It is widely agreed that insecticides play havoc with ecological systems. Many biologists are convinced that there must be diverse ecosystems for the biological health of the world. The widely heralded ecological crisis reflects a growing public concern about the gross disturbance of the balance of nature.

Persistent insecticides are perhaps the worst polluters of the global ecosystem. They are applied widely and abundantly, and they concentrate and remain in living systems. By contrast, most pollutants are concentrated near cities, and their threat to the ecosystem is more local.

Insecticides are usually intended for specific insect pests. However, many forms of life are destroyed by their action, from pollinating insects to birds. As a result there are fewer species and, by ecological law, there is greater instability in the ecosystem.

Predator insects which normally control the population of unwanted insects are hardest hit by insecticides because of biological magnification in their food chain. As predators are removed, population outbreaks of the unchecked pest may occur. Quite often minor pests, with the natural predators gone, erupt to the status of major pests. Other ecological complications, too numerous to mention here, occur frequently.

Persistent insecticides move up the food chain, becoming increasingly concentrated. Damage can occur at any link in this chain. Phytoplankton, at the bottom of the chain and the ultimate source of food for most ocean life, are adversely affected by a few parts per billion DDT. However, the greatest accumulations and the most serious damage occur at the top of the food chain, with predatory birds, fish, and mammals.

Birds have been most seriously affected. Bird populations have been reduced or locally eliminated in many parts of the world. In this country, for instance, the bald eagle no longer nests in the northeastern United States. The peregrine falcon (duck hawk) has been exterminated in the East and severely reduced elsewhere in North America, as shown in Figure 10-11. Earlier, robin populations in the midwest were decimated by DDT used in the multistate campaign against Dutch elm disease. The DDT on fallen elm leaves was passed to robins in lethal amounts through the common earthworm, a major link in the robin's food chain.

Most bird damage is not caused by direct poisoning. Rather, sublethal amounts of organochlorines cause severe changes in the reproductive systems. Worst, they lead to thin-shelled eggs in many species, including the bald eagle. These fragile eggs do not hatch successfully.

It is thought that the thinning of eggshells is caused by the increased production of liver enzymes, as noted earlier. These enzymes have many functions, including the metabolism (chemical break-

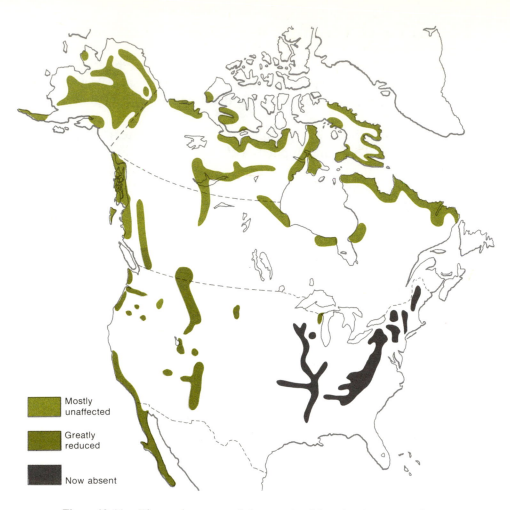

Mostly unaffected

Greatly reduced

Now absent

Figure 10-11. The nesting range of the peregrine falcon has been severely reduced as a result of widespread insecticide contamination.

down) of the female sex hormone, estrogen. Increased enzyme production lowers the estrogen concentration. Estrogen has a vital role in reproduction in birds and mammals, including humans. In birds, estrogen controls the transfer of calcium from bone to egg-shell. When estrogen levels are reduced, thin-shelled, unhatchable eggs result.

DDT is the principal but not sole villain in ecological damage. The ban on DDT in this country will reduce this damage. However, DDT will remain an ecological contaminant for years beyond its withdrawal. First, there is a worldwide circulation of this substance, as we have noted, which brings DDT from its source to distant parts of the world. Second, DDT presently in the food chain will only slowly be eliminated. It is believed by some scientists that DDT concentrations at the top of the food chain could continue to rise for years after a general halt in the application of DDT.

Insect Resistance. Over two hundred pest insects have developed a resistance to one or more of the insecticides. They are genetic variants of the parent insect population. Genes control enzyme production, as shown in Figure 10-10. Resistant insects have special genes which promote the manufacture of enzymes to chemically alter and remove the insecticide from the insect body.

Ridding the body of foreign chemicals is an age-old trick of living systems. If the molecule is oil-like and resides in the fatty tissues, as most insecticides do, a polar (water-loving) group may be attached through enzyme action. The reaction below is typical, and it may occur with some of the insecticides that contain benzene rings.

$$2\,R-\!\!\left\langle\!\!\begin{array}{c}\end{array}\!\!\right\rangle + O_2 \rightarrow 2\,R-\!\!\left\langle\!\!\begin{array}{c}\end{array}\!\!\right\rangle-OH$$

<table>
<tr><td>oil-like molecule,
insoluble in water</td><td>now slightly
water soluble</td></tr>
</table>

With the hydroxyl (—OH) group attached, the molecule is more water soluble and can therefore enter the blood, which is mostly water, and be carried off and eliminated. Other polar groups sometimes added by enzymes are hydrogen sulfate ($-OSO_3H$), amino ($-NH_2$), and various sugarlike groups.

Some enzymes work by hydrolyzing the foreign molecule, a process we showed earlier (Figure 10-3) for parathion. Such is the fate of most organophosphorus compounds in living systems.

An enzyme which can modify and remove one toxic molecule often can modify and remove related molecules by the same chemical mechanism. Thus resistance to several organochlorine compounds often develops simultaneously. Likewise insects resistant to one organophosphorus compound very frequently are resistant to the others. In this way insects can develop resistance to entire classes of insecticides, leaving few effective controls. For instance, some populations of bollworms and spider mites, both cotton pests, can no longer be controlled with available insecticides. Bollworm resistance to DDT alone increased in one area 30,000 times in 5 years. Resistance to other insecticides increased simultaneously but less dramatically.

Insect resistance has grown to a problem of major proportions in the public health field as well as in world agriculture. Organochlorines and organophosphorus compounds are rarely used now for mosquito control in the south because of resistance. It is reported that some mosquitoes in California can no longer be controlled by chemical insecticides. More significantly, the worldwide malaria program is starting to show fractures. Mosquitoes in Africa and Central America, particularly, are developing significant resistance to both organochlorine and organophosphorus compounds.

It has been said that we are losing the game of chemical warfare with insects. True or not, there are some discouraging advantages on their side. A short life cycle makes possible the quick development of genetic resistance. This resistance covers whole classes of insecticides at once, making it increasingly difficult to synthesize effective new

Spraying over the Congo. The incessant rain of insecticides used to control malarial mosquitoes in the tropics has begun to backfire, generating resistant mosquitoes. (*Photo courtesy United Nations.*)

compounds at a pace fast enough to keep ahead. Meanwhile, the human adaptation to this new chemical environment is virtually nonexistent. First, the generation-to-generation turnover is too slow in man. Second, human mortality due to insecticides is, fortunately, too low to provide selectivity, which is necessary for such evolutionary changes.

Man, compared to insects, is extremely sluggish in adaptability. Any gross alteration in world environment—chemical, thermal, or radioactive—will certainly be more readily survived by the lowly insect than by man.

OPTIONS TO CHEMICAL INSECT CONTROL

Few would dispute the record of human benefits from chemical insecticides. Malaria programs are a prime example, with a substantial fraction of world malaria now under control. However, human welfare does not require massive and universal applications of insecticide to every crawling creature. This approach serves only to damage the ecology, increase insect resistance, and endanger human health. We should start off by being much more selective and restrained in using insecticides.

In the long run, the present method is doomed by insect resistance. It would be better to seek an alternative now than under crisis conditions later on.

Sound approaches to insect control will be relatively expensive, measured in narrow economic terms. They may be cheap if we consider long-term ecological and health implications.

Insect Control in Transition.   Insect control is changing rapidly. For three decades the battle against insects was led by DDT and other persistent, cumulative, widely toxic chemicals. Now DDT is restricted in several European countries and the United States. The emerging trends in insect control avoid some of the environmental snags of DDT and its organochlorine relatives. Nonetheless, many of the proposed substitutes still cause environmental problems. They also encounter serious economic hurdles. Here we shall discuss some of the promises and pitfalls of various options to the departing king— DDT.

Chemical Control.   Chemists are now so skillful at making new molecules that many chemical options exist. We have discussed already chemicals that are nonpersistent, that degrade in a few weeks' time. Nonpersistent insecticides of the organophosphorus and organochlorine types are being used more and more. Methyl parathion, an organophosphorus insecticide (Figure 10-2), is replacing DDT for many uses. These nonpersistent chemicals, because of their rapid breakdown, do not accumulate in living organisms and do not magnify in food chains. But most nonpersistent insecticides are highly toxic to living organisms, including man. These insecticides must be applied frequently, and they quickly generate insect resistance. Still, they are a marked improvement for the environment.

Ideally, insecticides should be *selective*—they should destroy only a single target insect. Here the main hurdle is economic. The research, development, and testing of a marketable insecticide costs $4 million. Chemical companies have steered away from investing this much money in a narrow-market chemical good only for one kind of insect. There is no chance of striking another profit-maker like DDT in selective insecticides, and the investment is simply not worthwhile. Perhaps this area should receive strong government support.

Another promising approach involves the use of chemicals that are natural to insects. Synthetic versions of these chemicals, applied at certain times and places, can destroy insects by interference with natural functioning. For instance, a substance called *juvenile hormone* is needed by immature insects for normal growth. However, it must disappear before they can become adults. If juvenile hormone is applied artificially as an insecticide, it arrests the insect's adult development and propagation without much affecting noninsect species. Insect resistance to juvenile hormones will probably be slow in emerging: insects must be careful in combatting a chemical that is essential to their own life cycle.

In a similar vein, very small amounts of *synthetic sex attractants* can draw insects great distances to traps. The Mediterranean fruit fly was eradicated in Florida by this method. It may also lead to control of the gypsy moth and the elm bark beetle, where the earlier DDT campaigns failed.

Biological Control.   Methods based on biological insect destruction provide yet another direction with minimal side effects. Here one releases and encourages predators, parasites, or pathogens (disease microorganisms) that prey on the pest species. Already over 20 successful applications of such control have been made in this country. The earliest was 80 years ago when the vedalia, a lady beetle, was imported from Australia to prey on a citrus scale insect. Control was quickly gained over this ravaging pest that had seriously threatened the California citrus industry. The sequel to this success story is also instructive. With the introduction of DDT to the citrus groves, the vedalia population was decimated. The subsequent resurgence of the citrus scale caused damage worse than any since vedalia was imported.

Agricultural Control.   A new, more flexible approach to agriculture could alleviate much of the present problem. Vast acres of single crops (monocultures) are in accord with the age of automation, but they bring about maximum ecological instability and thus susceptibility to pests. With some increase in costs, interplanting and crop rotation can help keep pest insects off balance. Also, crop planting can be timed in some cases to miss peak seasonal infestations. Thoughtful management of water, fertilizers, and soil can all be helpful in discouraging insect pests. Crops that are resistant to insects are under active development.

Integrated Control.   Maximum flexibility is provided by *integrated control*. Here one manipulates the biological, agricultural, and chemical variables all at once to gain insect control with minimum side effects. This approach requires a level of ecological sophistication that is rarely used at present.

It is doubtful that agriculture would be irreparably hurt by a careful and flexible reduction in insecticides. Before 1945 the toll of insect devastation averaged little more than 10 percent of farm crops. It is still very near that figure today. Undoubtedly, many modern agricultural practices would need modification to maintain this low level with reduced insecticide usage.

The years ahead will profoundly challenge our wisdom in dealing with insects, for an exploding human population will demand more food—less insect damage to crops—but a fragile environment will not tolerate endless applications of cheap, persistent insecticides to gain that food. We must be quick in developing suitable options, or we may suddenly find ourselves residents of a world that is both hungry and ecologically sterile.

Exercises

1.   Classify each of the following insecticides as (a) organochlorine, (b) organophosphorus, or (c) carbamate.

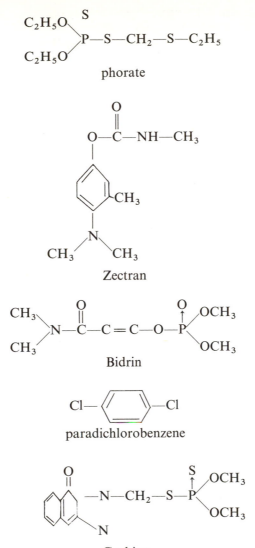

phorate

Zectran

Bidrin

paradichlorobenzene

Guthion

2. In Chapter 7 we discussed the merits of taxing sulfur dioxide (SO_2), a severe air pollutant. Discuss the good and bad points of taxing DDT—which also causes environmental damage—instead of applying an outright ban on this insecticide.

3. Using the structural formulas shown in the text, calculate the molecular weights of DDT, methoxychlor, and aldrin.

4. Calculate the molecular weights of lindane, malathion, and DDVP.

5. If the total pesticide produced in the United States in 2 years—2 billion lb—were put in the oceans and stirred, what would the pesticide concentration of the 1.5×10^{18} tons of ocean water be? Would this level endanger the phytoplankton, whose photosynthesis is reduced by 10 parts per billion (ppb) of some pesticides? (The productive coastal regions—as well as inland lakes—are more limited in extent and thus much more susceptible to insecticide damage.)

6. Predict and draw the structure of the breakdown product (molecules) resulting from the hydrolysis of mevinphos. (Hint: The hydrolysis process is much like that shown in Figure 10-3 for parathion.)

7. One hundred lb of aldrin sprayed on a corn field is eventually converted to dieldrin through the reaction shown in Figure 10-4. How many pounds of dieldrin are thus created?

8. Coho salmon from Lake Michigan commonly contained 5 to 20 ppm DDT in 1970, before DDT was banned from Michigan farm lands. How many milligrams of DDT would you consume eating 0.5 kg of Coho salmon at 10 ppm DDT?

9. Fatty tissues constitute about 20 percent of a person's weight. Assuming 10 ppm DDT storage in your fatty tissues, estimate your own allotment of DDT in milligrams. This amount can be visualized by comparison with an aspirin tablet, which weighs about 300 mg.

10. How many mg of DDT are there in a 70 kg Eskimo whose fatty tissues, constituting 20 percent of his body weight, contain 3 ppm DDT?

11. The water of a Long Island salt marsh was estimated to have 0.00005 ppm of DDT and its DDE and DDD relatives [George F. Woodwell *et al.*, *Science 156*: 821 (1967)]. Plankton had an insecticide concentration 800 times greater than the concentration in the water. Shrimp contained a concentration four times higher than that of the plankton. Needlefish contained 12 times more DDT and its relatives than did shrimp, and ring-billed gulls were found to have concentrations 35 times greater than those of the needlefish. What was the total magnification in insecticide concentration in going from water to ring-billed gulls?

12. Shrimp mortality is 50 percent in water containing 0.4 ppb (parts per billion) lindane, endrin, or heptachlor. One acre of water 10 ft deep (a total of 10 acre-ft) weighs about 13,500 tons or 27 million lb. How many pounds of one of the above insecticides dissolved in the water would create this toxic condition for shrimp? Could 1/2 lb/acre applied over such water damage the shrimp?

13. Use the LD_{50} table to estimate the lethal dose of parathion in g for a person weighing 60 kg (132 lb).

14. Estimate the dose of dieldrin ($LD_{50} = 50$) in milligrams that would be lethal to a person weighing 50 kg (110 lb).

15. Juvenile hormone analogs are so potent to insects that a dose of 1 nanogram (10^{-9} g) will disable a 0.02 g insect. What is the disabling dose in mg/kg? Compare it with the LD_{50} value for several conventional insecticides to obtain relative potencies.

Glossary

Acetylcholine   substance that transmits nerve impulses from one nerve cell to another.

Biological magnification   repeated concentration of a substance in food by each organism in a food chain.

Carbamate insecticides   insecticides that contain the atomic core

$$\text{N}-\overset{\overset{\text{O}}{\|}}{\text{C}}-\text{O}-.$$

Chlorinated hydrocarbon insecticides   insecticides containing chlorine atoms bonded directly to carbon.

Food chain   sequence in which a plant is eaten by an animal, which in turn is eaten by a predator, and so on.

Fungicide   a substance that kills fungi.

Herbicide   a substance that kills plants.

Hydrolysis   reaction with water.

Insecticide   a substance that kills insects.

Integrated control   manipulation of biological, agricultural and chemical variables all at once to gain insect control with minimum side effects.

Juvenile hormone   substance needed for normal growth in immature insects, but which prevents adult development and propagation.

Nonpersistent insecticide: insecticide that remains active from 1 to 12 weeks before degrading.

Organochlorine insecticides   chlorinated hydrocarbon insecticides.

Organophosphorus insecticides   insecticides containing phosphorus and carbon, usually having a core of P bonded to three O atoms and one S atom.

Permanent insecticides   insecticides that remain in soil indefinitely, including arsenic and lead compounds.

Persistent insecticides   insecticides that retain their activity from 2 to 20 or more years.

Pesticide   a substance that is used to kill pests.

Photochemical reactions   reactions caused or catalyzed by light energy.

Rodenticide   a substance that kills rodents.

Selective   affecting only one or several target organism.

Synergism   cooperative action of two agents so that the total effect is greater than the sum of the effects of each one alone.

Synthetic sex attractants man-made substances that make use of mating responses to attract insects to traps.

Additional Reading

Starred selections are most suitable for the nontechnical reader.

*Carson, Rachel, *Silent Spring*. Boston: Houghton Mifflin, 1962.

Edwards, Clive A., *Persistent Pesticides in the Environment*. Cleveland, Ohio: CRC Press, 1970.

*Giddings, J. Calvin, and Manus B. Monroe, eds., *Our Chemical Environment*. San Francisco: Canfield, 1972.

Gould, Robert F., ed., *Organic Pesticides in the Environment*, Advances in Chemistry Series, 60. Washington, D.C.: American Chemical Society, 1966.

*Graham, Frank, Jr., *Since Silent Spring*. Boston: Houghton Mifflin, 1970.

O'Brien, R. D., *Insecticides, Action and Metabolism*. New York: Academic Press, 1967.

Report of the Secretary's Commission on Pesticides and Their Relationship to Environmental Health, Parts I and II, U.S. Department of Health, Education, and Welfare, December 1969.

*Whiteside, Thomas, *What Are Our Herbicides Doing to Us? Defoliation*. New York: Ballantine/Friends of the Earth, 1970.

Periodic table of elements:

1 H
2 He
3 Li · 4 Be
5 B · 6 C · 7 N · 8 O · 9 F · 10 Ne
11 Na · 12 Mg
13 Al · 14 Si · 15 P · 16 S · 17 Cl · 18 Ar
19 K · 20 Ca · 21 Sc · 22 Ti · 23 V · 24 Cr · 25 Mn · 26 Fe · 27 Co · 28 Ni · 29 Cu · 30 Zn · 31 Ga · 32 Ge · 33 As · 34 Se · 35 Br · 36 Kr
37 Rb · 38 Sr · 39 Y · 40 Zr · 41 Nb · 42 Mo · 43 Tc · 44 Ru · 45 Rh · 46 Pd · 47 Ag · 48 Cd · 49 In · 50 Sn · 51 Sb · 52 Te · 53 I · 54 Xe
55 Cs · 56 Ba · 57 La · 72 Hf · 73 Ta · 74 W · 75 Re · 76 Os · 77 Ir · 78 Pt · 79 Au · 80 Hg · 81 Tl · 82 Pb · 83 Bi · 84 Po · 85 At · 86 Rn
87 Fr · 88 Ra · 89 Ac · 104 Ku · 105 Ha · 106 · 107

58 Ce · 59 Pr · 60 Nd · 61 Pm · 62 Sm · 63 Eu · 64 Gd · 65 Tb · 66 Dy · 67 Ho · 68 Er · 69 Tm · 70 Yb · 71 Lu
90 Th · 91 Pa · 92 U · 93 Np · 94 Pu · 95 Am · 96 Cm · 97 Bk · 98 Cf · 99 Es · 100 Fm · 101 Md · 102 No · 103 Lr

HYDROGEN, KRYPTON, STRONTIUM, IODINE, CESIUM, URANIUM, PLUTONIUM
These seven elements, shown in white boxes, are of overriding importance to our nuclear-age environment. Eleven others, shown in light green boxes, are of considerable significance.

11 NUCLEAR ENERGY, RADIOACTIVITY, AND ENVIRONMENT

INTRODUCTION

The awesome force of nuclear weapons, the promise of abundant electric power, and the threat of radioactive fallout are three related environmental factors of the nuclear age destined to influence man's future. Each factor is enmeshed in controversy because each is associated directly or indirectly with risks and benefits on an enormous scale. The scientific basis of these factors and the controversies surrounding them will be explored in this chapter.

The importance of the nuclear age derives ultimately from the properties of the tiny nucleus of the atom. To this time we have dealt only with chemical properties of atoms, which reflect the activity of electrons in outer shells. Deep in the center of the atom the positive nucleus is far removed from chemical events. Occasionally this nucleus undergoes a major rearrangement in which energetic nuclear particles and rays are thrown off. This cast-off radiation disrupts the outer electrons of nearby atoms and in this way causes profound chemical changes. When these chemical changes involve atoms in the molecules of living systems, the machinery of life is damaged or destroyed.

The emission of nuclear radiation is called *radioactivity*. Only a limited number of elements are normally *radioactive*—inclined to spew out nuclear particles and rays. Natural radioactive events occur only infrequently in the lifetime of but a few nuclei, so that the resultant chemical effects are ordinarily brief and isolated. For this reason we can ignore the chemical effects of radioactivity throughout most of chemistry. However, radioactivity is induced artificially within nuclear reactors and nuclear weapons, causing in some cases sufficiently high concentration of disintegrating nuclei that chemical and biological effects are indeed significant.

Foremost among the basic properties of nuclei that promise to shape human events is their high energy. The energies that bind protons and neutrons together in a nucleus are a million times greater than the chemical-bond energies that hold atoms together in molecules. Only recently has man learned how to tap this enormous storehouse of energy.

The heralded nuclear age is the child of this high-intensity energy. This is most apparent in nuclear weapons, where some of the vast energy of nuclei can be released suddenly to create an explosion of awesome proportions. It is also apparent in the nuclear-power industry, where a single pound of uranium releases nuclear energy to create as much electric power as 1400 tons of coal.

The energy basis of the nuclear age is less obvious when we come to the topic of radioactivity and radioactive pollution. It is less obvious, but no less significant. In radioactivity, fragments are hurled from the nucleus with tremendous energies, powered by the energy of the nucleus. Radioactive contamination is feared precisely because it involves high energy nuclear radiation that can literally rip molecules apart. Recall from Chapter 2 that molecules cannot remain intact when bombarded by high-energy electromagnetic radiation. By the same token, high-energy fragments from a nucleus will alter the bonding arrangements of atoms in molecules. Since life depends on the integrity of some rather large and intricate molecules, such as DNA and the enzymes, it is highly susceptible to damage by energetic radioactive emissions. We shall examine the nature of this damage as well as the possible sources of radioactivity which may produce it.

THE NONENVIRONMENTAL NUCLEUS

This chapter will focus on the effect of nuclear energy and radioactivity on the environment. Both pluses and minuses to human welfare are obvious. However, a proper perspective of the nuclear age must take into account the significant nonenvironmental uses of the nucleus.

Radioactive materials have contributed enormously to knowledge in many fields. Their nuclear disintegration, picked up by sensitive instruments, allows them to be tracked, so that new knowledge on molecular pathways is unfolded. Advances in medicine, chemistry, biology, agriculture, ecology, archeology, geology, engineering, and many other fields have been truly spectacular.

Radioactivity, we shall learn later, can cause cancer under some conditions. Very large doses, however, destroy cancer cells. Radioactive emitters are therefore frequently inserted into tumors with moderate success in controlling cancer. Radioactive substances are also important in medical diagnosis.

Among many other uses of radioactivity, one deserves special mention. Radiation has been used to sterilize male populations of certain insects which, when released, mate with females to produce infertile eggs. Insect control by this method has been particularly effective for a major pest in Florida, the screwworm, doing away with the need for ecologically damaging insecticides. In this case an "indoor" use of radioactivity has had ecological impact. However, this impact is minor on the scale of nuclear power, nuclear weapons, and environmental radioactivity, subjects for which we shall now lay a scientific foundation.

ISOTOPES

The nucleus of an atom, as we learned in Chapter 2, is small but massive. Its diameter is 100,000 times smaller than that of the atom within which it is centered. Yet this mere speck in the already small world of atoms contains well over 99.9 percent of the mass of the atom. This is so because the nucleus consists of protons and neutrons, heaviest of the common fundamental particles of nature. The planetary electrons are light by comparison and contribute virtually nothing to the mass of the atom.

Recall further from Chapter 2 that each element has a fixed number of protons in the atomic nucleus; this is the *atomic number*, Z. However, nuclei with Z protons may have neutron contents that vary from nucleus to nucleus. These different neutron combinations constitute the *isotopes* of the element. For instance, one natural isotope of uranium has 143 neutrons and another has 146 neutrons. They are both isotopes of the same element, uranium, because each has 92 protons, the atomic number of uranium.

The chemical properties of different isotopes of an element are virtually identical. Chemical properties, recall, are determined by the number of electrons in the electron orbitals of atoms. The orbital electrons and nuclear protons of an atom exist in equal numbers. Therefore once the element is determined by the number of nuclear protons, Z, the electron content and hence chemical properties are unvarying whether neutrons are added or not.

By contrast, nuclear properties, particularly radioactivity, depend on the number of neutrons as well as protons in the nucleus. Therefore in studying all nuclear phenomena, we must specify the particular isotopes involved.

Isotopes of an element are distinguished from one another by their *mass number, A*. Mass number is the sum of the number of protons and neutrons in the nucleus. This number varies up and down depending on whether the nucleus has many or few neutrons. Consider again the two isotopes of uranium, with 143 and 146 neutrons, respectively. Since each has 92 protons, the mass numbers are

isotope 1

isotope 2

number of protons $= 92$

number of neutrons $= 143$

mass number, $A = 235$

number of protons $= 92$

number of neutrons $= 146$

mass number, $A = 238$

The popular designation for these isotopes incorporates the mass numbers: uranium-235 (U-235) and uranium-238 (U-238).[1] These symbols are recognized worldwide because uranium in these two isotopic forms is truly the mother element of the nuclear age. U-235 was the explosive ingredient in the first weapon of nuclear warfare, dropped over Hiroshima on a fateful August day in 1945. U-235 is also the present energy source of nuclear power. U-238 is the isotope used to synthesize plutonium-239, the principal ingredient of nuclear weapons around the world.

In the scientific literature, isotope symbols include both mass number and atomic number. They appear as superscripts and subscripts, as follows:

$$\underset{\text{Atomic No.}}{\overset{\text{Mass No.}}{}}\mathbf{X}$$

By this designation, uranium-235 appears as

$$^{235}_{92}\mathbf{U}$$

It may seem redundant to show both the symbol of the element (U) and its atomic number (92). Nonetheless, when we begin to examine the nuclear reactions which produce the significant isotopes, we shall

---

[1] If the mass number A and atomic number Z of an isotope are given, the number of neutrons can be obtained by subtraction: number of neutrons $= A - Z$. The reason is that A represents total protons and neutrons, and if we subtract the Z protons, we are left with neutrons. For instance, with U-235, the number of neutrons $= 235 - 92 = 143$, the same number as given above for U-235. (We could deduce the same equation from the definition of A, namely $A =$ number of neutrons $+ Z$, which becomes $A - Z =$ number of neutrons, when Z is subtracted from each side.)

be happy to have the atomic number with the isotope symbol in keeping track of protons and thus of nuclear charge.

Most of the 90 elements found on earth (two others are found in microscopic amounts) have two or more natural isotopes. Tin has the greatest number, 10 in all, including $^{112}_{50}Sn$, $^{114}_{50}Sn$, $^{115}_{50}Sn$, $^{116}_{50}Sn$, $^{117}_{50}Sn$, $^{118}_{50}Sn$, $^{119}_{50}Sn$, $^{120}_{50}Sn$, $^{122}_{50}Sn$, and $^{124}_{50}Sn$. Tin is followed by xenon and cadmium, with nine and eight isotopes, respectively. Only 20 or so of the elements are found in a single isotopic form. Altogether there are nearly 300 natural isotopes for these 90 elements, an average of just over three each.

Isotopes of an element usually occur in nature in grossly unequal amounts. For instance, uranium atoms from natural ore deposits are an overwhelming 99.3 percent $^{238}_{92}U$ and a mere 0.7 percent $^{235}_{92}U$. A third isotope, $^{234}_{92}U$, is present at the negligible level of 0.005 percent. These relative amounts are called the *abundance ratios*. Some abundance ratios, most with nuclear-age significance, are listed in Table 11-1.

Table 11-1. NATURAL ABUNDANCE RATIO OF SOME ISOTOPES WITH NUCLEAR-AGE SIGNIFICANCE

Element	Abundance Ratio[a]		Element	Abundance Ratio[a]	
Hydrogen (H)	99.98%	^1_1H	Oxygen (O)	99.8%	$^{16}_8O$
	0.02%	^2_1H		0.2%	$^{18}_8O$
	(trace)	^3_1H		(trace)	$^{17}_8O$
Lithium (Li)	92.6%	^7_3Li	Thorium (Th)	100%	$^{232}_{90}Th$
	7.4%	^6_3Li			
			Uranium (U)	99.3%	$^{238}_{92}U$
Carbon (C)	98.9%	$^{12}_6C$		0.7%	$^{235}_{92}U$
	1.1%	$^{13}_6C$		(trace)	$^{234}_{92}U$
	(trace)	$^{14}_6C$			

[a]The abundance ratio is the percentage of the nuclei of the element having the specified isotopic form.

Many new isotopes have been made since man learned how to alter atomic nuclei. Over a thousand isotopes not previously present on earth have been synthesized in scientific laboratories. A great number of these are significant in chemical, biological, and medical research.

Many isotopes are radioactive and many are not. The isotopes that occur naturally tend to be *stable* (not radioactive): otherwise most of them would have disintegrated and disappeared long ago. However, there are exceptions, as we shall note soon. The man-made isotopes are generally radioactive. Isotopes that are radioactive are called *radioisotopes*. Many radioisotopes are so unstable that they exist for only a fraction of a second after they are made.

For a synthetic isotope to achieve environmental significance, it must (1) be radioactive, (2) have a sufficient lifetime to be dispersed in the environment, (3) be an abundant product of some human activity in the nuclear field, and (4) have some chemical route to man. Of the thousand or more man-made isotopes, only a handful meet all these conditions and thus achieve environmental prominence. The major intruders are shown in Table 11-2 in the standard isotopic notation.

Table 11-2. SOME MAN-MADE RADIOISOTOPES THAT POSE AN ENVIRONMENTAL THREAT

Element	Isotope
Plutonium	$^{239}_{94}Pu$
Iodine	$^{131}_{53}I$
Cesium	$^{137}_{55}Cs$
Strontium	$^{90}_{38}Sr$
Carbon	$^{14}_{6}C$
Hydrogen	$^{3}_{1}H$
Krypton	$^{85}_{36}Kr$
Zirconium	$^{95}_{40}Zr$

NUCLEAR INSTABILITY

Whether a nucleus is stable or prone to nuclear change depends on the forces binding the nucleus together. Scientists have tried to understand these forces and their associated energies since the nucleus was discovered by Rutherford in 1911. A fair understanding has been gained, although some puzzles remain about the ultimate nature and origin of nuclear forces and particles.

Basically the nucleus is bound together by the same kind of quantum mechanical forces that hold atoms together in molecules. These *nuclear forces* "glue" the protons and neutrons to one another in the nucleus. As atomic number increases, protons and neutrons gain more neighbors with which to bond and hence become more tightly secured into the nucleus.

Opposing the quantum mechanical forces of attraction in the nucleus are the electrostatic repulsion forces between the positive nuclear protons. These forces reach out beyond near neighbors, farther than the attractive forces go. Thus as the nucleus gets bigger the repulsive forces become dominant by being able to act disruptively across the entire nucleus. Eventually, with increasing nuclear size, the repulsion between protons will literally split the nucleus apart.

This analysis suggests that as atomic number increases, the nucleus at first becomes more stable because of the increasing nuclear forces and later becomes unstable because of overriding proton repulsion. Somewhere in the middle of the periodic table, binding forces must be at a maximum and stability must be greatest. If this analysis is correct, fusion and fission can be explained simply as the tendency for very small and very large nuclei, respectively, to change their size toward the golden mean of maximum stability.

Scientists have indeed verified that the maximum binding energy is midway through the periodic table. Surprisingly, they have verified this by measuring the mass of the nucleus rather than its energy. Nuclear and atomic masses can be measured with great accuracy by use of a so-called mass spectrometer. Energy, by contrast, is difficult to measure directly. The connection between nuclear mass and nuclear binding energy is explained by the famous Einstein relationship between mass and energy.

In 1905 Einstein postulated that energy and mass are equivalent and are related by the universal equation

$$E = mc^2$$

where E stands for energy, m for mass, and c for the velocity of light, a constant quantity. Since c is very large, 3×10^{10} cm/sec, a very small amount of mass multiplied by the enormous quantity c^2 can provide a vast release of energy. This explains the abundant energy fueling the nuclear age.

When isolated protons and neutrons are joined together by nuclear forces to form an atomic nucleus, enormous energy is given off. As Einstein predicted, a small amount of mass is simultaneously lost. The result is that the nucleus weighs less than its protons and neutrons. This small deficiency in mass may be called the *mass defect*, although slightly different definitions exist.

Only by changes in mass defect can nuclear mass be translated into overt nuclear energy. Next we focus on this small but earth-shaking deficit in mass.

Formation of the helium nucleus from two protons and two neutrons provides a good example of the mass defect. On the standard atomic weight scale the mass defect is a mere 0.0302 atomic mass units (amu), less than 1 percent of the mass of the nucleus formed.[2] Small masses such as this have an enormous Einstein energy equivalent, as we have noted. This mass-energy may be given off as a powerful gamma ray (which we shall take up later). The hypothetical process of forming a helium nucleus is shown in Figure 11-1, next page.

In actual fact we do not go around forming nuclei from bare protons and neutrons and harvesting the emitted mass-defect energy. The difficulty of forcing a large number of protons and

---

[2] This result is obtained as follows:

Mass of 2 protons	$= 2 \times 1.0076$	$= 2.0152$ amu
Mass of 2 neutrons	$= 2 \times 1.0089$	$= 2.0178$ amu
Total		$= 4.0330$ amu

However, by mass spectrometry the helium nucleus has been found to have a mass slightly smaller, 4.0028 amu. The difference, $4.0330 - 4.0028 = 0.0302$ amu, is the mass defect, less than 1 percent of the total nuclear mass.

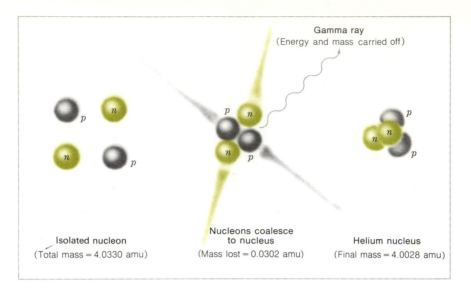

Gamma ray
(Energy and mass carried off)

Isolated nucleon	Nucleons coalesce to nucleus	Helium nucleus
(Total mass = 4.0330 amu)	(Mass lost = 0.0302 amu)	(Final mass = 4.0028 amu)

Figure 11-1. The mass defect of the helium nucleus is 0.0302 amu. This mass and a corresponding quantity of energy are lost upon formation of the nucleus.

neutrons together all at once rules this approach out. Instead we instigate simple nuclear transformations to take advantage of *differences* in mass defect. To illustrate how this can be achieved, Figure 11-2 is plotted to show the variation in mass defect from the beginning of the periodic system to its end.

Figure 11-2 shows that the mass defect reaches a broad maximum

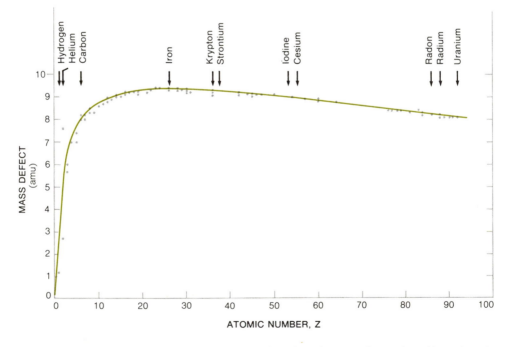

Figure 11-2. Mass defect per nucleon plotted against atomic number. (A nucleon is a proton or a neutron.)

NUCLEAR ENERGY, RADIOACTIVITY, AND ENVIRONMENT

just to the left of the center in the periodic system, near iron, $Z = 26$. Here the net nuclear forces are greatest, as we explained earlier. When nuclei near the ends of the periodic system are changed to those more central, the added mass defect becomes released energy.

There are clearly only two general ways to change nuclei toward a greater mass defect. First of all, we can take a very large nucleus, such as that of $^{235}_{92}U$, and split it into two intermediate-sized nuclei. This process is *fission*. Alternatively, we can take two light nuclei and join them together. This method is *fusion*. The increase in mass defect with each process is shown by the arrows in Figure 11-3. The energy accompanying this mass-defect increase may be released in nuclear explosions or, in the case of fission, also in electric-power production. Later we shall discuss the scope and applications of fission and fusion.

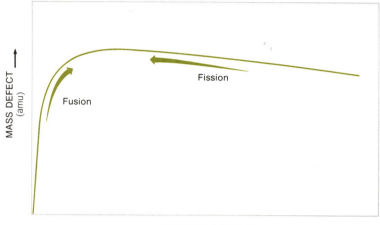

Figure 11-3. The mass defect can be increased and a corresponding amount of energy released in both fission (the splitting of nuclei) and fusion (the joining of nuclei).

Fission and fusion energy derive from broad, general changes in mass defect across the periodic system of elements. However, on this mass-defect plot, if we include all isotopes, we find small sawtooth variations along with the broad trends that are responsible for fission and fusion. These small variations are responsible for the phenomenon of radioactivity. Here the nucleus undergoes only minor change to obtain an advantageous binding energy and increased stability.

Unlike the major changes occurring in fusion and fission, radioactivity can happen to a susceptible nucleus anywhere, at any time, without the prompting of other nuclear particles. A radioactive event in the nucleus is beyond human intervention, a fact of importance to our efforts to control environmental radioactivity.

The sawtooth variations in mass defect are due largely to another quantum mechanical phenomenon: the tendency to form stable shells of protons and neutrons. Just as a shell structure arranges electrons external to the nucleus and governs chemical stability, protons and neutrons in the nucleus also form shells with variations in nuclear stability. The level of stability depends on the degree of shell completion. Specifically, certain shell structures with an even number of protons and neutrons are most stable, resembling the chemical stability of closed electron shells. Nuclei with an odd number of either protons or neutrons are less stable. These "odd" isotopes are more often radioactive than are the "even" isotopes, and they are less common in nature, as shown in Table 11-3. Apparently when the nuclei of the universe were formed, by some primitive and poorly understood process, the more stable isotopes were produced in the greatest quantities. The most abundant elements in the earth's crust, for instance, have an even number of protons: oxygen ($Z = 8$) and silicon ($Z = 14$). The most abundant isotopes of these two elements also have an even number of neutrons: oxygen is 99.8 percent $^{16}_{8}O$ (8 neutrons) and silicon is 92.3 percent $^{28}_{14}Si$ (14 neutrons). Only 10 isotopes having an odd number of both protons and neutrons occur in nature, as shown in Table 11-3. Of them, only four are nonradioactive, and these isotopes are all near the beginning of the periodic system: $^{2}_{1}H$, $^{6}_{3}Li$, $^{10}_{5}B$, $^{14}_{7}N$. The fact that there are few totally "odd" nuclei has also a chemical analog— the rarity of molecules with an odd number of electrons. Only three such odd molecules have been observed (including the pollutants NO and NO_2) out of more than a million molecules known to exist.

Table 11-3. THE NUMBER OF NATURAL ISOTOPES WITH EVEN AND ODD NUMBERS OF PROTONS AND NEUTRONS

Number of Isotopes	Protons	Neutrons
177	Even	Even
58	Even	Odd
54	Odd	Even
10	Odd	Odd

Many of the original nuclei with less stable nuclear shells have long since decayed away by radioactivity, and others are in the process of doing so. With the occurrence of nuclear fission and fusion, which man has learned to induce, most of the new nuclei formed are also radioactive. In this way radioactivity is invariably associated with efforts to extract energy from nuclear reactions, and it will pose some level of threat to the environment as long as nuclear power and nuclear weapons are a part of civilization.

RADIOACTIVITY

The use of radioactivity is nature's most common way of transforming nuclei to a more stable form with greater binding energy.

The emitted particles and rays are the carriers of the excess energy.

Natural radioactivity was first observed quite by accident by Henri Becquerel in 1896. He discovered that uranium compounds emitted a penetrating radiation that would darken a photographic plate without exposure to light. In an enthusiastic and dedicated study of the phenomenon, Mme Marja Curie with her husband Pierre discovered two new elements, both more radioactive than uranium. They are polonium (Po) and radium (Ra). Others carried on similar work, and within 15 years of Becquerel's discovery, about three dozen radioactive substances had been discovered. At present, over fifty naturally occurring radioisotopes are known.

Three major kinds of radiation (radioactive emissions) are given off by radioactive nuclei (Figure 11-4). They are *alpha particles*,

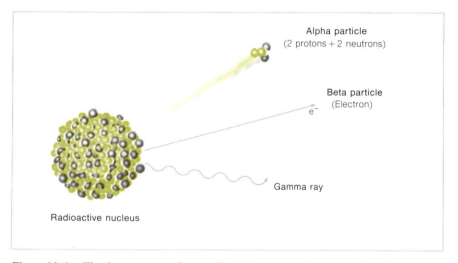

Figure 11-4. The three common forms of radiation emitted when a radioactive nucleus decays.

beta particles, and *gamma rays*. In a given radioactive event, a nucleus usually loses just one particle, alpha or beta, and the particle frequently is accompanied by a gamma ray. The characteristics of each of these forms of radiation is as follows.

1. Alpha (α) Particles. These particles are composed of two protons and two neutrons and are identical to the nucleus of the helium atom. They are by far the most massive of the three radiations, but the slowest and least penetrating. They travel, at maximum, at 10 percent of the speed of light (still very fast) and can be stopped by a piece of paper or a thickness of skin. Therefore penetration into living systems from the outside is small. Damage occurs only when alpha-emitting isotopes are ingested or inhaled, and thus brought into intimate contact with internal tissue and organs.

As an emitted alpha particle penetrates tissue, its positive charge attracts and pulls away the electrons that belong to atoms within the tissue. This disruption causes the adverse chemical effects we mentioned earlier. Because an alpha particle is doubly charged (two protons) and relatively slow moving, its effectiveness at stripping off electrons is very high. Even though the alpha particle's distance of penetration is slight, it causes considerable chemical havoc along its short path.

2. Beta (β) Particles. This particle is simply an electron emitted by the nucleus, but its energy and velocity are so high that it can be very destructive. It may travel up to 90 percent of the speed of light. The penetration of tissue can be 1 cm or more.

A beta particle streaking through tissue occasionally passes very close to an orbital electron. The repulsion of the two negative particles may force the orbital electron into another orbital or kick it out of the atom altogether. Again the chemistry is disturbed because of the simple forces acting between charged particles, one atomic (the electron) and one nuclear (an alpha or beta particle).

3. Gamma (γ) Rays. The gamma ray is radiation of distinctly different character. It is identical to energetic x rays, but it has a strictly nuclear origin. It is thus a form of electromagnetic radiation, like visible light itself, and travels at the same velocity. Gamma rays, like x rays, have a much shorter wavelength than light, a far greater energy, and an impressive penetrating power. They can pass right through a human being.

Electromagnetic radiation, as its name implies, is associated with traveling electric and magnetic fields (recall Chapter 2). The passing electric field can eject electrons from atoms, if enough energy is available. Gamma rays are extremely energetic, enough to cause many such ionizations as they penetrate tissue or other matter. Ordinary visible light lacks the energy to disturb electrons in most materials, and it is thus essentially harmless.

Several other forms of radioactivity occur, but not often enough to have general environmental impact. The less common radioactive fragments include neutrons and positrons, the latter being a kind of positive electron. Neutrons, of course, are common in fission, a point we shall get to soon. The properties of the various forms of radiation are summarized in Table 11-4.

Table 11-4. PROPERTIES OF VARIOUS FORMS OF NUCLEAR RADIATION

	Type	Symbols	Mass Number	Charge	Approximate Tissue Penetration (cm)	Makeup
Most common	Alpha particle	α, $_2^4\text{He}$	4	$+2$	0.01	2 protons + 2 neutrons
	Beta particle	β, $_{-1}^0\beta$, β^-, $_{-1}^0\text{e}$, e^-	0	-1	1	1 electron
	Gamma ray	γ	0	0	100	electromagnetic radiation
	Neutron	n, $_0^1\text{n}$	1	0	10	neutron
	Positron	β^+, $_1^0\text{e}$, e^+	0	$+1$	1	positive electron

HALF-LIVES

Like the chemical reaction of molecules, the radioactive breakup of nuclei requires a finite time to occur. And then the many nuclei do not change all at once. Instead they disintegrate randomly, first one and then another, until over a period of time most of the original nuclei are gone. The time required for half of the original nuclei to disappear is called the *half-life*. Each radioisotope has a characteristic half-life. They range from millionths of a second to a time greater than the age of the earth, several billion years. Some half-lives of isotopes of environmental significance are listed in Table 11-5, along with the radiation emitted by that isotope.

Table 11-5. HALF-LIVES AND THE TYPE OF RADIATION EMITTED FROM SOME RADIOACTIVE ISOTOPES OF ENVIRONMENTAL SIGNIFICANCE

	Isotope	Half-Life	Radiation Emitted
Natural radioisotopes	$^{238}_{92}U$	4.5×10^9 years	α, γ
	$^{235}_{92}U$	7.1×10^8 years	α, γ
	$^{232}_{90}Th$	1.4×10^{10} years	α, γ
	$^{40}_{19}K$	1.3×10^9 years	α, γ
	$^{226}_{88}Ra$	1600 years	α, γ
	$^{14}_{6}C$	5600 years	β
Radioisotopes from human activities	$^{239}_{94}Pu$	24,000 years	α, γ
	$^{137}_{55}Cs$	30 years	β, γ
	$^{90}_{38}Sr$	28 years	β
	$^{131}_{53}I$	8 days	β, γ

If one starts with a given amount of radioisotope, one half of it will undergo nuclear change in a single half-life. Half of that remaining will disintegrate in the next half-life, leaving a quarter of the original radioisotope behind. In three half-lives, only one half of this amount will remain, one eighth of the starting amount. Such decreases follow Table 11-6.

Table 11-6. ATTRITION OF RADIOISOTOPE WITH ELAPSED HALF-LIVES

Number of Half-Life Periods Elapsed	Fraction of Original Radioisotope Remaining
1	$\frac{1}{2} = 0.5$
2	$\frac{1}{4} = 0.25$
3	$\frac{1}{8} = 0.125$
4	$\frac{1}{16} = 0.062$
5	$\frac{1}{32} = 0.031$
6	$\frac{1}{64} = 0.016$
7	$\frac{1}{128} = 0.008$
8	$\frac{1}{256} = 0.004$
9	$\frac{1}{512} = 0.002$
10	$\frac{1}{1024} = 0.001$

More graphically, the amount of radioisotope can be plotted against time, as in Figure 11-5. The curve is known as a *decay curve*, sometimes as an *exponential decay curve*, because *"exponential"* describes the property of being either halved or doubled in fixed time intervals. Population growth curves are sometimes called "exponential" because populations often tend to double in a fixed time (the *doubling time*). With radioactivity, exponential refers to the halving of the quantity of isotope with the passage of each half-life period.

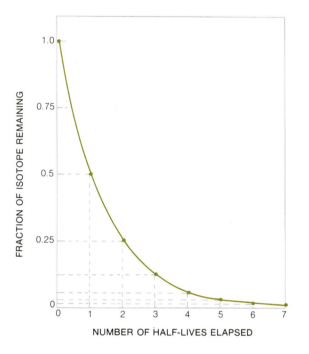

Figure 11-5. Exponential decay curve for radioisotopes. Notice that the amount of isotope approaches zero rather rapidly in the first few half-lives. However, it does not actually reach zero, because halving the amount of isotope with each passing half-life always insures that a small amount is left behind.

As an isotope is eaten away by its own radioactivity, there is proportionally less emitted radiation. However, a residue of radioactivity still remains after many half-lives, as Table 11-6 shows.[3]

NUCLEAR TRANSFORMATIONS

The alchemists of ages past dreamed often of turning one chemical element into another, thus producing gold and the like from elements less precious. As the atomic theory became entrenched, this dream

---

[3] For example, Table 11-6 shows that after five half-lives, 1/32 of the original isotope will remain. Accordingly, the intensity of its radioactivity will also be reduced to 1/32 of the original. For strontium (Sr)-90, half-life 28 years, this reduction will take 5 × 28 years = 140 years. For plutonium (Pu)-239, half-life 24,000 years, a reduction to 1/32 will require 5 × 24,000 years = 120,000 years.

was shattered, for the atom assumed the role of the one stable and indivisible unit of matter. However, the study of radioactivity early in this century brought with it a realization that atoms do change from one kind to another during their occasional nuclear events. Alas, the transformation is not efficient enough to produce gold in quantity (although traces of synthetic radioisotope gold (Au)-198 are used in tumor control). But changes much more momentous have occurred, such as the production of plutonium atoms from uranium atoms, in amounts great enough to change the complexion of civilization.

In most nuclear events—fission, fusion, and radioactivity— new elements form as old disappear. This change happens as the nucleus gains or loses positive charge and thus has its atomic number altered. For instance, in alpha decay, an alpha particle carries two protons (as well as two neutrons) away from the nucleus, thereby reducing the atomic number of the nucleus by two. The new element formed thus appears two places ahead of its parent in the periodic table.

Similarly, in beta decay, the departure of a negative charge leaves the nucleus with one additional positive charge. An element is born with an atomic number one unit higher.

The transformation of elements is represented by a nuclear equation, just as chemical transformations are represented by chemical equations. The rules for balancing the two are similar. By learning to manipulate nuclear equations, we often can predict the products of nuclear reactions.

Consider a very simple nuclear reaction, that in which an isolated neutron (n) undergoes beta decay, leaving a proton (p) behind. This event, having a half-life of 12 min, is represented by the equation

$$\,^{1}_{0}n \longrightarrow \,^{1}_{1}p^{+} + \,^{0}_{-1}\beta$$

| Neutron | Proton | Beta particles |

Note that superscripts as well as subscripts, when summed on the right, equal the left-hand totals:

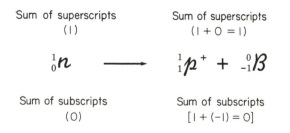

Sum of superscripts (1)

$$\,^{1}_{0}n \longrightarrow$$

Sum of superscripts $(1 + 0 = 1)$

$$\,^{1}_{1}p^{+} + \,^{0}_{-1}\beta$$

Sum of subscripts (0)

Sum of subscripts $[1 + (-1) = 0]$

This equation is balanced because it satisfies two rules: mass number (sum of superscripts) and atomic number (sum of subscripts) do not change in nuclear processes.

Beta decay within an atomic nucleus is much like the beta decay of the bare neutron we have shown: there is a loss of a neutron and a gain of a proton in the nucleus. However, neutrons within a nucleus are usually not subject to beta decay, because the nuclear binding forces stabilize the neutron. Only in radioactive isotopes are the stabilizing forces weak enough so that beta decay may occur. An example is strontium (Sr)-90, a bone-seeking isotope found in atomic fallout.

The beta decay of Sr-90 is represented by the equation

$$^{90}_{38}\text{Sr} \quad \rightarrow \quad {}^{90}_{39}\text{Y} \quad + \quad {}^{0}_{-1}\beta$$

strontium-90   yttrium-90   beta particle

Notice that the new isotope formed, yttrium (Y)-90, has an atomic number (39) one higher than that of the parent Sr-90 and a mass number (90) the same. Such is always the case in beta decay. Otherwise, considering the mass and charge of the beta particle, the equation could not possibly balance.

The isotope Y-90, produced in the beta decay of Sr-90, undergoes a beta decay of its own. The identity of the isotope produced in this decay is deduced as follows.

The beta decay of Y-90 to unknown isotope ${}^{A}_{Z}\text{X}$ can be represented by the nuclear equation

$$^{90}_{39}\text{Y} \rightarrow {}^{A}_{Z}\text{X} + {}^{0}_{-1}\beta$$

The equation balances only if superscripts balance from left to right ($90 = A + 0$, giving $A = 90$) and if subscripts also balance ($39 = Z - 1$, giving $Z = 40$). Element 40 is zirconium (Zr). The unknown isotope is thus ${}^{90}_{40}\text{Zr}$. (Less formally, we could arrive at the same answer for any beta decay by keeping A constant and increasing atomic number Z by one.)

A gamma ray is emitted in most beta decay processes (although not in the beta decay of Sr-90), and it adds to the radiation damage. It is not necessary to show the gamma ray in the nuclear equation since it has zero mass number and zero charge and thus contributes nothing to either total. However, it is biologically and energetically important, and it may therefore appear in the equation, as we show here in the beta decay of another important fallout radioisotope, iodine (I)-131.

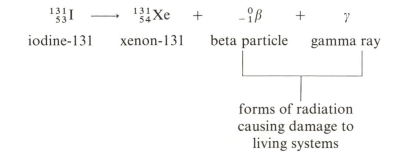

$$^{131}_{53}\text{I} \quad \longrightarrow \quad {}^{131}_{54}\text{Xe} \quad + \quad {}^{0}_{-1}\beta \quad + \quad \gamma$$

iodine-131   xenon-131   beta particle   gamma ray

forms of radiation
causing damage to
living systems

Alpha decay may similarly be represented by a nuclear equation. For instance, most of the U-238 on earth is gradually being eaten away by alpha decay, although some is lost through spontaneous fission. The alpha decay is represented by

$$^{238}_{92}U \rightarrow \quad ^{234}_{90}Th \quad + \quad ^{4}_{2}\alpha \quad + \quad \gamma$$

U-238 thorium-234 alpha particle gamma ray

The sums of superscripts ($238 = 234 + 4$) and subscripts ($92 = 90 + 2$) both remain unchanged, as required. As we mentioned before, since the alpha particle carries off two charge units and four mass units, the new nucleus is always two less than its parent in atomic number and four less in mass number. In this manner, we can easily predict the product of alpha decay,[4] in this case thorium (Th)-234.

The isotope Th-234 is itself significant, for it is the first step in a decay chain (soon to be discussed) that produces much of this earth's natural radioactivity.

The process of beta and alpha decay can be represented on a rectangular grid, with mass number increasing upward and atomic number plotted to the right. The alpha decay of U-238 thus appears as

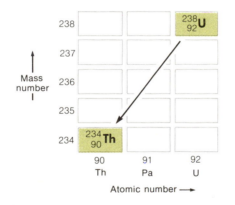

The daughter isotope, Th-234, is four layers down, because its mass number is four less, and two boxes to the left, because its atomic number is two lower. The whole process is indicated by the heavy diagonal arrow, which runs from the top right rectangle to the lower left rectangle.

Beta decay is represented by a single displacement right in the grid, because the atomic number increases by one. Thus the beta

---

[4] For instance, the alpha decay of plutonium-239 or $^{239}_{94}Pu$ gives an isotope $^{A}_{Z}X$ with $A = 239 - 4 = 235$ and $Z = 94 - 2 = 92$. Element 92, of course, is uranium. The product is thus $^{235}_{92}U$ or U-235. We can get the same result by finding the A and Z values necessary to balance the nuclear equation

$$^{239}_{94}Pu \rightarrow ^{A}_{Z}X + ^{4}_{2}\alpha$$

decay of strontium-90 represented on part of a rectangular grid looks like this:

Here the usual superscript and subscript symbols are omitted because mass and atomic number are given by the grid coordinates. Thus a glance at the grid makes it clear that the isotopes involved are $^{90}_{38}$Sr and $^{90}_{39}$Y.

The rectangular-grid picture of nuclear changes is the best way of showing the complicated nuclear history of some important, long-lived isotopes found in nature, including uranium (U)-238. The U-238 isotope is fairly abundant on earth. It decays slowly, with a half-life of 4.5 billion years, and thereby starts an entire sequence of nuclear disintegrations, many of which are prominent in natural radioactivity. This decay chain, or *nuclear disintegration series*, is shown in the rectangular grid of Figure 11-6. The sequence of nuclear events is shown by the arrows, starting in the upper right-hand corner with U-238. It ends with lead (Pb)-206, a stable isotope.

The grid in Figure 11-6 shows some familiar isotopes, including

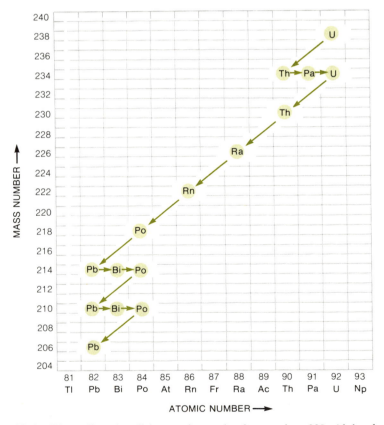

Figure 11-6. The radioactive disintegration series for uranium-238. Alpha decay is represented by long diagonal arrows ($\swarrow$); beta decay by short horizontal arrows ($\rightarrow$).

radium (Ra)-226, the isotope discovered by the Curies, and polonium (Po)-214, whose alpha particles led Rutherford to the discovery of the nucleus. In addition, the isotopes U-238, Ra-226, radon (Rn)-222, lead (Pb)-210, and Po-210 are responsible for much of the earth's background radiation.

Each intermediate isotope in the disintegration series has its own unique half-life. These times vary widely, ranging from 1.6×10^{-4} sec for polonium (Po)-214 to 2.5×10^5 years for uranium (U)-234.

Altogether there are three radioactive disintegration series in nature. They begin with U-238, U-235, and thorium (Th)-232. Each series, like the U-238 sequence just discussed, goes through many intermediate radioisotopes before terminating with a stable isotopic species.

RADIATION DOSAGE

Radiation causes biological damage. The amount of damage is directly related to dosage. Any realistic discussion of radiation hazards must therefore include dosage and dosage units. These dosage units, mainly "curies," "rads," and "rems," quite often appear in today's newspapers and magazines as the public becomes more involved with its radioactive environment. Here we define these terms.

1. Curie. The *curie* (Ci) is the level of radioactivity caused by 3.7×10^{10} radioactive disintegrations per sec. The odd number comes from historical usage, in which radium (Ra) was used as a standard of radioactivity. One g of radium has 3.7×10^{10} disintegrations per sec, a single curie of radioactivity. A curie is a lot of radioactivity, 37 billion nuclear events per sec, so the more moderate units millicurie (10^{-3} Ci), microcurie (10^{-6} Ci), and picocurie (10^{-12} Ci) are often used.[5]

The curie is a basic unit of radioactivity, but unfortunately, equal curies produced by different kinds of isotopes do not have equal biological effects. The violence of the nuclear event differs from one kind of isotope to another, and thus the type and energy of the radioactive fragments differ and the chemical-biological damage is variable. In other words, the damage from a curie coming from radioactive substances A and B may be quite different. If we wish to assess bio-

---

[5] Later we shall learn that 10 μg (10^{-5} g) of radium (Ra)-226 retained in man causes cancer in 50 percent of those exposed. Since 1 g of Ra-226 produces 1 Ci, 10μg produces 10 microcuries (10^{-5} Ci). The number of nuclear disintegrations per sec in a man so exposed is calculated from the definition 1 Ci $= 3.7 \times 10^{10}$ disintegrations/sec. Thus:

$$10^{-5} \text{ Ci} \times \frac{3.7 \times 10^{10} \text{ disintegrations/sec}}{1 \text{ Ci}} = 3.7 \times 10^5 \text{ disintegrations/sec}$$

This result shows that such an exposed man has 370,000 internal nuclear disintegrations every second.

logical damage, a unit which is more comparable between isotopes (still not entirely so, as we shall see) is the rad.

2. Rad. This unit expresses the amount of energy actually absorbed from radioactive fragments by tissue or other material. One rad signifies the transfer of 0.0024 calories (= 0.01 joules, a metric expression of energy) of radioactive energy to a kilogram (kg) of material. One rad, a mere fraction of a calorie in a kg of tissue, is not much energy; converted to heat it would raise the temperature of tissue no more than a few millionths of a degree. But because it is concentrated in energetic radioactive projectiles that can disrupt vital molecules, this small dose of highly selective energy is devastatingly effective.

3. Rem. The rem is much like the rad; often they are identical. One rem is, in fact, 1 rad multiplied by an adjustment factor called the *relative biological equivalent (RBE)*, or sometimes the *quality factor (QF)*.

$$\text{number of rems} = \text{number of rads} \times \text{RBE}$$

The RBE factor accounts for the occasional differences in biological damage caused by different kinds of radioactive particles having the same energy. For instance, an alpha particle released in tissue is ten times more damaging than a beta particle of the same energy. The RBE for an alpha particle is 10 and for a beta particle is 1, which reflects their unequal destructiveness. Various RBE's are shown in Table 11-7.[6] When RBE = 1, as for beta particles and gamma rays, the rem and the rad are indeed equal. For most purposes in this chapter, they can be assumed the same.

The rem is a more accurate reflection of biological damage, and it is thus commonly used in technical descriptions of actual and permissible radioactive doses.

A fourth unit of radiation intensity used extensively in the past and still used occasionally is the roentgen.

Table 11-7.   RELATIVE BIOLOGICAL EQUIVALENTS (RBEs) FOR VARIOUS FORMS OF RADIATION

	Radiation Type	RBE
Most common radiation	Gamma rays (or x rays)	1
	Beta particles	1
	Alpha particles	10
	Fast neutrons	10
	Slow (thermal) neutrons	5

---

[6] Here we show how RBE values convert a given rad dose to a dose expressed in rems. For example, 5 rads of exposure by alpha particles, for which Table 11-7 shows RBE = no. rems/no. rads = 10, gives a rem exposure of

$$5 \text{ rads} \times \frac{10 \text{ rems}}{1 \text{ rad}} = 50 \text{ rems}$$

On the other hand a 5 rad dose due to beta particles, for which RBE = 1, gives, in rems

$$5 \text{ rads} \times \frac{1 \text{ rem}}{1 \text{ rad}} = 5 \text{ rems}$$

4. Roentgen. This unit is a measure of exposure to gamma and x radiation. One roentgen is an amount of radiation that will ionize enough molecules in a kilogram of dry air to create a fixed electrical charge, equivalent to 2.6×10^{15} ion pairs. It turns out that when this amount of gamma or x radiation—one roentgen—bombards soft tissue or water, approximately one rad (or one rem) of dosage is accumulated. Therefore the roentgen, the rad, and the rem are roughly equivalent for gamma and x rays.

Many devices have been developed for the measurement of radiation exposure. Best known is the Geiger counter, in which a high voltage is applied to the opposite walls of a small chamber. When radiation enters the chamber and causes the ionization of molecules, the charged ions and electrons released are drawn to the chamber walls by the applied voltage. Thus a small pulse of electrical current flows. The counting of these pulses gives directly the frequency of radioactive disintegrations in a given region.

DOSAGE AND BIOLOGICAL DAMAGE

The radiation dosage at which human injury begins is unknown. As with other pollutants, there are chronic (long-term) effects of low doses which would be difficult to unravel even with extensive and dangerous experiments on man himself. Acute (short-term) effects are more directly observable, but they are still clouded by a lack of human data and insufficient knowledge of how biological damage occurs. Nonetheless, animal experiments have provided an important framework of information, confirmed in broad outline by the hapless victims of nuclear accidents and nuclear warfare (Hiroshima and Nagasaki).

Acute Exposure. As a frame of reference, we should know the acute exposure which will cause death. Table 11-8 shows the dose of x rays that is fatal within 30 days of exposure to 50 percent of various mammal populations. This is expressed in terms of LD_{50}, the "lethal dose to 50 percent."

Table 11-8. LD_{50} VALUES FOR MAMMALS[a]

Mammal	LD_{50} (rads or rems)
Rabbit	800
Rat	750
Mouse	600
Monkey	550
Guinea pig	400
Dog	350
Goat	300
Pig	250
Sheep	250

[a]The dose in rads or rems, applied over the whole body, that is fatal to half of each animal population.

Since man has biochemical machinery similar to that of other mammals, he presumably belongs in this range. Roughly, then, a sudden exposure of 500 or more rads (or rems) would be expected to cause death in man. Based on information gleaned from the Hiroshima and Nagasaki disasters and on a few reactor accidents, death is caused by blood disorders, gastrointestinal failure, and damage to the central nervous system.

Man and his fellow mammals are weaklings compared to most of the simpler life forms in resisting radiation effects. Some insects can withstand over 100,000 rads, and microorganisms even more. Clearly man will be one of the first victims of his own mistake should radiation exposure get out of hand.

Long-Term Exposure.   Low but persistent exposure to radiation seems to have a number of major long-term consequences. Among them are (1) premature aging, (2) teratogenic (injury to the unborn) effects, (3) carcinogenic (cancer-producing) effects, and (4) mutagenic (mutation-producing) effects.

It has been recognized for at least three decades that radiation exposure causes many changes normally associated with aging. Such changes occur throughout the system: changes in the density of fine blood vessels, in connective tissue, the formation of eye cataracts, even the graying of hair. Mortality rates are increased. A study of mortality statistics among laboratory animals suggests that man may lose 17 days of life for each rad of radiation exposure. Radiologists, working constantly with x rays, were found in another study to die an average of 5 years younger than other physicians.

Radiation injury to an unborn fetus occurs with even less exposure. The fetus is roughly five times more sensitive to radiation than adults. Children, too, are more sensitive. The effects are broad, with damage to the brain, skeleton, eyes, and other organs.

It has long been known that radiation causes cancer: Thomas Edison's assistant, Clarence Dally, died from it in 1904 because of intensive exposure to x rays in Edison's laboratory. Only with Dally's death did Edison curtail his own exposure to x rays.

Skin cancer, bone cancer, thyroid cancer, leukemia, and other cancers are products of radiation exposure. Leukemia is probably most significant. After studying leukemia in subjects with undue x ray exposure, Hiroshima survivors, and others, E. B. Lewis concluded that for a 1 rad dose, one to two cases of radiation-induced leukemia would occur each year in a human population of 1 million. This leukemia incidence might seem small, but is presumably repeated for each and every year for 20 years after the exposure. Other cancers add to this total. Thus, far short of the (acute) dose that causes quick death, a few individuals here and there will die from exposures many times smaller. In fact, it is now thought by some that even very tiny doses will cause a proportional level of leukemia and other cancer: for example, 0.01 rad (instead of 1 rad) is expected to cause a leukemia incidence of 0.01 to 0.02 (instead of 1 to 2) per million per year (one or two cases per 100 million per year). Thus even the most minute dose may entail some additional risk.

The theory (Figure 11-7) that risk is proportional to exposure even down to very low exposures—the *linear theory*—is accepted by some experts, but not all. Opposed to this is the *threshold theory*, which states that below a certain dosage, no injury occurs. Currently, many radiobiologists believe that a *sigmoid* (S-shaped) curve is more realistic for cancer induction. The whole controversy surfaced with charges by John Gofman and Arthur Tamplin that, because of a presumed linear response, present radiation standards entail risks that are altogether too high. Criticism of their employer, the Atomic Energy Commission (AEC), has had widespread newspaper coverage and political repercussions. We shall return to this controversy in our study of nuclear power plants.

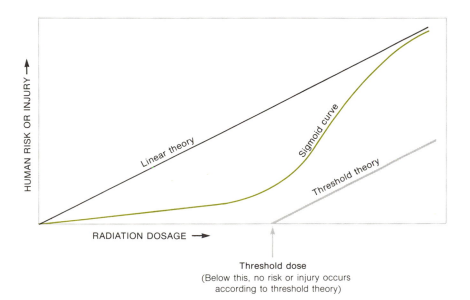

Figure 11-7. Threshold and linear theories. The major difference between them is that at low doses (characteristic of environmental contamination), the injury or risk incurred by radiation is finite (linear theory) or zero (threshold theory). The sigmoid curve, currently favored, leads to an intermediate conclusion: the risk at low doses is finite but extremely small.

Last but not least on the ledger of radiation effects is genetic damage. The genetic risk is much more difficult to study than cancer because genetic damage may not surface for many generations. Genetic damage—damage to the gene structure—may cause death, or simply a variety of defects such as cleft palate. From the limited data available, it has been estimated that following the exposure of a population of 1 million to 1 rad, 200 offspring of that 1 million will suffer lethal genetic changes (many deaths are prenatal deaths) and an indefinite number of other defects. The damage will continue at only a slightly decreasing rate from generation to generation, until perhaps 8000 genetic deaths occur over 50 generations. This number of mutations is small compared to those occurring naturally, but is nonetheless significant.

In summary, it takes about 500 rads to kill a person outright, but doses as small as a fraction of a rad may shorten life slightly and pick off a few individuals here and there, some many generations later.

Curies versus Rads. It may be difficult to see the connection between the biological effects of curies and of rads or rems. Indeed, the exact relationship is not simple.

Curies are defined as a level of radioactivity, 3.7×10^{10} nuclear bursts per second. The amount of a radioisotope stored or released at some point is often measured in terms of its curies of activity. A small fraction of the released curies will reach man and cause, over a period of time which depends on many factors, an exposure of so many rads or rems. The difficulties in connecting the two are that the rads or rems (measuring biological damage) from a curie may vary widely from isotope to isotope, and the period of exposure may vary.

The varying biological significance of a curie can best be explained by specific examples. For instance, a steady level of 10 microcuries (10^{-5} curies) of radium (Ra)-226 maintained in man will induce a 50 percent incidence of cancer. An observable increase in cancer is caused by a mere 1 microcurie of Ra-226. In swine, an observable increase is caused by 250 microcuries of strontium (Sr)-90. In contrast, experiments on mice suggest that the weak β emitter tritium (H-3) would need to be present at 70 million microcuries (70 Ci) to cause immediate death in man. Tritium is truly exceptional, but it does show the wide range in curies over which biological damage occurs.

We conclude that, with occasional exceptions, radioactivity of a few microcuries will endanger the life of man.

Relative Toxicity of Radioisotopes. Per unit of weight, radioactive substances are the most toxic materials known. Quite rightly, it is often said that radioisotopes are a million to a billion times more toxic than most chemical poisons. Ingestion of micrograms can be deadly. This matter is serious if there is a significant chance of exposure. Possible avenues of exposure will be explored in later parts of the chapter.

Table 11-9 compares directly the doses of various substances that have a given biological effect. Units of dosage are normalized to milligrams per kilogram of body weight so that the response of man and various animals can be compared.[7] Note that even the extreme potency of LSD is exceeded by most radioisotopes.

---

[7] If we assume that the injection of 0.0014 mg per kg of body weight of strontium (Sr)-90 will cause cancer in man as it does in mice, then a 70 kg man would run a risk of cancer with an amount of Sr-90:

$$70 \text{ kg body weight} \times \frac{0.0014 \text{ mg Sr-90}}{1 \text{ kg body weight}} = .098 \text{ mg Sr-90}$$

a quantity less than a tenth of a milligram.

Table 11-9. DOSES OF VARIOUS SUBSTANCES HAVING THE STATED BIOLOGICAL EFFECT

Substance	Injected mg per kg Body Weight	Effect Caused by Dose
Strontium (Sr)-90[a]	0.0014	14% incidence of bone cancer in mice
	0.00068	57% incidence of bone cancer in beagles
Radium (Ra)-226[a]	0.003	30% incidence of bone cancer in man
	0.01	50% incidence of bone cancer in man
	0.00034	38% incidence of bone cancer in beagles
Plutonium (Pu)-239[a]	0.00026	33% incidence of bone cancer in beagles
	0.00080	69% incidence of bone cancer in beagles
LSD	0.0036	Hallucinogenic dose
Methyl mercury[b]	10	100% death in mice
Parathion[c]	8	50% death in mammals

[a] Data for radioactive substances provided by Professor Charles W. Mays of the Radiobiology Division, University of Utah.

[b] A deadly water pollutant.

[c] One of the most toxic insecticides.

MOLECULAR ORIGIN OF BIOLOGICAL DAMAGE

The ultimate sources of radiation injuries are unwanted changes in the molecules of life caused by impinging radiation. However, much radiation sickness has not yet been tracked to its molecular origins. Here we give a brief summary of some molecules involved and an example of molecular damage known to be caused in living systems by radiation.

Many large molecules carry out the intricate chemical processes upon which life is based. Proteins, enzymes, nucleic acids, carbohydrates, lipids, and others are involved in the sensitive chemical machinery. All must function like clockwork, where a failing part can throw the whole apparatus off. This delicate system is an easy target for molecule-breaking radiation. It is no wonder that the lethal dose of some radioisotopes, such as strontium (Sr)-90, is measured in millionths of a gram.

Typical of the important molecules that can be damaged is DNA (deoxyribonucleic acid). DNA is the carrier of genetic information, the pattern which guides the growth of individuals generation after generation. Damage to DNA will thus endanger individuals long into the future.

DNA is made up of two molecular strands, which gently twist around one another in the famed "double helix" configuration. Forming the backbone of each strand are sugar and phosphate groups (Figure 11-8). Attached to the strand are side groups—nitrogen-containing bases called purines and pyrimidines. More detail of its complex structure need not concern us here. What do concern us

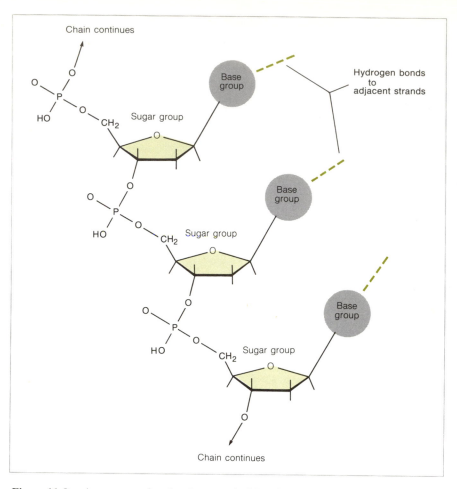

Figure 11-8. A segment of molecular strand of DNA. Two such chains spiral around each other to form the double helix.

are the kinds of chemical changes that radiation causes in this giant molecule. Many different forms of damage have been found:

1. One of the molecular strands may be fractured.
2. Both strands may be fractured.
3. The strands may bond chemically to each other (cross-linking).
4. The strands may bond to an outside molecule.
5. Hydrogen bonds holding the strands together may be ruptured.
6. A base side group may be torn off.
7. The chemical nature of the base may be changed.

Each change involves a rearrangement of atoms within the molecule, usually leading to the formation of a new chemical bond. Each disturbs the delicate structure of the DNA molecule so that its proper functioning may no longer be possible. Thus each may be the source of a genetic defect.

The multiple possibilities of damage to DNA are typical of the susceptibility of molecules to radiation. The great energy of radiation can tear almost any molecule apart at any number of places. Thus radiation damage within a living system involves a most complex set of chemical processes within a very large number of essential molecules.

CHEMICAL FACTORS IN HUMAN EXPOSURE TO RADIOISOTOPES

Nuclei are encased in shells of electrons. The movement of a nucleus through the environment depends entirely on the movement of its "electron casing." If the outer electrons unite by a chemical bond to form a molecule, that nucleus is then forced to follow the molecule wherever it goes. The properties of molecules that we have studied—size, polarity, chemical reactivity, and so on—determine whether the molecule will be airborne, dissolve appreciably in water, attach to falling dust particles, or be absorbed in the various tissues of man.

We see then that the hazard potential of a radioisotope depends on two quite different properties. First are the nuclear properties that we have been discussing, which determine when the nucleus is apt to disintegrate and the kind of radioactive fragments that will be given off. Second and equally important are the chemical properties characteristic of outer electron shells, which determine where the nucleus will go and thus where it is most likely to be when its radioactive burst takes place.

Chemical factors are too varied and complicated to follow in detail. They may determine how a radioisotope enters the environment, how rapidly it falls to earth, its absorption by plants and soil, its transmission to man, and its retention and excretion by man. These factors, in combination, fix human exposure. Figure 11-9 shows the interrelationship of some chemical pathways.

An example of chemical factors at work is provided by the inert-gas radioisotopes, especially those with atomic numbers 18 (argon), 36 (krypton), 54 (xenon), and 86 (radon). Since their outer electron shells are closed, atoms of these elements are not restrained by chemical bonds to other atoms, and they thus escape freely to the atmosphere.

One inert-gas radioisotope of importance is krypton (Kr)-85. We shall learn that it is produced by fission in nuclear power plants. When the nuclear fuel is being reprocessed in a special plant to purge it of fission radioisotopes, Kr-85 escapes to the atmosphere. No simple way has been found to capture Kr-85, because it cannot be chemically immobilized, as most of its sister fission products are. Once in the atmosphere, it exposes human skin and lungs. It does not expose internal organs appreciably because, again, its closed outer

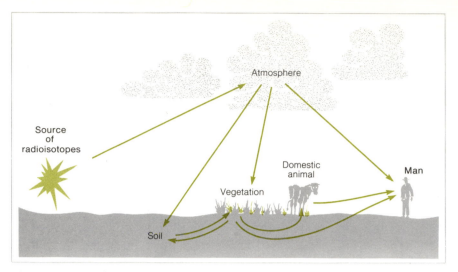

Figure 11-9. Various pathways followed by radioisotopes from their source to man. Each pathway is governed largely by chemical factors.

shell does not allow it to interact enough with molecules in body fluids to encourage it to dissolve. Both escape and exposure depend upon electron-chemical structure.

For similar reasons, the principal radioisotopes escaping directly from nuclear power plants are the inert gases Kr-85, xenon (Xe)-133, and Xe-135. A proposed gas-cooled reactor would release argon (Ar)-41. Also, in underground nuclear blasts to stimulate natural-gas production, Kr-85 remains mixed with the natural gas rather than settling out as do most other radioactive bomb products.

Uranium in soil is slightly radioactive, and it contributes some background radiation. However, it is locked firmly into the soil by chemical bonds. When the radioactive decay series of U-238 (discussed earlier) arrives at the first inert gas, radon (Rn)-222, which has no chemical bonds to immobilize it, escape to the atmosphere may occur (Figure 11-10). The above-ground location of Rn-222 encourages a greater exposure than would occur if it were buried. Its followers in the series, now unearthed, also contribute more to human exposure than they would have if they had remained underground. Lead (Pb)-210 will be mentioned in this connection later.

Some other radioisotopes, while not chemically inert, tend to form gaseous molecules that escape readily. Iodine (I)-131 is one of them, and in the chemical form of methyl iodide (CH_3I) and diatomic iodine (I_2) it is a major contaminant from reactor accidents. Hydrogen (H) not only forms gas or vapor molecules in the form of H_2 or H_2O, but the hydrogen atom is so small (that is, its electron "casing" is so small) that it can seep through metal retainers in nuclear reactors. Thus some H-3 (tritium) escapes from nuclear power plants.

One of the most feared radioisotopes is strontium (Sr)-90. Strontium is in the alkaline earth family of metals, as is calcium (Figure 11-11). Calcium enters the body to form bones and carry on

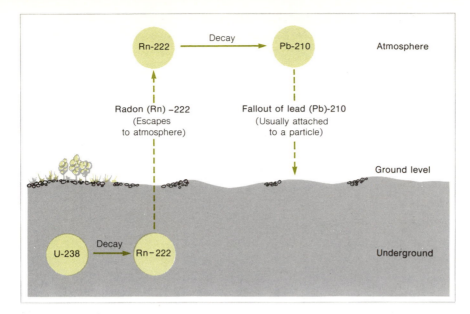

Figure 11-10. The critical role of radon (Rn)-222 in bringing the radioactive decay series of U-238 above ground, where exposure is enhanced. Rn-222 has this role only because it is chemically inert. (Note: Each heavy arrow indicating "decay" represents several steps in the decay chain. For simplicity the intermediates are not shown.)

other physiological functions. Sr-90 follows its chemical sister, calcium, into the body and to the bone. There it chemically deposits with calcium and becomes a source of bone cancer, leukemia, and other bone- and marrow-related disorders.

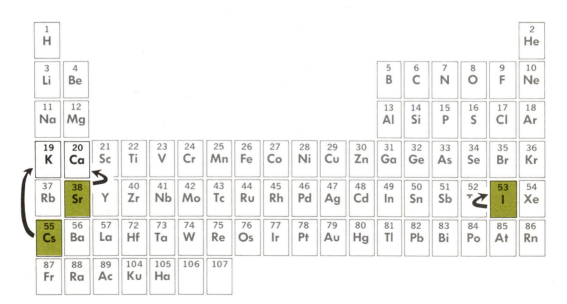

Figure 11-11. Important radioisotopes of the shaded elements follow chemically related elements (or, in the case of iodine, a nonradioactive form of the same element) into the body and to specific organs. The location and severity of exposure thus depends on chemical similarity in families of the periodic table.

Another element chemically related to calcium is radium. The radioisotope Ra-226 therefore goes to the bone also. Other significant radioisotopes seeking bone (based on more subtle chemical reasons) are plutonium (Pu)-239, zirconium (Zr)-95, and lead (Pb)-210.

Just as strontium is a bone seeker because of its chemical nature, ingested iodine (I) is directed to the thyroid by chemical forces. There it is used in the body's manufacture of the iodine-containing hormone thyroxin. Radioactive iodine (I)-131 concentrates in the thyroid just as chemically identical nonradioactive iodine does. There its nuclear disintegration can cause thyroid cancer. Largely for this reason, I-131 is one of the most serious radioactive contaminants entering the environment.

Cesium (Cs) is an alkali metal in the same chemical family as potassium, important to cells throughout the body. Therefore radioactive cesium (Cs)-137 follows chemical routes into and through the body as potassium does and exposes almost the entire body.

Human exposure is also influenced by the ease with which a radioisotope is absorbed into the bloodstream and tissues and by its speed of excretion. A radioisotope that is absorbed readily and excreted slowly will accumulate in the body and pose a greater threat than will one with opposite behavior. These factors also depend on chemical nature. A summary for several isotopes appears in Table 11-10.

The chemical pathways of radioisotopes in the body are only part of the story, as we have noted. Pathways that a radioisotope takes before it enters the body are equally significant. For instance, cesium (Cs) which falls to earth becomes chemically bound to soils. Therefore it is less readily absorbed from soil by plants than, for instance, Sr-90. The amount of Cs-137 passed to man by plants is thereby relatively reduced.

Biological concentration and *biological magnification* are processes in which living organisms extract certain chemicals from the environment. These substances become concentrated in the organism, and this high concentration frequently passes up the food chain, sometimes being magnified at each step. Certain radioisotopes in water become concentrated in marine life, which, if eaten by man, intensifies

Table 11-10. SUMMARY OF CHEMICALLY CONTROLLED ABSORPTION, EXCRETION, AND CONCENTRATION FOR SEVERAL RADIOISOTOPES IN MAN

Radioisotope	Half-Life	Ease of Absorption	Speed of Excretion	Place of Concentration
Strontium (Sr)-90	28 years	Good	Slow	Bone
Iodine (I)-131	8 days	Good	Months	Thyroid
Cesium (Cs)-137	30 years	Good	Months	All tissue
Carbon (C)-14	5600 years	Good	Slow	All tissue
Zirconium (Zr)-95	65 days	Fair	Slow	Bone

his exposure. Cs-137 in fresh water, for example, becomes concentrated by a factor of 1000 and sometimes up to 9000 in fish and 2000 to 6000 in waterfowl. Again, the chemical factors which govern this concentration influence the exposure of the ultimate consumer, man.

EXPOSURE TO NATURAL RADIOACTIVITY

For perspective, it is essential to realize that radioactivity has always existed on earth, and with each passing second something like a thousand nuclear disintegrations occur within each of us. We exist in a sea of radioactivity. We must recognize the magnitude and origin of this background radioactivity, for it is widely held that synthetic radioactivity is not excessive if it does not greatly exceed nature's own allotment. Others argue that we should have a more direct concern with the potential damage of the added radioactivity, for there is relatively little we can do about the natural form (except perhaps to stay out of granite buildings and seek lower elevations). It has been here since the origin of the earth, and perhaps, along with UV radiation, it helped to promote the molecular rearrangement from which life itself took root over 3 billion years ago.

Since radioactive isotopes disintegrate almost completely to other isotopic species in a few half-lives, it might seem puzzling that any natural radioactivity remains on earth. Nonetheless, 66 radioisotopes exist naturally in the earth's crust, oceans, and atmosphere. They originate in three ways:

1. Long-Lived Radioisotopes. Some radioisotopes that were apparently common on earth when it was born over 4 billion years ago have sufficiently long half-lives, near a billion years or more, to remain in quantity. The uranium isotopes U-235 and U-238 are the best-known examples.

2. Derivative Radioisotopes. Several of the long-lived radioisotopes decay to isotopes that are also radioactive. They were shown earlier as components in the radioactive decay series. Within these series an entire sequence or chain of derivative radioisotopes all stem ultimately from a single long-lived parent.

As shown earlier, U-238 is the parent radioisotope in a chain containing over a dozen other radioisotopes. All are now present on earth, despite the fact that some have short half-lives, because a new supply constantly passes down the chain from the decaying U-238. Radium (Ra)-226, half-life 1622 years, and radon (Rn)-222, half-life 3.8 days, are the best known examples of radioactive progeny of U-238.

3. Cosmic-Ray Radioisotopes. Some radioisotopes are created with the aid of cosmic rays. Cosmic rays bombard the upper atmosphere, their extreme energy shattering the nuclei of atmospheric gases, sending showers of neutrons and other nuclear particles toward earth. The neutrons start nuclear reactions in other nuclei in the manner we saw earlier, forming radioisotopes. For instance, radioactive carbon (C)-14 is formed by cosmic-ray neutrons bombarding atmospheric nitrogen (N) by the nuclear reactions

$$^{14}_{7}N + ^{1}_{0}n \rightarrow ^{15}_{7}N$$

$$^{15}_{7}N \rightarrow ^{14}_{6}C + ^{1}_{1}proton$$

Although C-14 decays with a half-life of 5600 years, its steady renewal by cosmic-ray neutrons means that a trace always exists in the atmosphere.

When atmospheric C-14 becomes incorporated with other carbon in living material which later dies, the C-14 gradually decays. Its declining ratio relative to normal C-12 tells the age of the object. This ratio is the basis of radiocarbon dating. Its discovery by chemist Willard Libby in 1950 has revolutionized archaeological dating.[8]

Table 11-11 summarizes some major natural radioisotopes in the above three categories.

Table 11-11. SOME IMPORTANT NATURAL RADIOISOTOPES, GROUPED ACCORDING TO ORIGIN

	Radioisotope	Half-Life	Radioactive Emission
Long-lived radioisotopes	Uranium (U)-238	4.5×10^9 years	α, γ
	Uranium (U)-235	7.1×10^8 years	α, γ
	Thorium (Th)-232	1.4×10^{10} years	α, γ
	Potassium (K)-40	1.3×10^9 years	β, γ
	Rubidium (Rb)-87	4.7×10^{10} years	β
Derivative radioisotopes	Radium (Ra)-226	1622 years	α, γ
	Radium (Rn)-228	5.8 years	β
	Radon (Rn)-222	3.8 days	α, γ
	Radon (Rn)-220	55 seconds	α, γ
	Polonium (Po)-210	138 days	α, γ
	Lead (Pb)-210	21 years	β, γ
Cosmic-ray radioisotopes	Hydrogen (H)-3 (tritium)	12 years	β
	Carbon (C)-14	5600 years	β
	Phosphorus (P)-32	14 days	β
	Sulfur (S)-35	87 days	β

[8] Enough C-14 exists in living matter to provide approximately 16 β decays in 1 g of carbon per min. An object buried one C-14 half-life ago, 5600 years, would have its β decay reduced to 8 disintegrations/min, and an object dead 2×5600 years = 11,200 years would show activity of only 4 disintegrations/min. Thus carbon from a cypress beam in the tomb of Egyptian Pharaoh Sneferu exhibited just over 8 decays/min average, suggesting an age somewhat less than the 5600 year half-life, calculated by more precise means to be 4800 years old. Archaeological records suggest an age of 4600 years, in good agreement. Most sites cannot be dated by archaeological means alone; in these cases C-14 dating is invaluable.

Radiation Exposure. All of the natural radioisotopes combined with the direct and indirect effects of cosmic rays lead to an average worldwide radiation exposure of slightly over 0.1 rads (100 millirads) per year. Exposure varies with location; statewide averages are shown in Figure 11-12. Exposure increases about 30 millirads per mile of elevation because cosmic-ray activity is greatest near the upper atmosphere. The level of exposure also varies depending on the amount of uranium and other radioisotopes in the local soil; in one Brazilian village, exposures as high as 12 rads (12,000 millirads) per year are encountered.

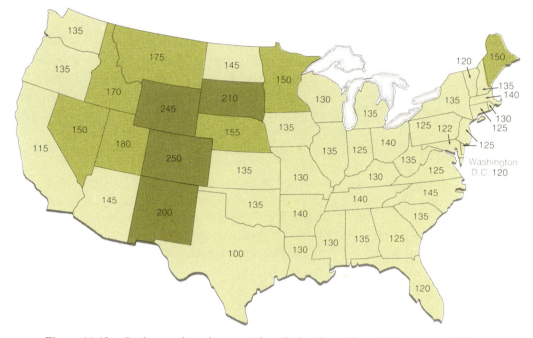

Figure 11-12. Background environmental radiation in various states, expressed in millirads. Radiation from granite or brick structures may add 50 percent to exposure. The contribution of cosmic rays to the total radiation dose increases with elevation, giving the mountain states the highest exposures.

Natural radiation exposure is best divided into three categories: cosmic rays, radioisotopes external to the body whose gamma rays penetrate the body (alpha and beta particles penetrate too weakly for this), and radioisotopes which enter the body, there to disintegrate with the release of alpha, beta, and gamma radiation. The average human exposure near sea level in these three categories is shown in Table 11-12.

The principal external radioisotopes are U-238, thorium (Th)-232, radium (Ra)-226, and potassium (K)-40. They are abundant in soil and rock. What is more important, they are components of building materials such as concrete and granite blocks, from which vantage point they expose occupants to their incessant gamma emissions.

One radioisotope, potassium (K)-40, is responsible for about

Table 11-12. SOURCES OF NATURAL RADIATION EXPOSURE NEAR SEA LEVEL

Source	Average Dose (millirads)
Cosmic rays	35
External radioisotopes	60
Internal radioisotopes	25
Total	120

90 percent of general internal exposure. Carbon (C)-14 is second. These radioisotopes enter the body by following the chemical pathways of nonradioactive potassium and carbon, two of the most important elements of living systems.

Certain parts of the body receive a high dose from other sources. Radon (Rn)-222, the inert gas in the U-238 decay chain mentioned earlier, escapes into the atmosphere from uranium-bearing minerals, there to offer a high exposure to inhaling lungs. Lung tissue may, through this mechanism, selectively receive over 100 millirads per year. Radium (Ra)-226, the immediate precursor of Rn-222 in the series, has an affinity for bone and teeth, giving added exposures of 1 to 2 millirads per year.

Special chemical routes of exposure abound. For instance, atmospheric Rn-222 decays through several short-lived daughters to reach lead (Pb)-210, half-life 21 years. Pb-210 fallout (Figure 11-10) may contaminate plants or soil. Enough is gathered by the big leaf of the tobacco plant to give an estimated 40 millirads per year to the lung tissue of a pack-a-day smoker.

Diagnostic X Rays. Although it is somewhat removed from the subject of natural radioactivity, another common source of radiation exists that is not related to the nuclear industry. It is the use of x rays in medical diagnosis. This radiation contributes 30 or more millirads per year average to human exposure.

Zero Radioactive Pollution. Again it is most clear that zero pollution is an impractical goal for our world. Nature (bless her!) rules it out unequivocally. Even if we roamed naked in the world, all traces of civilization gone, we would inject real amounts of carbon (C)-14 into the air with the carbon dioxide (CO_2) in each exhaled breath.

We should be careful, however, not to use the inevitability of pollution as an excuse to pollute with abandon. We must carefully weigh the impact of different levels of pollution, then act in such a way that a livable world is ensured. What this means in terms of man-made radioactivity will be explored in various sections to follow.

NUCLEAR FISSION

Fission is quite literally the "splitting of atoms." It is a special kind of nuclear transformation in which a heavy nucleus breaks into two or more medium-sized daughter nuclei. Protons and neutrons in the daughter nuclei are bound more tightly than in the parent, as we explained earlier, and therefore fission is always accompanied by the release of a large quantity of energy.

Fission occurs when very heavy nuclei are struck with some projectile—a neutron, an alpha particle, a proton, or a gamma ray. Man can therefore induce and control fission by bombarding fissionable materials with these projectiles. In contrast, the occurrence of radioactivity is virtually uncontrolled.

Neutrons are the only projectiles abundant and inexpensive enough to provide a net economical energy release from fission. The reason is that, once started, fission provides its own neutrons. As the nucleus splits, these extra neutrons fly out, soon to collide with other fissionable nuclei and induce even more fissions. Thus the *chain reaction* can expand, with each fission releasing an increasing number of neutrons to induce still further fission. In this way the fissionable material may be used up and a great energy release achieved.

Since nature provides its own neutrons for fission, the fission process would hardly seem controllable by man. Yet by arrangements we shall describe later, man can see to it that most of the neutrons pass out of the fissionable material and are lost, thus terminating the chain reaction.

Fission Products. The fissionable nucleus is like a pane of glass, able to shatter in many ways. Each leads to a different division of protons and neutrons among daughter nuclei. Several fission routes for U-235, producing the precursors to strontium (Sr)-90, iodine (I)-131, and cesium (Cs)-137, are shown on next page by balanced nuclear equations.[9] The first of these fission routes is also illustrated in Figure 11-13. The final steps leading to these environmental radioisotopes will be shown in the next section.

$$\boxed{\text{becomes strontium-90}}$$
$$\uparrow$$
$$^{1}_{0}n \;+\; ^{235}_{92}U \;\rightarrow\; ^{90}_{35}Br \;+\; ^{143}_{57}La \;+\; 3^{1}_{0}n$$

neutron　uranium-235　bromine-90　lanthanum-143　3 neutrons

---

[9] If we know one fission isotope and the number of neutrons released, we can determine the other isotope by balancing the nuclear equation. Thus krypton (Kr)-93, released with 2 neutrons, is represented by

$$^{1}_{0}n \;+\; ^{235}_{92}U \;\rightarrow\; ^{93}_{36}Kr \;+\; ^{A}_{Z}X \;+\; 2^{1}_{0}n$$

1 neutron　　uranium-235　　krypton-93　　?　　2 neutrons

Unknown isotope X must have an atomic number Z such that added to krypton's 36 it gives uranium's 92: $Z + 36 = 92$. Only atomic number $Z = 56$ satisfies this condition. The periodic table shows element 56 to be barium (Ba). The mass number A of the barium isotope must be such that added to krypton's 93 and the two neutrons' 2 it gives uranium's 235 plus the neutron's 1: $A + 93 + 2 = 235 + 1$. Hence A must be 141. In this way we find that the isotope released with Kr-85 and two neutrons is barium-141 or $^{141}_{56}Ba$.

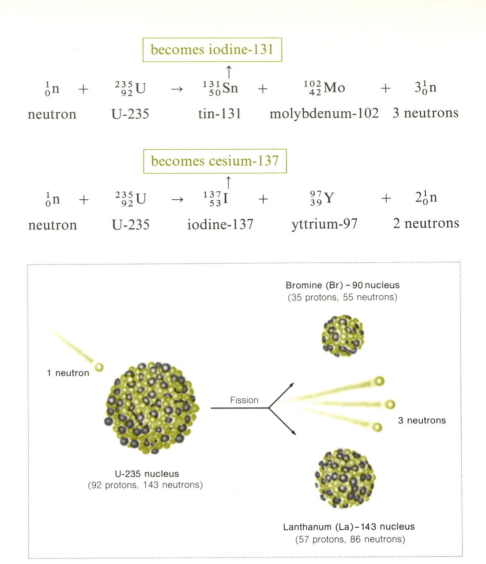

$$\text{becomes iodine-131}$$
$$\uparrow$$
$$^1_0\text{n} + ^{235}_{92}\text{U} \rightarrow ^{131}_{50}\text{Sn} + ^{102}_{42}\text{Mo} + 3^1_0\text{n}$$
$$\text{neutron} \quad\quad \text{U-235} \quad\quad \text{tin-131} \quad \text{molybdenum-102} \quad \text{3 neutrons}$$

$$\text{becomes cesium-137}$$
$$\uparrow$$
$$^1_0\text{n} + ^{235}_{92}\text{U} \rightarrow ^{137}_{53}\text{I} + ^{97}_{39}\text{Y} + 2^1_0\text{n}$$
$$\text{neutron} \quad\quad \text{U-235} \quad\quad \text{iodine-137} \quad\quad \text{yttrium-97} \quad\quad \text{2 neutrons}$$

Bromine (Br) – 90 nucleus
(35 protons, 55 neutrons)

1 neutron

Fission

3 neutrons

U-235 nucleus
(92 protons, 143 neutrons)

Lanthanum (La) – 143 nucleus
(57 protons, 86 neutrons)

Figure 11-13. One routes for the fission of U-235. Note that the same number of protons (92) are found left and right, and the same number of neutrons (144). The three neutrons set free by fission may go on to instigate fission in other U-235 nuclei, thus starting a chain reaction.

A hundred or so other fission reactions could be written, some leading to other environmental radioisotopes.

Two of the above fission reactions produce three neutrons and one of them two neutrons. All of the many fission reactions of U-235 combined produce an average of 2.4 neutrons. The fission of plutonium (Pu)-239 produces an average of 2.9 neutrons. For U-233, this number is 2.5 neutrons. Any isotope whose fission produces an average of more than one neutron can support an expanding chain reaction; in theory such an isotope can be fuel for nuclear weapons or nuclear power plants.

Imagine an isotope whose fission produces two neutrons. These two neutrons can produce two fissions; the four neutrons released from two fissions can produce four fissions, or eight neutrons

(Figure 11-14). By stages, the chain reaction can double and double again from 1 fission to 2, 4, 8, 16, 32, 64, 128, . . . to near infinity. If it grows uncontrolled, a bomb burst results. If controlled by the removal of some of the growing number of neutrons, it produces a steady flow of energy, possibly for power production.

A number of the heavy synthetic elements undergo fission. Few of them can be used because (1) they are not available in large quantities or (2) they are intensely radioactive, losing their fission activity and energy in an uncontrolled, dangerous manner. The three

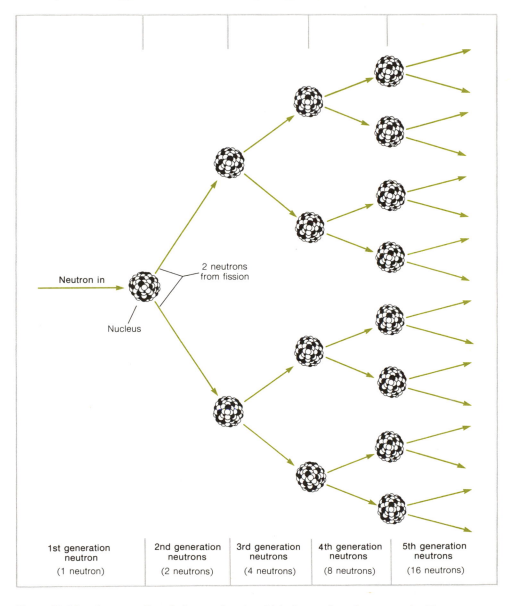

1st generation neutron	2nd generation neutrons	3rd generation neutrons	4th generation neutrons	5th generation neutrons
(1 neutron)	(2 neutrons)	(4 neutrons)	(8 neutrons)	(16 neutrons)

Figure 11-14. An expanding chain reaction, in which the number of neutrons doubles with every fission, from 1, 2, 4, 8, 16, 32 to trillions. The fission of some nuclei releases 3 neutrons, giving a tripling at each stage, from 1, 3, 9, 27, 81, 243, . . . to trillions of neutrons. Nuclear processes are so fast that the trillions of neutrons can be generated from one in less than a millionth of a second, giving a nuclear explosion.

useful and potentially useful fissionable isotopes are U-235, plutonium (Pu)-239, and U-233. We shall discuss them later.

BETA DECAY TO ENVIRONMENTAL RADIOISOTOPES FOLLOWING FISSION

The daughter nuclei produced in fission are almost always radioactive. The reason is that nuclear instability at high atomic numbers is caused by proton-proton repulsion, a factor which disturbs the common one-to-one neutron-proton balance.

The ratio of neutrons to protons in stable elements varies across the periodic system. This variation is illustrated in Figure 11-15, a plot of the number of neutrons against atomic number, Z. At low and medium Z values, neutron and proton shells have roughly equal binding stability and therefore fill up at about the same rate. Stable nuclei in this part of the periodic system therefore tend to have an equal number of protons and neutrons. The oxygen isotope $^{16}_{8}O$ (8 neutrons, 8 protons), the most abundant isotope in the earth's crust, is typical.

Further along in the periodic system the mutual repulsion of protons weakens proton binding in the nucleus, and thus neutrons

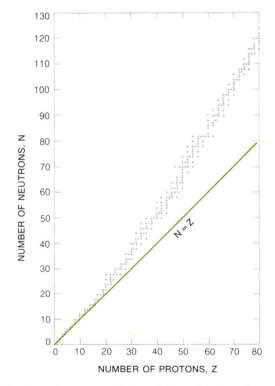

Figure 11-15. Number of neutrons, N, in stable nuclei plotted as a function of the number of protons, Z. A point on the straight line would represent a nucleus with an equal number of protons and neutrons. The points veer upwards from this line, indicating an increasing neutron excess as Z becomes large.

are favored. Consequently, nuclei of the heavy elements have high neutron-proton ratios. For instance, uranium-235 ($^{235}_{92}$U) has 143 neutrons and 92 protons, a ratio of about three to two. Plutonium-239 ($^{239}_{94}$Pu) has a similar ratio.

When the fission of an isotope like U-235 occurs, a few of the excess neutrons fly off to continue the chain reaction, but most of them end up in the daughter nuclei. These nuclei, being roughly half of the U-235 nucleus, are in the central part of the periodic table, where neutrons and protons are favored in roughly equal numbers. We are not surprised, then, that processes begin to take place in these newly formed fission daughters that will reduce their considerable excess of neutrons over protons. The nuclear process that best does so is beta decay, in which a nuclear neutron becomes a nuclear proton plus a cast-off electron (beta particle). Thus in a single beta decay, the neutron count is reduced one and the proton count is raised one. Typically, several successive beta decays occur to bring the nucleus close enough to a one-to-one ratio to achieve stability.

It is clear that the primary radioactivity from man-induced fission, whether power production or nuclear bombs, is beta activity. Gamma rays often accompany beta emission, so they are also important. Very little alpha activity is present, although a few alpha particles are thrown off as the nuclei rupture.

Long-Lived Environmental Radioisotopes. Generally the first few beta decays of a fission daughter occur very rapidly, within seconds or minutes. Their beta and gamma radioactivity is therefore confined near the site of fission (power plant or bomb) and does little damage at points distant from the site. After several fast beta emissions, a radioisotope of long half-life is often produced, one that has time before decay to spread widely into the environment, where its radioactivity can cause considerable damage.

The production of significant environmental radioisotopes can be illustrated by bone-seeking strontium (Sr)-90. To start with, fission produces a considerable amount of bromine-90, as we have seen. In three quick steps, bromine (Br)-90 beta decays to krypton (Kr)-90, which beta decays to rubidium (Rb)-90, which then yields strontium (Sr)-90. The sequence of nuclear reactions and their half-lives are

$$^{90}_{35}\text{Br} \rightarrow \,^{90}_{36}\text{Kr} + \,^{0}_{-1}\beta \qquad \text{(1.4 sec)}$$

$$^{90}_{36}\text{Kr} \rightarrow \,^{90}_{37}\text{Rb} + \,^{0}_{-1}\beta \qquad \text{(33 sec)}$$

$$^{90}_{37}\text{Rb} \rightarrow \,^{90}_{38}\text{Sr} + \,^{0}_{-1}\beta \qquad \text{(2.7 min)}$$

Then, as shown earlier, Sr-90 decays with a 28-year half-life to yttrium (Y)-90:

$$\boxed{^{90}_{38}\text{Sr} \rightarrow \,^{90}_{39}\text{Y} + \,^{0}_{-1}\beta} \qquad \text{(28 years)}$$

BETA DECAY TO ENVIRONMENTAL RADIOISOTOPES

Since Sr-90, but not its precursors, has a lifetime sufficient to circulate widely in the environment before decay, only its beta emission, of those shown, is an environmental threat. For this reason its beta emission is set apart by a box.

After one more beta decay,[10] from yttrium-90, a stable isotope is reached, zirconium (Zr)-90 or $^{90}_{40}$Zr.

The other prominent environmental isotopes have a similar history. Iodine (I)-131 is formed from tin (Sn)-131 through the following sequence of reactions, terminating in stable xenon (Xe)-131.

$$^{131}_{50}\text{Sn} \rightarrow \, ^{131}_{51}\text{Sb} + \, ^{0}_{-1}\beta \qquad \text{(3.4 min)}$$

$$^{131}_{51}\text{Sb} \rightarrow \, ^{131}_{52}\text{Te} + \, ^{0}_{-1}\beta \qquad \text{(23 min)}$$

$$^{131}_{52}\text{Te} \rightarrow \, ^{131}_{53}\text{I} \; + \, ^{0}_{-1}\beta \qquad \text{(24 min)}$$

$$\boxed{^{131}_{53}\text{I} \; \rightarrow \, ^{131}_{54}\text{Xe} + \, ^{0}_{-1}\beta} \qquad \text{(8 days)}$$

Cesium (Cs)-137 begins as the fission daughter iodine (I)-137 and ends as stable barium (Ba)-137.

$$^{137}_{53}\text{I} \; \rightarrow \, ^{137}_{54}\text{Xe} + \, ^{0}_{-1}\beta \qquad \text{(24 sec)}$$

$$^{137}_{54}\text{Xe} \rightarrow \, ^{137}_{55}\text{Cs} + \, ^{0}_{-1}\beta \qquad \text{(3.9 min)}$$

$$\boxed{^{137}_{55}\text{Cs} \; \rightarrow \, ^{137}_{56}\text{Ba} + \, ^{0}_{-1}\beta} \qquad \text{(30 years)}$$

These three beta sequences (and there are many more) give, with each step, an element one unit higher in the periodic system. This progression is best seen in a rectangular-grid diagram, where the steps along the atomic-number scale are readily apparent:

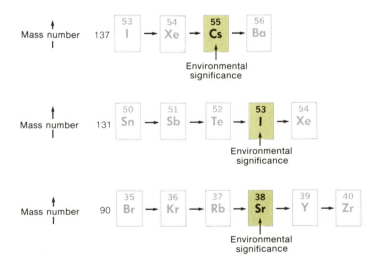

[10]This beta-decay sequence starts with $^{90}_{35}$Br and ends with $^{90}_{40}$Zr. From beginning to end, neutrons have decreased from $90 - 35 = 55$ in number to $90 - 40 = 50$, while protons have increased from 35 to 40. The neutron-proton ratio has thereby decreased from $(55/35) = 1.57$ for original fission product $^{90}_{35}$Br to $(50/40) = 1.25$ for the final product, stable $^{90}_{40}$Zr.

Whether a given long-lived radioisotope is important depends on whether its chemical nature encourages transmission to and absorption by the body and upon whether it is produced in abundance. The chemical route into the body is a complicated subject. There are many links in the chain between source and man, as we have noted. The level of production of each radioisotope from fission is, by contrast, perfectly known; values are shown in Table 11-13. This table gives the *yield*, the number of long-lived radioactive nuclei produced for each 100 fission processes. Yields for U-235 are those whose fission is induced by slow neutrons (so called thermal neutrons), and they are most applicable to nuclear power plants. However, nuclear weapons based on fast-neutron fission produce a similar yield. The yields for plutonium (Pu)-239 are for fast-neutron fission, and thus they approximate the long-lived radioactivity produced by many nuclear weapons.

Table 11-13. **YIELD OF SOME LONG-LIVED RADIOISOTOPES THAT MAY ENTER THE ENVIRONMENT AFTER THE FISSION OF URANIUM (U)-235 OR PLUTONIUM (Pu)-239**

| | Radioisotope | Half-Life | Yield (per hundred fissions) | |
			U-235	Pu-239
Most significant	Strontium (Sr)-90	28 years	5.8	2.2
	Iodine (I)-131	8 days	3.1	3.8
	Cesium (Cs)-137	30 years	6.1	5.2
	Strontium (Sr)-89	51 days	4.8	1.9
	Zirconium (Zr)-95	65 days	6.2	5.9
	Krypton (Kr)-85	10.3 years	0.3	—
	Ruthenium (Ru)-103	40 days	3.0	—
	Barium (Ba)-140	13 days	6.4	5.7
	Cerium (Ce)-141	33 days	6.0	5.2

Clearly in the fission process the seething energy of the nucleus leads to many kinds of nuclear fracture, thus explaining the great diversity of radioisotopes produced. In a few cases, smaller particles are thrown off during fission. Thus for every 300 fissions (yield, 0.3 percent), an alpha particle is emitted, giving the fission process a small and localized alpha activity. Also, once in every 10,000 fissions (yield, 0.01 percent) of U-235, a tritium nucleus is produced. Tritium is a radioactive isotope of hydrogen, and one of the few radioisotopes that has a distinct name. Technically it is hydrogen (H)-3, or ^3_1H. It is a weak beta emitter (half-life 12.3 years) whose beta particles are used to light up luminous chemicals on watch dials. It is also an environmental contaminant.

PLUTONIUM

Plutonium (Pu), element 94, is virtually absent from nature, as are all elements with atomic number greater than uranium's 92. The synthesis of artificial plutonium was but one step in the extension of the periodic table from element 92 to its present upper bound of 105. It was a most profound step in its significance to mankind, for plutonium-239 is a long-lived (half-life 24,000 years) isotope that is highly susceptible to chain-reaction fission. When its nucleus splits, an average of nearly three neutrons are released, more than from any other long-lived isotope and more than enough to continue the chain reaction with explosive vigor. It is, sadly, ideal bomb material.

The synthesis and detection of elements beyond 92 is one of the outstanding feats of science of the past third of a century. For most such elements, ingenious nuclear transformations were instigated by use of the nuclear fragments produced in giant accelerators. Plutonium itself, however, is produced more simply, and by means that are altogether logical.

We found earlier that neutron-rich nuclei tend to beta decay, producing elements of increased atomic number. It might be expected, then, that if a neutron-rich uranium nucleus could be produced, it would beta decay to elements with new highs in atomic number. Following this logic, Enrico Fermi first tried to produce new elements in 1934. Although he was unable to detect the desired product, this scheme has since been used to produce plutonium in abundance.

The starting material is uranium (U)-238, the natural isotope of uranium that has the most neutrons. When U-238 is bombarded with neutrons, its nucleus captures one additional neutron, and the nuclear isotope U-239 is formed.

$$^{238}_{92}\text{U} + ^{1}_{0}\text{n} \rightarrow ^{239}_{92}\text{U}$$

U-239, now overly rich in neutrons, undergoes two successive beta decays, forming first neptunium (Np)-239 then plutonium (Pu)-239.

$$^{239}_{92}\text{U} \rightarrow ^{239}_{93}\text{Np} + ^{0}_{-1}\beta \qquad \text{(24 min)}$$

$$^{239}_{92}\text{Np} \rightarrow ^{239}_{92}\text{Pu} + ^{0}_{-1}\beta \qquad \text{(2.4 days)}$$

Through this sequence of reactions, in which U-238 is the fuel and U-235 fission provides the neutrons, hundreds of tons of plutonium have been produced in so-called nuclear reactors. Much of it now resides in nuclear warheads aimed at the cities of the world.

FISSION BOMBS

The full impact of the nuclear age was thrust on the world on August 6, 1945. On that date a fission bomb was detonated over Hiroshima, Japan, killing 70,000 people.

Before that time, on December 2, 1942, an unheralded event of equal significance took place. A nuclear reactor in Chicago with Enrico Fermi in charge accomplished a controlled and gradual release of fission energy, as if to anticipate a new kind of world with nuclear energy at its technological foundation. But it was wartime, and all the quiet developments of atomic energy were secret until the bomb burst suddenly onto the world.

Bombs and reactors both tap the enormous energy of the nucleus; only the suddenness of the energy release is different. We discuss bombs here and reactors in the next section. However, they are not totally independent. The explosive agent in most fission bombs—plutonium (Pu)-239—is made in reactors. Also, certain fission mixtures used for reactor fuel can theoretically be diverted to fission bombs. Likewise, bomb material can become reactor fuel.

The building of a fission bomb (often called "atomic bomb") hinges on the concept of *critical mass*. We learned earlier that a chain reaction can grow rapidly from one neutron causing one fission to 2, 4, 8, 16, . . . , and very soon to explosive trillions of neutrons causing trillions of energy-releasing fissions. A chain reaction like this one, unchecked, becomes a violent explosion.

Neutrons released in a small piece of fissionable material may escape to the outside before they have an opportunity to induce fission (Figure 11-16). If enough neutrons are lost this way, the chain reaction dies and there is no explosion. Such a condition is called *subcritical*. At a certain larger mass, the *critical mass*, the chain

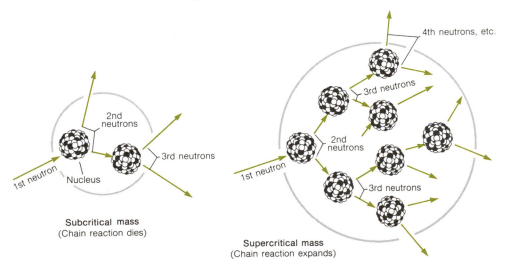

Figure 11-16. The chain reaction initiated by a neutron in a small, "subcritical" mass of fissionable material dies because too many neutrons escape unproductively across its boundaries. Few neutrons escape the larger, "supercritical" mass, and they cause the chain reaction to expand, usually with explosive force.

neither grows nor dies. At a bigger, *supercritical* mass, only a few neutrons escape to the boundaries, and there are more than enough left to expand the chain into an explosion.

The critical mass—the balance point between explosion and nonexplosion—varies for different fissionable materials. It is roughly 3 kg for Pu-239 and 10 kg for U-235. It varies somewhat, depending on shape.

The principle of a fission bomb is amazingly simple. Two or more small, subcritical pieces of fissionable material, each nonexplosive by itself, are brought together into a supercritical mass, which immediately explodes.

While the principle of a fission bomb is simple, its engineering is somewhat more difficult, because the subcritical masses must be brought together with great speed to avoid a premature "fizzle." The Hiroshima bomb was a simple "gun-tube" device (Figure 11-17) in which one subcritical mass of U-235 was hurled into another at the opposite end of a cannon tube by a charge of chemical explosive. Even though the explosive heart of the Hiroshima bomb weighed only a few kg (somewhat above the U-235 critical mass), the entire device with gun tube and explosives was a rather monstrous 5 tons. Such was the power in the few kilograms of fissionable U-235 that an explosion equivalent to 14,000 tons (14 kilotons) of TNT rocked Hiroshima.

The second Japanese bomb, dropped over Nagasaki three days later, was a more sophisticated "implosion" device, in which a hollow sphere of Pu-239 was crushed to a supercritical density by chemical explosives.

Since Hiroshima and Nagasaki, bomb design has vastly im-

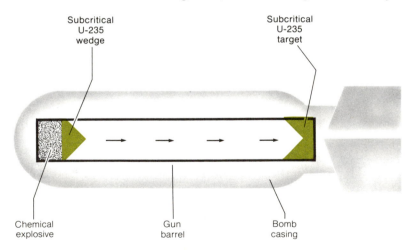

Figure 11-17. Configuration of "gun-tube" fission bomb, the type that destroyed Hiroshima. The high explosive fires the U-235 wedge into the U-235 target. Upon impact, the total U-235 mass becomes supercritical and a powerful explosion occurs.

proved, and greater explosive power has been obtained for a given bomb size. At one extreme, giant fusion bombs (H-bombs) were designed with a vastly increased explosion power, but this is another story, to be told after we have studied fusion. The opposite extreme may have equal significance—the development of smaller and smaller bombs with a Hiroshima punch. Such bombs have been incorporated into artillery shells, bazookas, and other battlefront weapons, but not yet used. Nuclear expert Ralph Lapp claims that such a bomb can now be packaged in a hand-carried case. Such a weapon could easily be concealed, and thus may be subject to theft and smuggling at some future date when fissionable materials are more commonly available.

ENVIRONMENTAL THREAT OF NUCLEAR WEAPONS

The velocity of neutrons from a fissioned nucleus is incredibly high, and they cause fission so rapidly that in less than a millionth of a second the chain reaction can escalate to explosive proportions. So great is the sudden burst of energy that the fissionable core of the bomb is heated to tens of millions of degrees. Bomb casing and all surrounding materials are vaporized and hurled away. These hot, expanding materials form the fireball which, in an air detonation, bursts out to 600 ft in diameter in slightly over a hundredth of a second. This burst causes intense shock waves in the surrounding air, destructive for many miles. Gamma rays from the nuclear reactions and other electromagnetic radiation in the visible, ultraviolet, and infrared ranges are emitted at damaging levels. The fireball residue, very hot and thus lighter than air, rises like a giant balloon, pulling a column of rushing air after it which contains debris and dust sucked up from the ground. The fireball and air column together form a *mushroom-shaped cloud*, now a symbol of the nuclear age.

Radioactive Fallout. Within the mushroom-shaped cloud are vaporized, radioactive fission products. Upon cooling, many of these substances attach by chemical or intermolecular forces to dust particles; others join water droplets. Dust and rain come eventually to earth as *radioactive fallout*. Fallout may begin locally a few hours after detonation, but some of the radioactive materials from the most powerful explosions are carried in the stratosphere for years, and they finally come down all across the face of the earth.

The most dangerous fallout isotopes are the long-lived fission products listed in Table 11-14, including strontium (Sr)-90, iodine (I)-131, cesium (Cs)-137, strontium (Sr)-89, and zirconium (Zr)-95,

Table 11-14. APPROXIMATE LIFETIME EXPOSURE OF THE AVERAGE WORLD CITIZEN TO SOME FALLOUT RADIOISOTOPES FROM FISSION BOMB TESTS, 1954–1965

Radioisotope	Organ	Dose (millirads)
Cesium (Cs)-137	All organs (external radiation)	25
	Gonads	15
	Marrow	15
	Cells lining bone	15
Strontium (Sr)-90	Gonads	0
	Marrow	78
	Cells lining bone	156
Carbon (C)-14	Gonads	13
	Marrow	13
	Cells lining bone	20

plus a large residue of unfissioned plutonium (Pu)-239.[11] Of these isotopes, I-131 provides the greatest exposure in the first few weeks following the nuclear blast.

Some radioactivity is also induced as neutrons impinge on surrounding materials. We have already shown how cosmic-ray neutrons convert atmospheric nitrogen (N)-14 to radioactive carbon (C)-14. Bomb neutrons do the same thing, and C-14 thus becomes an important environmental contaminant. Other neutron-induced radioisotopes are calcium (Ca)-45 and sodium (Na)-24.

A 1966 United Nations report assessed the probable dose to the average person from various radioisotopes produced by nuclear tests from 1954 through 1965. Because of chemical factors, various organs (gonads, bone marrow, cells lining bone) are exposed unequally, as Table 11-14 shows. In total, these doses represent 100 to 200 millirads—not a frightening amount. However, the exposure is not equally distributed: some individuals in the arctic are committed to much higher doses because of their unique chemical food chain. Also, C-14 is long-lived enough that many successive generations will suffer a similar exposure, which will give cumulative genetic damage. These difficulties, of course, pale in comparison to those that would result from general nuclear war.

---

[11]The amounts of various fission isotopes can be roughly estimated. A typical fission explosion may involve 20 kg of Pu-239. Only about 10 percent (2 kg) of the Pu-239 is fissioned: the other 90 percent (18 kg) is swept away by the intense forces of explosion before fission can occur. Thus around 18 kg of deadly Pu-239 are released. The other prominent fission products occur at yields of about 2 to 6 percent (two to six product nuclei are released for each 100 Pu-239 nuclei fissioned). These product nuclei average only one half the mass of the Pu-239 nucleus. Therefore 2 to 6 nuclei in 100 have a mass of only (2 to 6 percent)/2 = 1 to 3 percent relative to parent Pu-239. For instance I-131, with a yield of 3.8 percent from Pu-239, would be created in the approximate amount 3.8 percent/2 = 1.9 percent of the mass of fissioned Pu-239. This ratio gives

$$2 \text{ kg Pu-239} \times \frac{1.9 \text{ parts I-131}}{100 \text{ parts Pu-239}} = 0.038 \text{ kg I-131}$$

By similar reasoning the other fission products, at 2 to 6 percent yield, would be created in amounts varying from 0.02 to 0.06 kg, or 20 to 60 g. (The radioactivity from these isotopes would range from 10^3 to 10^7 Ci.)

Proliferation. One of the direct threats to man and his environment is the proliferation of nuclear weapons. The danger to the world if nuclear bombs become widespread in the arsenals of man was well summarized by the late President John F. Kennedy in 1963.

I ask you to stop and think what it would mean to have nuclear weapons in so many hands, in the hands of countries large and small, stable and unstable, responsible and irresponsible, scattered throughout the world. There would be no rest for anyone then, no stability, no real security . . . .

What is the chance that widespread proliferation will occur? Unfortunately, it may be high. A fission bomb is based on simple principles, as we have seen, and in all likelihood it is relatively simple to make compared to many modern devices. Exact details of construction are top secret, so there would be some air of uncertainty about the effort. However, one nuclear expert stated, without contradiction from his colleagues, that he himself could construct a gun-tube fission bomb given the necessary amount of plutonium. The hurdle would be in getting the plutonium. For nations as well as for bomb-building extremists in smaller groups, the difficulty of obtaining high-grade fissionable fuel—Pu-239, U-235, or U-233—has severely limited nuclear weapons development. Any action that would circulate these nuclear fuels worldwide carries with it a threat of the first magnitude. Such a threat exists in an expanding nuclear-power industry, as we shall discuss.

Fissionable isotopes are difficult to produce because of a combination of circumstances. Of the uranium isotopes U-235 and U-238, only U-235 is fissionable, but it is always mixed with a great preponderance (140 parts to 1) of U-238. Since the chemical properties of isotopes are virtually identical, there is no simple chemical method for separating U-235 from its chemical twin, U-238. One of them cannot be dissolved selectively away by, for instance, an acid, because they both dissolve equally well in acid. However, combined in the gaseous molecule uranium hexafluoride (UF_6), the two isotopes diffuse through air or any other gas at an ever so slightly different rate. Because they diffuse unequally, the light isotope gets a trifle ahead of the other and becomes slightly enriched in that region. Based on this principle, and with miles of pipe and hundreds of porous diffusion barriers, a giant gaseous-diffusion plant was first built at Oak Ridge to separate bomb-quality U-235 from U-238. This plant was a staggering technological undertaking, far more intricate and expensive than building the bomb itself. It clearly cannot be undertaken by technology-poor nations.

Because isotope separation is so difficult, wartime experts also focused on making the fissionable, synthetic element plutonium in bomb quantities. Pu-239, we learned earlier, can be made from U-238 bombarded by neutrons. At that time an intense neutron source was not available. The chain reaction of U-235 produces abundant neutrons, but pure U-235 was not available. However scientists discovered that if highly energetic fission neutrons were slowed down to thermal velocities (the velocity of nuclei and atoms under ordinary thermal agitation), a nonexplosive chain reaction could be sustained in U-235 even when it was admixed with U-238. Accordingly, the first nuclear reactor (or atomic pile) was built in wartime secrecy in a converted squash court at the University of Chicago. Graphite, a form of carbon, was used to slow the neutrons.

December 2, 1942 was a turning point of the nuclear age. On that date nuclear forces were first harnessed with 1 watt of flickering energy and a few resultant Pu-239 nuclei. A production-scale reactor at Hanford, Washington started producing serious amounts of energy and Pu-239 soon thereafter. Plutonium, being a different element with unlike chemical properties, was easily separated from the reactor uranium. Enough was produced to charge several fission bombs by August 1945, one of them destined to destroy Nagasaki, Japan. The plutonium arsenal has grown steadily since that time, stockpiles in this country now being of the order of 1000 tons, enough to build a half-million bombs of Hiroshima strength.[12]

The energy produced in the first reactors was discarded in the headlong rush to generate bomb material. The situation is now quite the opposite, with large reactors being constructed in a steady stream to quench the energy thirst of mankind.

ELECTRIC-POWER GENERATION BY FISSION

An energy crisis looms ahead for mankind because of a rapid growth in demand coupled with a decline in available fuels. Electric-power use in the United States is doubling every 10 years. This rate is another of the exponential growth trends (recall that exponential means doubling in a fixed time) which must level off as man faces the realities of a finite world.

Many people now believe that nuclear energy must be widely

---

[12] According to Ralph Lapp, 4 kilotons of TNT explosive yield can be obtained from each pound of Pu-239 in modern weapons. The total explosive yield from 1000 tons (2×10^6 lb) is

$$2 \times 10^6 \text{ lb Pu-239} \times \frac{4 \text{ kilotons TNT}}{1 \text{ lb Pu-239}} = 8 \times 10^6 \text{ kilotons TNT}$$

or 8 billion tons of TNT equivalent. The Hiroshima bomb released 14 kiloton TNT energy. The number of such Hiroshima explosions theoretically available is

$$\frac{8 \times 10^6 \text{ kilotons TNT}}{14 \text{ kilotons TNT/bomb}} = 570,000 \text{ bombs}$$

Actually, the high efficiency cited above cannot be obtained in "small" Hiroshima-type bombs, but it could be reached with somewhat fewer bombs of greater power releasing the same total energy.

harnessed to fill at least temporarily the impending gap between fuel supply and booming demand. They also feel that the environmental risks of nuclear power are smaller than those stemming from coal-burning power plants.

Accordingly, nuclear power is in a rapid initial growth phase in this country and elsewhere. The first significant commercial power facility was the Shippingport 90-megawatt plant started up near Pittsburgh in 1957. At present, over one hundred nuclear power plants are in operation or in some stage of construction or planning. They are concentrated in the eastern half of the country, as Table 11-15 (p. 438) shows. Nuclear energy is expected to provide over half of the total power generated in the United States by the year 2000. In the world as a whole, nuclear-power production may grow a hundred-fold beyond 1970 levels before the century ends.

Principles of Power Production. Nuclear power plants are based on the same concept as fossil-fuel plants. The difference is that nuclear fuel replaces fossil fuel as a source of heat. The major elements of a nuclear-power unit are shown in Figure 11-18. This design is the simplest power unit, the so-called *boiling-water reactor (BWR)*. Somewhat more complicated in design is the *pressurized-water reactor (PWR)*, but it works on the same principles.

The nuclear fuel is contained in the reactor core. Here the energy of fission appears in the form of heat. Water circulating through the reactor and over this hot core is turned to steam. The steam is led to a turbine, which drives the electrical generator. The spent steam is

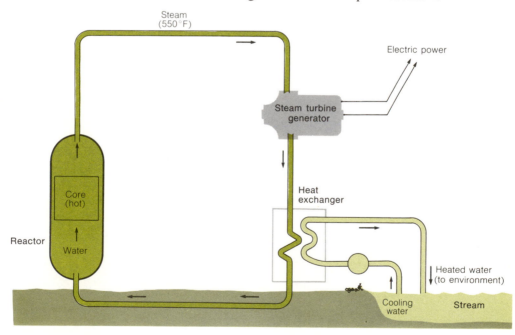

Figure 11-18. The essential elements of a nuclear power plant. A fossil-fuel plant is similar, except that the reactor core is replaced by burning fuel to provide heat for making steam.

Table 11-15.   LOCATION OF NUCLEAR PLANTS IN VARIOUS STATES[a]

State	Location	Date of Startup	State	Location	Date of Startup
Ala.	Browns Ferry	1970	Neb.	Fort Calhoun	1971
	Browns Ferry	1971		Brownsville	1972
	Browns Ferry	1972	N. H.	Seabrook	1974
Ark.	Dardanelle Lake	1972	N. J.	Toms River	1968
Calif.	Humboldt Bay	1963		Toms River	1973
	San Clemente	1967		Artificial Island	1971
	Corral Canyon	1973		Artificial Island	1973
	Diablo Canyon No. 1	1971	N. Y.	Indian Point No. 1	1962
	Diablo Canyon No. 2	1974		Indian Point No. 2	1969
	Sacramento County	1973		Indian Point No. 3	1971
Colo.	Platteville	1971		Scriba	1968
Conn.	Haddam Neck	1967		Easton	1971
	Waterford No. 1	1969		Rochester	1969
	Waterford No. 2	1974		Shoreham	1973
Fla.	Turkey Point No. 3	1970		Lansing	1973
	Turkey Point No. 4	1971		Nine Mile Point	1973
	Red Level	1972	N. C.	Southport	1973
	Hutchinson Island	1973		Southport	1974
Ga.	Baxley	1973	Ohio	Oak Harbor	1974
Ill.	Morris No. 1	1959	Oreg.	Rainier	1974
	Morris No. 2	1968	Pa.	Peach Bottom No. 1	1966
	Morris No. 3	1969		Peach Bottom No. 2	1971
	Zion No. 1	1972		Peach Bottom No. 3	1973
	Zion No. 2	1973		Shippingport No. 1	1957
	Quad Cities No. 1	1970		Shippingport No. 2	1973
	Quad Cities No. 2	1971		Three Mile Island	1971
Ind.	Burns Harbor	1970s	S. C.	Hartsville	1970
Iowa	Cedar Rapids	1973		Lake Keowee No. 1	1971
Maine	Wiscasset	1972		Lake Keowee No. 2	1972
Md.	Lusby	1973		Lake Keowee No. 3	1973
	Lusby	1974	Tenn.	Daisy	1973
Mass.	Rowe	1960		Daisy	1973
	Plymouth	1971	Vt.	Vernon	1970
Mich.	Big Rock Point	1962	Va.	Hog Island	1971
	South Haven	1969		Hog Island	1972
	Lagoona Beach	1963		Louisa County	1974
	Lagoona Beach	1974	Wash.	Richland	1966
	Bridgman	1972	Wisc.	Genoa	1967
	Bridgman	1973		Two Creeks No. 1	1970
	Midland	1974		Two Creeks No. 2	1971
	Midland	1975		Carlton	1972
Minn.	Elk River	1962			
	Monticello	1970			
	Red Wing No. 1	1972			
	Red Wing No. 2	1974			

[a]Some startup dates are tentative.

condensed to water as it passes over tubes of cooling water supplied from the outside. It then circulates back to the reactor to repeat the cycle.

One effect of a power-generating plant—fossil fuel or nuclear—is that the outside cooling water is warmed by the condensing steam. In fact, its function is to carry off the excess heat—usually to a lake or river—so that the steam will be able to condense. Unfortunately, more energy is lost into the cooling water than is converted to electricity. Nuclear plants presently operate at about 32 percent efficiency, which means that the ratio of waste heat to electrical energy is somewhat over two to one.[13] Fossil-fuel plants are only slightly better; at an efficiency of about 40 percent, they waste 1.5 times as much energy as is converted to electricity. When we realize that these plants generate an enormous amount of electrical energy, then we are not surprised that the waste heat, even greater in amount, heats and considerably disturbs the local environment. This environmental heating is called *thermal pollution*, as we noted in Chapter 1.

We now examine the reactor core in more detail. It is the heart of the reactor, the site at which energy, plutonium, and radioactive fission products are generated.

The core is made up of thousands of *fuel rods*, sealed metal tubes

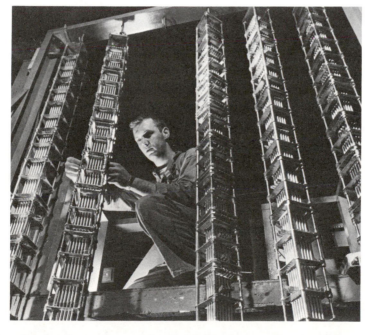

Bundles of fuel rods are prepared for installation in a reactor core. (*Photo courtesy Westinghouse Atomic Power Division.*)

---

[13] An efficiency of 32 percent means that for every 100 units of nuclear heat (thermal) energy produced, only 32 units become electrical energy and the remaining 68 units enter the cooling water as thermal waste. The ratio of waste heat to electrical energy is thus 68/32, over 2 to 1. The 40 percent efficiency of coal-fired plants means that there are $100-40 = 60$ units of waste heat to 40 units of electrical energy for a ratio of $60/40 = 1.5$.

containing pellets of uranium dioxide (UO_2). The pellets are enriched in U-235, containing 2 to 4 percent in contrast to the natural U-235 abundance of 0.7 percent (Table 11-1). The remainder of the core uranium is U-238. Altogether, some 100 tons of uranium is used as fuel, several tons of it being U-235. This amount dwarfs the critical mass—about 10 kg of U-235—needed to build minimal fission bombs.

At intervals the movable blades of *control rods* are inserted between the fuel rods. These rods contain boron or cadmium, the nuclei of which absorb neutrons. If the chain reaction shows any signs of getting out of hand, the control rods are thrust in to soak up the proliferating neutrons and extinguish fissioning.

A high density of neutrons sustains the chain reaction. They penetrate into the water which is circulating around the fuel rods. When neutrons collide with a very light nucleus, such as the single proton constituting the nucleus of hydrogen in water (H_2O), they lose energy. In this way the circulating water serves as a *moderator* to the high initial velocity of the neutrons. In the original Chicago reactor, graphite served as the moderator. The neutrons slow down until they reach thermal velocity, typical of atomic motion at that temperature. These *thermal neutrons* are used in modern reactors because they are more effective in initiating fission than are fast neutrons.

Radioisotopes in Nuclear Power Plants.   Radioisotopes are produced in two ways in a nuclear reactor, just as they are in the explosion of a fission bomb. First, slight amounts of neutron-induced isotopes cobalt (Co)-58, Co-60, manganese (Mn)-54, iron (Fe)-59, chromium (Cr)-51, and zinc (Zn)-65 are formed. Second and most important, radioactive fissions products are produced directly in large quantities.

The fission daughters in a reactor core are both abundant and highly radioactive. Most of them are trapped in the uranium dioxide pellets, but a few break free. A second barrier to escape is the metal cladding of the fuel rod. However, cladding leaks occur occasionally during operation, freeing small amounts of radioactive products into the circulating steam and water. Also, the cladding is not entirely free of "tramp uranium," small amounts adhering to the outside of the cladding and some contained in the cladding as impurity. Much of the tramp uranium is already external to the cladding, and it therefore releases fission products directly to the steam system.

The steam driving the turbine is circulated time and again, with no environmental contact. However, gases formed in this steam must be tapped off for proper operation. Such gases are later released to the atmosphere. (Some plants are now being designed to collect and store the tapped gases.) These gases contain those fission products which by their chemical nature are gaseous, including the noble gases krypton (Kr)-85, xenon (Xe)-133, Xe-135, and slight amounts of iodine (I)-131 and tritium, or H-3. Their release carries radioactivity from the nuclear power plant to the environment. Unfortunately, this release is only one small part of the environmental threat of nuclear power, as the following section shows.

ENVIRONMENTAL IMPACT OF NUCLEAR POWER

The production of nuclear power involves a chain of steps, from the mining and preparation of fissionable fuel to the disposal of radioactive wastes. These steps and their environmental impact are outlined in Figure 11-19.

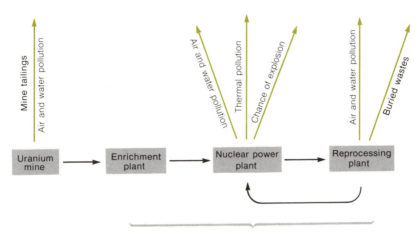

Figure 11-19. The principal steps in the production and use of fissionable fuel for power production. The primary environmental threats at each stage are shown.

We have noted already that nuclear-power stations release thermal pollution and some radioactive gases. The curies of radioactivity released to the atmosphere by several nuclear power plants, including three of the worst "offenders," are shown in Table 11-16. These releases vary greatly, depending on power-plant design and

Table 11-16. ANNUAL RELEASE OF RADIOACTIVE PRODUCTS TO THE ATMOSPHERE BY SEVERAL NUCLEAR POWER PLANTS[a]

Power Plant	Curies/Year
Shippingport	0.0015
Yankee	1.5
Connecticut Yankee	1.9
San Onofre	4.4
Indian Point	39
Big Rock Point	204,000
Dresden I	215,000
Humboldt Bay	900,000

[a] Figure given is average yearly release for 1967–1968.

unpredictable defects in fuel-rod cladding. Pressurized-water reactors, the first five in the table, are obviously "cleaner" than are boiling-water reactors, the bottom three.

Releases to the water environment by these plants are generally smaller than those to air, varying from 0.03 to 9 Ci per year.

Regulation of Radioactive Releases.   The release of radioactive isotopes from nuclear power plants is limited by Atomic Energy Commission (AEC) regulations. These regulations are based on guidelines for maximum human exposure set forth by the Federal Radiation Council (FRC). The FRC is a panel of high-level government officials established by executive order in 1959. Its guidelines are based in large part on recommendations from two scientific bodies, the National Council on Radiation Protection and Measurements (NCRP) and the International Commission on Radiation Protection (ICRP).

In brief, FRC guidelines state that the public at large should receive a radiation exposure no larger than 170 millirems (approximately 170 millirads) per year from man-made radioisotopes. Maximum public exposure—perhaps someone living near a nuclear power plant—is limited to 500 millirems per year. For reference, note that the permitted average public exposure of 170 millirems per year is in the neighborhood of natural radiation exposure (slightly over 100 millirads or millirems), and it is some 3000 times lower than the acute dose that would lead to quick death.

The AEC enforces these exposure maxima by regulations which limit the radioactivity in air and water released from the plant site. *Maximum permissible concentrations (MPC)* are set for individual radioisotopes, as shown in Table 11-17. These levels are fixed such that dosage will not exceed the guideline maximum of 500 millirems per year for a "standard" man drinking $2\,\ell$ of such water or breathing $20,000\,\ell$ of such air.[14]

Critics of the AEC contend, first, that a yearly dose of 170 millirems is too great for the public at large. Since even small doses such as this may cause occasional cancer, the total number of radiation-caused cancers (also genetic defects) when the entire population of the United States is exposed would be considerable. John Gofman and Arthur Tamplin now estimate 100,000 extra cancer deaths per year

---

[14]A man breathing $20,000\,\ell$ of air daily contaminated with, for example, the MPC of 0.03 picocuries (3×10^{-14} Ci) per liter for Sr-90 would inhale:

$$\frac{3 \times 10^{-4}\,\text{Ci}}{\ell} \times \frac{2 \times 10^4\,\ell}{\text{day}} = 6 \times 10^{-10}\,\frac{\text{Ci}}{\text{day}}$$

or, for a year:

$$\frac{6 \times 10^{-10}\,\text{Ci}}{\text{day}} \times \frac{365\,\text{days}}{\text{year}} = 0.22 \times 10^{-6}\,\frac{\text{curies}}{\text{year}}$$

or 0.22 microcuries per year. About 250 microcuries body burden will induce leukemia in swine and presumably also in humans. These numbers are not exactly comparable for several reasons, but it is clear that allowed intake is well below the point where damage becomes obvious.

Table 11-17. MAXIMUM PERMISSIBLE CONCENTRATION (MPC) FOR SOME IMPORTANT RADIOISOTOPES IN WATER AND AIR

Radioisotope	MPC for Water (picocuries[a]/ℓ)	MPC for Air (picocuries/ℓ)
Strontium (Sr)-90	100	0.03
Cesium (Cs)-137	20,000	2.0
Iodine (I)-131	—	0.1
Plutonium (Pu)-239	—	0.00006

[a] 1 picocurie = 10^{-12} Ci.

for 170 millirems exposure nationwide. They have persistently recommended a tenfold reduction in man-made radiation limits.

Atomic-energy proponents answer, first, that Gofman and Tamplin overestimate radiation damage by using the linear theory of damage-dose proportionality. Second, they contend that actual releases of radioactivity are far below AEC maxima. Only two or three plants—those at the bottom of Table 11-17—presently release radiation greater than 1 percent of the allowed limit.[15] Perhaps to meet critics on this point, the AEC proposed in 1971 a hundredfold reduction in the permissible release of radioisotopes.

Critics also contend that radiation intake may far exceed the amount that a person gains by simply drinking water and breathing air. Both physical and biological magnification (each a function of chemical factors) can occur for radioisotopes as they do for other pollutants, such as mercury. For instance, fish concentrate cesium (Cs)-137 to a level 1000 times greater than that in the surrounding water. Tamplin calculates that a man eating 1 lb of fish a week caught in water acceptable to the AEC would receive a dosage of 15 rems (15,000 millirems) per year, 90 times the dose acceptable for the population at large. The concentration of iodine in milk is even worse: a child drinking 1 qt a day could receive a phenomenal 2500 rems a year because of iodine fallout from "acceptable" air onto pasture lands. The AEC has started to allow for these concentration factors—particularly for iodine (I)-131 in milk. All in all, radiation standards are becoming more stringent as more factors of this complex subject are understood. In the 1950s, some MPC's were dropped by 1000 to 10,000-fold. There is comfort in seeing these permissible levels inch down, but at the same time there is a little apprehension that the trend may continue to a point where today's standards are unacceptable by the presumably more rational criteria of the future.

Long-Term Dangers. While the dispute over everyday radiation release continues, a threat of far more serious proportions is casting

---

[15] For example, the maximum release permitted for the Humboldt Bay plant, calculated according to local meteorological and other data to give a 500 millirem per year exposure at the plant boundary, is 1.8 million Ci per year. Actual releases in 1967–1968 were about 0.9 million Ci per year, according to Table 11-18. Thus this plant, the worst one, operated at that time at $(0.9/1.8) \times 100 = 50\%$ of its maximum permissible level.

its shadow on the future. This threat stems, very simply, from the enormous quantities of radioactive and fissionable materials associated with nuclear-power production. Should an accident or improper diversion of a mere fraction of this vast and deadly storehouse occur, mankind would suffer enormously.

The amount of fissionable materials to be consumed yearly in producing power is shown in Figure 11-20. Clearly, by the end of the century many hundred tons will be consumed every year, enough, if diverted, to produce about 100,000 fission boms.[16]

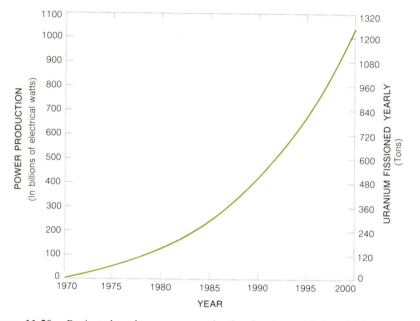

Figure 11-20. Projected nuclear-power production for the remainder of this century is shown on the left-hand axis. On the right is shown the tons of fuel to be fissioned yearly in producing this power.

Each ton fissioned produces nearly 1 ton of radioactive fission products. About 15 percent of this amount, 300 lb, ends up as long-lived fission radioisotopes, with half-lives from 8 days to thousands of years. By the turn of the century, nearly 200 tons of such fission products will be produced each year.[17] Recall that these substances

---

[16] The 1200 tons or 1200 × 2000 = 2,400,000 lb of fissionable material expected to generate power in the year 2000 can be converted to kilograms kg as follows:

$$2.4 \times 10^6 \text{ lbs} \times \frac{1 \text{ kg}}{2.2 \text{ lb}} = 1.1 \times 10^6 \text{ kg}$$

Assuming that a minimal bomb can be made from 10 kg (10 kg/bomb), the number of such bombs theoretically available from this fuel would be

$$1.1 \times 10^6 \text{ kg} \times \frac{1 \text{ bomb}}{10 \text{ kg}} = 110,000 \text{ bombs}$$

[17] Long-lived fission products (llfp) at 15 percent of the 1200 tons fuel/year gives:

$$\frac{1200 \text{ tons fuel}}{\text{year}} \times \frac{15 \text{ parts llfp}}{100 \text{ parts fuel}} = \frac{180 \text{ ton llfp}}{\text{year}}$$

are dangerous at the microgram level. Literally trillions of doses capable of being fatal will be produced yearly as the nuclear industry gains momentum.

In terms of single power stations, a moderately large 1000 megawatt plant will contain 7 billion Ci of radioactivity in its core after 6 months of operation. Two million Ci of it will be strontium (Sr)-90. Together the power plants of year 2000 will produce 10^{10} Ci of Sr-90 every year. Again, fatal doses are measured in millionths of a curie.

Present-day nuclear reactors cannot explode as a fission bomb would, despite their heavy loading with fissionable U-235. The reason is that average U-235 enrichment is only 2 to 4 percent. The bulk of uranium is U-238, which absorbs some neutrons. If the nuclear reaction started to get out of hand, heating the fuel, the increased thermal motion of the U-238 atoms and their nuclei would lead to the capture of more neutrons (the Doppler effect), and thus fission would be slowed back to normal. This effect acts as an automatic brake against total nuclear runaway.

Nonetheless, smaller energy excursions, chemical explosions, and fuel rod melting are conceivable in reactors. The blast would not devastate even a city block, but it might unloose enough of the internal radioactivity to immobilize the city.

The "hot" radioactive fission products under discussion build up during plant operation, most of them locked firmly within fuel rods where they are produced. After half a year or so, their buildup in the fuels rods interferes with power production. The fuel must then be removed and shipped to a special reprocessing plant where, entirely by remote control, the hot fission impurities are chemically separated from the uranium and plutonium. From here, two dangerous roads are traveled.

First, some of the reprocessed uranium is enriched to 90 percent in U-235. (It will eventually be assembled with unenriched fuel rods to give the average 2 to 4 percent of enrichment.) This is bomb material, yet it is shipped as a regular commercial product. On several occasions enough U-235 to make a number of atomic bombs has been misplaced in shipping and lost for days or weeks. Soon large amounts of plutonium (Pu)-239 will enter commercial traffic as a nuclear fuel. As the nuclear industry grows worldwide, the traffic in nuclear materials will increase and hijack opportunities will increase proportionally. Forged into bombs by unscrupulous hands, these materials could trigger an international disaster in a world already perched on the brink of a nuclear abyss.

Second, most of the millions of curies of radioactive fission wastes will be solidified and sent for burial in surface vaults or in salt mines, there to simmer in their own radiation for thousands of years. Only after the passage of many (as high as 20) half-lives for each long-lived fission fragment will this intensively radioactive material become truly safe. In the meantime, should some unusual geological event or human disturbance unearth these smouldering wastes, the environmental exposure would be extreme.

Most of the threats from nuclear-power production increase in proportion to the number and size of power plants. Everyday radiation release, the possibility of nuclear hijack, the amount of buried radioactive waste—these threats are all proportional to nuclear-power production. Nuclear-power demands, on the other hand, will increase enormously if world population continues to grow. We see that the threat of the nucleus—like so many other environmental threats—is linked intimately to the number of human inhabitants on spaceship earth.

One should be careful not to make nuclear energy a scapegoat for this prospect. If tens of billions of people should find themselves at once occupants of this planet, it would be better that nuclear energy could provide for them in a dangerous way than that they could not be provided for at all.

BREEDER REACTORS

The present generation of nuclear power reactors is rapidly consuming the available U-235 on earth. Uranium is not an abundant element—it averages only about 3 ppm in the earth's crust. The U-235 isotope is less than a hundredth part of this amount since its natural abundance is under 1 percent. M. King Hubbert estimates that, as a result of this scarcity, an acute shortage of uranium is likely to develop in the next 25 years. Certainly, then, present U-235-consuming reactors offer little in the way of a permanent solution to the impending energy crisis.

While U-235 is the only fissionable isotope of practical value in nature, there are two isotopes, each more abundant, which can be converted or "bred" to fissionable isotopes by nuclear transformations. These so called *fertile* isotopes are U-238 and thorium (Th)-232. We have already noted that U-238 is the abundant isotope of uranium, present in amounts over a hundredfold greater than is U-235. The isotope Th-232 is more abundant still, in the neighborhood of 8 ppm in the earth's crust. Provided these isotopes could be made fissionable in some practical device, nuclear fuel supplies could be enlarged by a factor of several hundred.

If U-238 and Th-232 were to be extracted from Conway granite, where their combined concentration is naturally enriched to 30 ppm, their conversion to fissionable isotopes would provide nuclear energy from each ton of granite equivalent to almost 100 tons of coal. In this way, through the mining of granite, mankind's every need could be met for hundreds of centuries.

The practical device for converting fertile isotopes is the breeder reactor. In a breeder reactor, neutrons must be carefully conserved and utilized, for they are essential for the nuclear transformation of fertile isotopes to fissionable ones.

We have already learned how neutrons can induce the nuclear change from U-238 to plutonium (Pu)-239, converting a "useless" isotope to a fissionable one. U-233 is another fissionable isotope that can be made if enough neutrons are available. The sequence of nuclear reactions—a neutron capture followed by two beta decays—is the same as that in which Pu-239 is produced, except that the starting material is Th-232.

$$^{232}_{90}\text{Th} + ^{1}_{0}\text{n} \rightarrow ^{233}_{90}\text{Th} \quad \text{(neutron capture)}$$

$$^{233}_{90}\text{Th} \rightarrow ^{233}_{91}\text{Pa} + ^{0}_{-1}\beta \quad \text{(beta decay)}$$

$$^{233}_{91}\text{Pa} \rightarrow ^{233}_{92}\text{U} + ^{0}_{-1}\beta \quad \text{(beta decay)}$$

The final beta decay yields the desired product, U-233.

In a breeder reactor, two nuclear conversions must occur: (1) one of the fissionable isotopes, U-233 or Pu-239, must be produced from a fertile isotope by neutron bombardment, and (2) the resultant U-233 or Pu-239 (or even starter U-235) must undergo neutron-induced fission to produce reactor energy. Clearly, each process requires one neutron per nucleus. A look at neutron economy tells us that, to sustain these twin processes, each fission must release two productive neutrons, one to create a new fissionable nucleus and another to initiate the subsequent fission. If nuclear power is to expand at all, slightly more than two neutrons are needed to produce the additional fuel. The breeder reactor works in just this way, producing more fissionable fuel than it consumes. An effective breeder, in fact, will double an initial amount of fissionable fuel in 7 to 10 years.

Since many neutrons are soaked up by nonfuel components of the reactor (such as the control rods), it is not easy to maintain anything above a two-neutron-per-fission economy. The special engineering of a breeder reactor is aimed at this objective.

Pu-239, which averages 2.9 neutrons per fission, is simpler to engineer into a breeder economy than is U-233, at 2.4 neutrons per fission. Thus development of a plutonium breeder has reached a more advanced stage, and it will undoubtedly be the first to produce commercial power, sometime around 1984.

The plutonium breeder now in development is called the *liquid metal fast breeder reactor (LMFBR)*. The term "liquid metal" refers to the medium, liquid sodium, that will carry heat from the reactor core to the turbine. As we noted earlier, water serves this function in present-day "light water" reactors.

The core of the plutonium breeder will contain fissionable Pu-239 in addition to fertile U-238. The U-238 helps prevent reactor runaway via the Doppler effect, just as it does in the present generation of nonbreeder reactors.

Surrounding the reactor core is an added "blanket" of fertile U-238, which produces additional Pu-239 with the aid of neutrons escaping the core. The neutrons themselves travel much faster than do "thermal" neutrons—hence the occasional descriptive name *fast breeder reactors.*

Present-day reactors also contain an abundance of U-238, as we already saw. While they are not engineered for high plutonium conversion, the neutron density in existing reactors is great enough so that roughly one fissionable plutonium nucleus is made for each two U-235 nuclei that fission. Recovered during reprocessing and placed back in the nuclear fuel cycle, this isotope by its fission already aids in conserving world supplies of U-235.

Because of high neutron density, neutron speed, and other unique factors, breeders tend to be less stable than other reactors against reactor runaway. Also, since breeder reactors work ultimately on fission energy, radioactive fission products are produced in large quantities. Runaway could release a considerable part of this large storehouse of radioactivity. Extraordinary engineering effort has gone into designing safe breeder reactors, but it is impossible yet to say how safe they will be in the long run.

Some nuclear experts, such as Alvin Weinberg, director of Oak Ridge, believe that breeder reactors show the only presently feasible pathway to an abundant energy supply for mankind's future. Weinberg feels that the long-term needs of 20 billion people (over five times the present population) can be supported easily by worldwide breeder reactors consuming 500 tons of thorium or uranium per day. Critics ask whether such an enormous quantity of potential bomb material, flowing daily to points across the face of the earth, could ever be made theft-proof or free from diversion by unstable governments or extremist organizations. They also ask whether 75 or so tons of hot fission wastes can be buried safely every day for centuries without an eventual catastrophic release.[18]

NUCLEAR FUSION

Fusion complements fission as a source of nuclear energy. As we said earlier, fusion energy appears when small nuclei are joined together to form larger nuclei which have greater binding energy and mass defect. The released energy is considerable.

---

[18] Assuming that 15% of all fission products (fp) have long-lived radioactivity, 500 tons fissioned per day would give

$$\frac{500 \text{ tons fp}}{1 \text{ day}} \times \frac{15 \text{ parts long-lived fp}}{100 \text{ parts fp}} = \frac{75 \text{ tons long-lived fp}}{1 \text{ day}}$$

Perhaps even more meaningful are the extreme radioactivity levels connected with this amount of waste. In one year, 270 billion Ci of strontium (Sr)-90 would be generated. Accumulated from year to year, 10 trillion Ci would eventually build up. Similar levels would exist for several other radioisotopes.

Fusion is much more difficult to initiate than is fission. For this reason, fusion weapons (the "hydrogen bomb") lagged well behind fission bombs in development, and power production by fusion is not even known for certain to be scientifically feasible. Fission power production, by contrast, is now routine.

The reason that fusion is difficult to initiate is very simple: the close approach of nuclei to one another necessary for fusion is resisted by the strong repulsive force caused by their positive charges.

Since no tools are fine enough to manipulate even atoms and molecules, the tiny nuclei must be forced into unwilling collisions by indirect means. Accelerator devices can make a few nuclei collide, but high temperature is the only known tool for obtaining nuclear collisions in abundance. Recall that high temperature means high atomic and, of course, nuclear velocities. When these velocities are great enough, two nuclei on a collision course will merge because their momentum will overcome positive repulsion (Figure 11-21). The *ignition temperature* is the minimum temperature that will cause enough fusion to produce a net energy release.

The induction of fusion by heat sounds simple until we look at the actual magnitude of ignition temperatures required. Temperatures in the many tens of millions of degrees must be reached for productive fusion! Fusion reactions are sometimes called *thermonuclear* reactions because of this prodigious temperature requirement. The problems associated with reaching such temperatures on earth will be discussed later.

The mass-defect curve in Figure 11-3 shows that the greatest energy can be gained by fusion of the lightest nuclei, those of hydro-

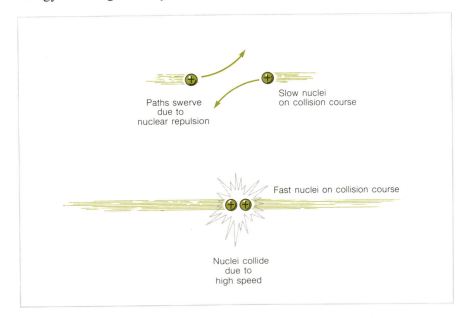

Figure 11-21. Slow nuclei on a collision course cannot collide and fuse because strong positive repulsion forces them to swerve apart. Fast nuclei will collide and fuse because their high momentum offsets repulsion.

gen. Also, it clearly would be easier to force together two hydrogen nuclei containing one positive charge each than, say, two carbon nuclei with six positive charges each. Thus almost all fusion processes use hydrogen for at least one of the fusing partners.

The most important fusion reaction on earth—the one occurring most frequently in hydrogen bombs and the one with the most immediate potential for power production—involves the combining of two hydrogen isotopes, nonradioactive deuterium (^2_1H) and radioactive tritium (^3_1H):

$$^2_1H \quad + \quad ^3_1H \quad \rightarrow \quad ^4_2He \quad + \quad ^1_0n$$

deuterium or   tritium or   helium-4   neutron
hydrogen-2   hydrogen-3

This transformation has the lowest ignition temperature of any fusion reaction—40 million °C, or 70 million °F.

Fusion is unnatural on this cold earth, but it powers the stars. The sun is a typical star, millions of degrees hot on the inside, and from fusion it derives an energy 50 trillion times larger than man's output of industrial energy. This energy is radiated into space, a small fraction (about one part in a billion) of which is intercepted by earth. Several fusion processes together are thought to yield this enormous energy, mostly from the conversion of hydrogen (H)-1 to helium (He)-4. One such fusion cycle is

$$^1_1H + ^1_1H \rightarrow ^2_1H + ^0_1positron$$

$$^1_1H + ^2_1H \rightarrow ^3_2He$$

$$^3_2H + ^3_2He \rightarrow ^4_2He + 2^1_1H$$

He-4   H-1 starts
product   fusion cycle
over again

FUSION BOMBS

Hydrogen, though not as abundant on earth as on the sun, is plentiful enough to revolutionize many facets of civilization with the harnessing of its energy of fusion. A new and more dangerous generation of nuclear weapons and a new and more promising source of controlled energy are two beckoning prospects of fusion. Both require the achievement of the stupendous ignition temperatures that are necessary for fusion. Only nuclear weaponry has become an accomplished fact. Efforts toward controlled energy production will be described in the next section.

We should not be surprised that a fusion explosion was achieved well ahead of any prospect for steady energy release by fusion. If the ignition temperature is reached at all, it would take man's full in-genuity to sustain it, for at such temperatures all matter—including the machine supporting the fusion—is turned to hot, structureless, ionized vapor. However, if the ignition temperature could be reached for but an instant, it would start a fusion explosion in which all nearby matter is to be vaporized anyway.

Accordingly, for explosion one needs only to reach for an instant the tens of millions of degrees ignition temperature. Even so, it might have been impossible if it were not for the fusion bomb's precursor—the fission bomb. Only in the center of an exploding fission device do temperatures on earth reach a magnitude sufficient to ignite fusion.

A fusion bomb—sometimes called a hydrogen bomb or a thermo-nuclear bomb—requires a fission-bomb heart. A blanket of fusion material surrounds the fission bomb and is ignited by the extreme temperature of the exploding fission bomb. There is essentially no limit to the explosive power of a fusion bomb because there is no such thing as a critical mass for fusion. The critical mass makes truly large fission bombs impractical because small, subcritical chunks must be widely separated to avoid premature explosion. However, hydrogen-containing fusion material can be assembled around a fission bomb in amounts limited only by the capability of delivering the device to its target. Consequently, enormous bombs have been built. The largest ever exploded was a Russian device tested in 1962. It released energy equivalent to 58,000,000 tons or 58 megatons of TNT. It thus contained the explosive power of over

Naval vessels are dwarfed by the mushroom cloud from a giant fusion bomb exploded at Bikini Atoll in 1946. (*Photo by Wide World Photos.*)

A Poseidon missile is launched from a submarine. This missile carries 10 separate nuclear weapons (the MIRV system), each with a 50,000 kiloton punch, giving a total explosive power equivalent to 0.5 megatons of TNT. (*Photo courtesy Lockheed Missiles and Space Company.*)

4000 Hiroshima fission bombs.[19] It would require a heavy bomber or a giant missile to deliver a weapon of such size. Missiles of the U.S. and U.S.S.R. are equipped with somewhat smaller weapons, in the vicinity of one or a few megatons.

The fusion blanket of a hydrogen bomb consists of the compound lithium hydride (LiH). Both the lithium and the hydrogen are special isotopes; Li-6 and H-2 (deuterium). Neutrons from both fission and fusion convert Li-6 to H-3 (tritium) in a nuclear transformation to be discussed in the next section. Deuterium and the released tritium then fuse through the process we outlined before:

$$\ce{^2_1H + ^3_1H -> ^4_2He + ^1_0n}$$

All of this occurs within a millionth of a second with the release of an enormous energy of fusion. More energy is sometimes obtained by the addition of still another blanket—this one of a fissionable material —to the bomb's exterior.

---

[19]The Hiroshima bomb had an explosive equivalent of 14,000 tons of TNT. The conversion factor from equivalent tons of TNT to Hiroshima bombs is therefore 1 Hir. bomb/14,000 tons TNT. The number of such bombs theoretically derivable from 58 megatons of TNT is therefore

$$58 \times 10^6 \text{ tons TNT} \times \frac{1 \text{ Hir. bomb}}{14 \times 10^3 \text{ tons TNT}} = 4100 \text{ Hir. bombs}$$

Significance of Fusion Bombs. The destructive potential of the world's hydrogen-bomb arsenals is clearly much greater than that of its fission bombs. However, in the age of overkill, the significance of this is dulled: a city can be destroyed and a life lost but once. Fusion bombs do play an important role in the balance of terror and in other matters connected with modern nuclear warfare, but it is a sad fact that so many lives are threatened by fission bombs alone that a substantial increase cannot occur. A war with either kind of device would unleash more human destruction than perhaps all the wars in the entire history of the human race.

Two classes of radioactivity are released with the explosion of fusion bombs. First is the type of radioactivity associated with the fission bomb trigger, already discussed. Second, large amounts of radioactive tritium are released with fusion. However, tritium is not as biologically damaging as most other radioisotopes, as we shall explain. Thus the threat of radioactivity from a fusion explosion is not much greater in magnitude nor much different in character from that released by a fission bomb. An exception occurs when an external fission blanket is added. These so called "dirty bombs" produce fission products in greater abundance than does a fission bomb by itself.

ELECTRIC-POWER GENERATION BY FUSION

Fusion is a potential source of controlled energy, a commodity so demanded by modern civilization. Fusion energy has the advantage that its basic fuel, deuterium, is abundant enough to last for eons and that radioactive contamination promises to be slight. The problem is that no one has devised a method for tapping fusion energy except by triggering nuclear explosions. It is not certain that a controlled supply of fusion power will ever be forthcoming because of the high, steady ignition temperatures required. Nonetheless, most experts are presently optimistic.

Fuel Supply. To start with, extensive fusion would require an abundant supply of the most promising fusion fuel, deuterium or hydrogen (H)-2. As shown in Table 11-1, the natural abundance of deuterium is 0.02 percent: hence roughly one part in five thousand of all hydrogen on earth is deuterium. On this basis, the natural deuterium supply of the world is close to 10^{14} tons.[20] This figure is

---

[20]Most hydrogen on earth is found in the ocean in water. The atomic weights of hydrogen (H) and oxygen (O) are 1.0 and 16; the mass breakdown of H_2O is (Chapter 4):

$$\begin{array}{lll} \text{mass of hydrogen} & = 2 \times 1.0 = & 2.0 \text{ amu} \\ \text{mass of oxygen} & = 1 \times 16 = & 16 \ \ \text{amu} \\ \hline \text{total mass (molecular weight)} & = 18 \ \ \text{amu} \end{array}$$

far greater than the estimated mass of fossil fuel reserves on earth (6.5×10^{12} tons). Moreover, recall that nuclear energy per atom is of the order of a million-fold greater than chemical energy, such as that released in fossil-fuel combustion. More precisely, a ton of deuterium can produce 10 million (10^7) times the energy of a ton of coal. (A ton of ocean water contains energy in its deuterium that is equivalent to over 400 tons of coal[21]—a virtual ocean of fuel.) Altogether, the world's supply of deuterium, over ten times larger in sheer mass than that of fossil fuels, will release more than 10×10^7 or 100 million (10^8) times as much energy as fossil fuels. A potential energy supply 10^8 times larger than that promised by all fossil fuels is so great that, if tapped, it could change the course of civilization for many millions of years.

Attempts to Reach Ignition Temperature. The primary difficulty in controlling fusion, as we have learned, is the heating of nuclear fuels to ignition temperatures in the tens of millions of degrees. They must be confined at that temperature for a short time so that fusion can proceed. They cannot be confined in an ordinary container because all materials known to man melt by 4000°C and vaporize before the temperature reaches 6000°C. These temperatures are pitifully small compared to the 40 million °C minimum for ignition.

To counter this, scientists in the 1940's proposed confining the fusion nuclei in a *magnetic bottle*. At these high temperatures, orbital electrons have been stripped away from the nuclei by the force of collisions. No neutral atoms remain. Instead, we have a gas composed of charged particles (nuclei and electrons)—a *plasma*. Now it is known that a charged particle cannot cross a strong magnetic force line—a line running from magnetic "north" to magnetic "south." Instead, it spirals around the line, as shown in Figure 11-22.

If now the magnetic force line can be bent around in a circle (Figure 11-23), charged particles will spiral endlessly around this

---

Water is therefore 2/18 or 1/9 H. The oceans contain 1.5×10^{18} metric tons of H_2O and therefore contain H in the amount

$$1.5 \times 10^{18} \text{ tons } H_2O \times \frac{1 \text{ part H}}{9 \text{ parts } H_2O} = 1.7 \times 10^{17} \text{ tons H}$$

The abundance of deuterium is 1 atom in 5000, but since deuterium is twice as heavy as ordinary hydrogen (its nucleus contains a neutron in addition to the proton), its mass relative to hydrogen is $2 \times 1/5000$ or 1 mass unit of deuterium in 2500 of hydrogen. Therefore in the oceans, 1.7×10^{17} tons of hydrogen are

$$1.7 \times 10^{17} \text{ tons H} \times \frac{1 \text{ part deuterium}}{2500 \text{ parts H}} = 6.8 \times 10^{13} \text{ tons deuterium.}$$

[21] To find the energy equivalent, in coal, of a ton of water, we must go through this sequence of known conversions: tons water → tons hydrogen → tons deuterium → tons coal. We have

$$1 \text{ ton } H_2O \times \frac{1 \text{ part H}}{9 \text{ parts } H_2O} \times \frac{1 \text{ part deuterium}}{2500 \text{ parts H}} \times \frac{10^7 \text{ parts coal}}{1 \text{ part deuterium}} = 440 \text{ tons coal}$$

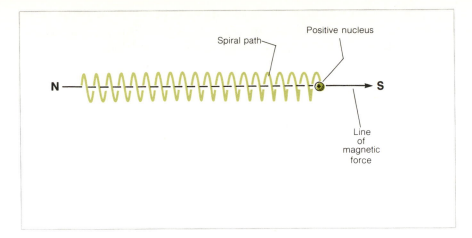

Figure 11-22. The spiraling of a nucleus around a magnetic force line.

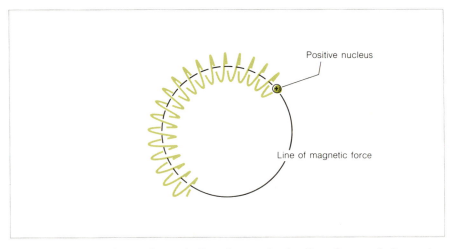

Figure 11-23. Positive nucleus spiraling along a circular line of magnetic force: the "toroidal" configuration.

circle—or at least long enough to fuse. Confined by this invisible magnetic bottle, they never touch solid matter. A circular magnetic field for this *toroidal* nuclear pathway can be induced by electrical currents, but to confine the hot plasmas of a fusion reaction requires magnetic fields of tremendous strength and stability.

The closest approach to practical fusion conditions has been achieved in a Russian device, the "Tokamak," which is based on toroidal magnetic confinement as we described it above. Russian workers have reached 5 million °C and have also approached the necessary ion density and confinement time to within a factor of about ten.

Fusion Reactors. In one of many possible designs, a power-generating fusion reactor of the future (conceivably before the year

2000) might have a large toroidal magnetic bottle 30 ft or more in diameter. Within it, a dilute deuterium and tritium plasma would fuse according to the nuclear reaction mentioned earlier:

$$\overset{2}{_1}H \quad + \quad \overset{3}{_1}H \quad \rightarrow \quad \overset{4}{_2}He \quad + \quad \overset{1}{_0}n$$

deuterium   tritium   helium-4   neutron

The neutron, uncharged and thus able to escape the magnetic bottle, would carry about 80 percent of the produced energy outside to a shield containing liquid lithium (Li). The molten lithium is heated by the bombardment, and its energy is transferred to steam for driving a turbine.

Lithium has another crucial role in this hypothetical reactor aside from receiving and transferring energy. With the aid of the impinging neutrons, it becomes the source of tritium. Both deuterium and tritium are needed for fuel in this particular reactor, as the preceding nuclear reaction shows, but tritium cannot be obtained in abundance from the oceans. Tritium, instead, can be produced by two different nuclear reactions between lithium and neutrons:

$$\overset{7}{_3}Li \quad + \quad \overset{1}{_0}n \quad \rightarrow \quad \overset{4}{_2}He \quad + \quad \overset{3}{_1}H \quad + \quad \overset{1}{_0}n$$

lithium-7   fast neutron   helium-4   tritium   slow neutron

$$\overset{6}{_3}Li \quad + \quad \overset{1}{_0}n \quad \rightarrow \quad \overset{4}{_2}He \quad + \quad \overset{3}{_1}H$$

lithium-6   slow neutron   helium-4   tritium

These reactions would guarantee an adequate supply of tritium.

Reduced Environmental Hazards of Fusion. Besides a near-infinite supply of energy, an attractive feature of fusion is that it would be a much less hazardous source of power than is fission. First of all, a fusion reaction cannot by itself get out of hand—instead nature is working very hard to keep the positive nuclei apart, and it will take all man's ingenuity to get them together for fusion. Only small quantities are in the magnetic bottle at any one time (unlike U-235 or Pu-239 in a reactor core), and any conceivable accident or failure of equipment would release them rather than cause an explosion.

The major radioactive contaminant from a fusion reactor would be tritium. It has been estimated that for a large power plant, about 10^8 curies of tritium would be in inventory, a small amount of which might escape through various mechanisms. A fission plant has a comparable amount of radioactive wastes, such as iodine (I)-131 and strontium (Sr)-90. A significant difference is that the tritium nucleus releases only slight energy in its radioactive decay, producing far fewer rads or rems. Thus per curie, it creates far less biological havoc than I-131 or Sr-90. Another important difference is that excess tritium is used up as part of the nuclear fuel, whereas the byproducts of fission are long-lived wastes to be discarded eventually to the environment.

The ultimate success or failure of fusion power may be largely determined by studies made in this decade. The outcome will have a considerable impact on the future environmental direction of mankind.

PLOWSHARE

Plowshare is an AEC program to utilize nuclear explosions for large earth-moving and mining projects. Shallow nuclear explosions in the ground create craters up to a mile in diameter. In the right place, this crater would be a new harbor. A continuous row of smaller craters would constitute a canal or a roadbed through steep mountains. Iron-ore deposits and oil shale would be reduced to rubble in preparation for mining and extraction. Explosions further down would fracture rock, creating storage areas for oil, gas, water, and garbage. Natural gas would seep out through such fractures in certain rock formations, thus enhancing gas production. Copper ores in the broken rock would be exposed, and copper could be leached out chemically. Certain rock formations that exist at high temperatures would be fractured so that geothermal energy could be extracted by circulating steam. These and other possibilities have been studied extensively by the AEC and private industry, using underground and cratering tests in the Nevada desert to guide them.

The largest Plowshare project ever to receive serious consideration was the digging of a new, sea-level Panama canal. A 57-mile route across Panama, cutting through hills a thousand feet high, would require a chain of nuclear explosives with total yield in excess of 150 megatons, equivalent to 10,000 Hiroshima bombs. Radioactivity has never been consistently retained in cratering experiments, so the release of considerable fission products would be expected. Of perhaps greater environmental significance, a sea-level canal would permit intermixing of Pacific and Atlantic ocean species, and large populations might be decimated through altered predator-prey relationships.

Very recently, planners began leaning toward excavation of the canal by chemical explosives. This change would relieve the radioactive risks, but it would be just as disturbing to the gross ecology of the oceans.

The first plowshare explosion in the U.S. took place 4000 ft under northern New Mexico in December, 1967. Dubbed Project Gasbuggy, the 26 kiloton explosive created a cavern full of rubble near gas-bearing rocks. Over a million cubic feet of gas a day leaked into the cavern initially. However the gas was so radioactive with gaseous tritium (H-3) and krypton (Kr)-85 that it could not enter commercial pipelines. Instead it had to be "flared"—burned in the open. This problem, encountered again in Project Rulison, a similar shot in Colorado in 1969, cast the future of nuclear-induced gas production in doubt. While Kr-85 and H-3 both decay with half-lives of about a decade, and in addition are swept out to a large extent with the first flared gas, residual radioactivity remains. Put into natural-gas pipelines, it would appear in home furnaces and stoves. The exposure level could be kept below present limits by mixing natural gas from several wells, but the margin is not great enough should new restrictions be put on the radioactive content of consumer products.

Plowshare has not yet proceeded far because the most economical explosions are large ones, in the megaton range, which shake the earth with the intensity of an earthquake. Small ones are economically marginal, especially if some of the product must be discarded because of radioactivity. Also, the project is facing a growing criticism from environmentalists concerned with radioactivity in consumer products and groundwater. For these various reasons, the future of Plowshare is now uncertain.

OVERVIEW OF RADIATION SOURCES

There are so many radiation sources that a summary helps give us perspective. Most sources contribute only slightly to man's exposure, but in some instances local exposures can be considerable. An overview will help the reader judge the present and potential importance of various local and worldwide sources of radiation exposure.

Recall that every individual is exposed to background radiation in the neighborhood of 100 millirads and that many are additionally exposed to medical x rays at a substantial fraction of that amount. Beyond this widespread radiation background, environmental contamination contributes a further increment to human exposure. Among the diverse sources are the following.

Bomb Fallout. This contribution is greatest immediately after periods of extensive atmospheric testing, which is now largely curtailed by the partial test-ban treaty of 1963. Extensive testing was done in 1957, 1958, 1961, and 1962. Local exposure was particularly heavy: according to environmental fallout expert Robert Pendleton, some children in Utah received thyroid doses of 14 rads because of the release of I-131 from atmospheric tests in Nevada in 1962. Exposure from earlier tests probably was higher. A residual environmental radioactivity of perhaps 5 to 15 millirads per year is a reminder that radioactivity, once released, does not disappear from earth except at its own leisurely pace, dictated by half-life. The major radioisotopes of bomb fallout are Sr-90, I-131, Cs-137, C-14, Pu-239, H-3 (tritium), Sr-89, and Zr-95.

Nuclear Power Plants. The release so far seems to have been slight, giving perhaps 5 millirads to someone near the plant site. The future is of more concern, because it is not certain that radioactive emissions can be kept small as this source of power expands vigorously—perhaps a hundredfold between 1970 and the end of the century (Figure 11-20). The principal radioisotopes emitted are the volatile ones, Kr-85, Xe-133, Xe-135, I-131, and H-3.

Reactor Accidents. Accidents disturbing the reactor core have occurred in the U.S., Canada, and Britain. Radioactivity released in the latter was considerable, including 20,000 Ci of I-131. Modern power reactors are much safer, but they are also more numerous. With so many reactors, accidents are a distinct possibility. An accident might release large quantities of the long-lived fission products ordinarily confined within the fuel rods: Sr-90, I-131, Cs-137, Sr-89, and H-3, along with Pu-239.

Fuel Reprocessing Plants. Here the built-up fission products are removed from uranium and plutonium, and some inadvertently escape in the process. Essentially all the noble-gas isotopes, including Kr-85, and all the tritium, H-3, escape through the fuel-processing stacks. Kr-85 remains in the atmosphere until decay. Although it is not now a problem, nuclear-power expansion could lead to worldwide exposures of 50 to 100 millirads per year within a century. H-3 is a little less serious because it chemically unites with oxygen (O) to form water (H_2O) and is "lost" in the vast oceans. Locally it can give measurable exposure—nearly 10 millirads, for instance, near Oak Ridge, Tennessee. Isotopes other than Kr-85 and H-3 which escape in fuel reprocessing are I-131, Ru-103, and Ru-106.

Plowshare Projects. Plowshare explosions are set off underground with the intention of moving or fracturing earth and rock. Fission products are generated at the point of explosion. If the explosion is shallow and vents to the atmosphere, these fission products will be released. If it is deep, they will be contained or possibly released to underground water supplies. In experiments stimulating natural gas production, the gaseous radioisotopes Kr-85 and H-3 will enter natural gas pipelines. Experience with Plowshare is so meager that a clear picture of the magnitude of the potential danger has not yet emerged.

Uranium Mining and Milling. Uranium ore and dust near mines, mills, and mine-tailing piles contains natural radioisotopes. The worst one is radon (Rn)-222, which, being a gas, enters and contaminates the nearby atmosphere. Mine tailings used as foundation material for houses in Grand Junction, Colorado, have caused human exposure and political furor. The problem is mainly local.

Fission Wastes. The fission products of nuclear reactors constitute by far the greatest inventory of concentrated radioactive materials on earth. Some of these wastes—separated from partly used uranium and plutonium fuel—are released during disposal. Included is Kr-85, released directly to the atmosphere. The more potent solid or liquid radioisotopes are stored or buried. For years the AEC planned to solidify future wastes for burial in a salt repository near Lyons, Kansas, but these plans are now in doubt. Burial

This vast excavation, hundreds of feet below the surface in Kansas salt, was prepared for the burial of nuclear wastes. Then the AEC changed its mind and is tentatively planning on surface storage. All waste storage methods entail risk, for these lethal wastes must be kept away from man for thousands of years, yet they cannot be insulated totally from future political and geological upheavals. (*Photo courtesy Oak Ridge National Laboratory*.)

near the surface is currently favored. Previously radioactive discards were buried as liquids in tanks, which occasionally leaked. While large losses have not yet occurred, the magnitude of the problem is indicated by the fission residue from plutonium production at Hanford, Washington. There, 140 giant tanks contain 50 million gal of liquid fission wastes. Their combined long-lived radioactivity is perhaps 10 billion Ci—equivalent to the expected release in a general nuclear war. These wastes are so radioactive that, despite decay, they will still be dangerous over five centuries from now. It seems almost inevitable that, as "nuclear graveyards" grow around the world, accidents and geological events such as earthquakes are going to release some of these wastes, with possible catastrophic results.

Radioisotopes in Space. A limited number of "atomic batteries," deriving energy from radioactivity to power satellites, are in use. These devices burn up and release their radioactivity on atmospheric reentry, so the exposure is slight. Nuclear rocket propulsion might pose a far greater threat to the future, but it is now a long way from practical realization.

Radioisotope Tracers. Radioisotopes provide a powerful tool for tracing important molecular pathways in the laboratory, industry, and the human body. The amount which escapes is small, and no substantial environmental contamination has so far resulted.

Coal-Burning Power Plants. Coal contains 1 to 2 ppm of both U-238 and Th-232. These and the products of their radioactive disintegration series are released with fly ash as the coal is burned. Major radioisotopes are Th-230, Th-228, Th-232, Ra-226, Ra-228, and U-238. Nuclear-power proponents take glee in the fact that coal-burning plants emit as much radioactivity as some nuclear plants of comparable size. However, human radiation exposure from coal plants is not now particularly significant.

Exercises

1. How many neutrons are in each nucleus of the following environmental isotopes? Strontium-90 ($^{90}_{38}Sr$), cesium-137 ($^{137}_{55}Cs$), iodine-131 ($^{131}_{53}I$), strontium-89 ($^{89}_{38}Sr$), plutonium-239 ($^{239}_{94}Pu$)

2. How many neutrons are in each nucleus of the following environmental isotopes? hydrogen-3, carbon-14, krypton-85, zirconium-95, radium-226.

3. A few radioactive substances such as plutonium (Pu)-239 are safe to hold for a brief period in a gloved hand, but deadly when ingested. From this fact, deduce whether the most significant radiation emitted is alpha, beta, or gamma. (Pu-239 emits two forms of radiation, but because of their relative energies, one is considerably more dangerous than the other. In this question we are interested in the more serious of the two.)

4. Each above-ground nuclear blast triggered by plutonium ($^{239}_{94}$Pu) may scatter kilogram quantities of plutonium into the environment. Pu-239 is one of the most deadly of radioactive materials because it collects in the bone near the blood-forming marrow, causing leukemia and other disorders. With a half-life of 24,000 years, it is retained in the environment for many centuries. How long, specifically, would be required for a 4 kg environmental injection of Pu-239 to decay to 125 g?

5. Eight kg of plutonium ($^{239}_{94}$Pu) are released in a nuclear explosion. How many kg will remain after 72,000 years? (The half-life of Pu-239 is 24,000 years.)

6. Eight g of an environmentally important isotope are released to the atmosphere. At a time 31.2 years later only 1 g of the isotope remains in the environment, the remainder having decayed. What is the half-life of this isotope? Identify the isotope from the various half-life listings in the text.

7. Krypton (Kr)-85 undergoes beta decay to a stable isotope. Balance the nuclear reaction for this beta decay and identify the isotope formed.

8. A zirconium (Zr)-95 nucleus undergoes two beta decays before a stable nucleus is formed. Identify the stable isotope.

9. Natural thorium (Th)-232 slowly undergoes alpha decay, thus initiating another radioactive decay series. Balance the nuclear equation for this alpha decay and identify the first daughter isotope in the series.

10. The fission byproduct tritium (H-3) from a single year's production of nuclear power (about 2×10^{10} watts) amounts to 2×10^5 Ci. How many nuclear disintegrations occur per sec in the tritium produced in just one year?

11. By the end of the century, it is anticipated that nuclear power will produce 10^{12} watts of electricity and 9000 kg of waste strontium (Sr)-90 per year. This 9000 kg produces about 1.3 billion curies of radioactivity. How many nuclear disintegrations would occur per sec in this amount of Sr-90?

12. In December 1970, an underground nuclear blast in Nevada vented radioactive materials to the atmosphere. On December 20 the air over Salt Lake City, Utah, several hundred miles northeast, was slightly radioactive, 200 picocuries/m^3. How many nuclear events per sec were occurring in a m^3 of that air?

13. The Atomic Energy Commission (AEC) has buried over 50 million gal of high-level radioactive wastes at Hanford, Washington, the site of reactors used to produce nuclear weapons fuel. It is estimated that these wastes contain 10^{10} Ci of radioactivity. Should these wastes escape and spread evenly over the surface of the earth, how many nuclear disintegrations would occur over each square mile of the earth's surface per sec? (The area of earth is about 2×10^8 mi^2. It is unlikely that all wastes would escape and equally unlikely that they would dilute uniformly into the world environment. Hence your answer would more likely apply to the possible danger to certain local regions of the earth.)

14. What weight of plutonium (Pu)-239 would cause a 33 percent risk of cancer in a 50 kg woman? Assume the dose rate in Table 11-9 to be applicable to humans.

15. Charcoal from the Lascoux Cave in France showed a count of 4 decays per min per g of carbon, as opposed to 16 for living material. Estimate when cave occupants left these charcoal remains.

16. Identify isotope X, which accompanies the release of zirconium (Zr)-99 and two neutrons in U-235 fission:

$$\ _{0}^{1}n + \ _{92}^{235}U \rightarrow \ _{40}^{99}Zr + X + 2\ _{0}^{1}n$$

17. What is the atomic number of the isotope released with tin (Sn)-132 in the fission of a U-235 nucleus?

18. Write the balanced nuclear equation for each beta decay in the cerium (Ce)-141 environmental series. This beta-decay sequence begins with xenon (Xe)-141 and ends with praseodymium (Pr)-141.

19. Calculate the change that occurs in the neutron-proton ratio in the I-131 beta-decay series, which begins with $_{50}^{131}Sn$ and ends with $_{54}^{131}Xe$.

20. In a typical fission explosion, fission occurs in about 10^{25} plutonium (Pu)-239 nuclei. How many cesium (Cs)-137 nuclei are produced in this explosion? (Use Table 11-14 to obtain the fission yield of Cs-137.)

21. Suppose U-238 naturally present at 2 ppm in a ton of coal were extracted from the coal and "burned" in a breeder reactor. Would this trace of U-238 provide more or less energy than the ton of coal it was imbedded in? By what factor? (Assume that 1 lb of U-238, like U-235, provides energy equivalent to 1400 tons of coal.)

22. If 1 part in 10,000 of fissionable breeder fuel were diverted to fission bombs, how many bombs would be built in one year if Weinberg's proposed consumption of 500 tons per day (180,000 tons per year) of fissionable fuel were to materialize? Assume that each bomb contains 10 kg of fissionable material.

23. The large Soviet SS-9 missile is capable of launching a 20 megaton fusion bomb across the ocean. To how many Hiroshima bombs (14 kilotons) would the explosion of such a warhead be equivalent?

24. Using the fact that the abundance of deuterium is 0.02 percent, calculate how many g of deuterium (fusion fuel) there are in each ton of ocean water.

25. Lake Michigan contains 5.4×10^{12} tons of water. If the fusion energy of the deuterium in Lake Michigan were used to replace worldwide fossil-fuel combustion at its present level, how many years would the lake's deuterium supply last? Present world consumption of fossil carbon is 5 billion tons per year. (Recall that the deuterium in 1 ton of water has an energy equivalent to 440 tons of coal, where coal and fossil carbon can be considered equivalent.)

26. Energy release in nuclear reactions is often stated as the number of "million electron volts" (MEV), the energy imparted to an electron by a million volts. The energies released per nuclear reaction for fusion and fission processes are shown below in parentheses.

fusion: $_{1}^{2}H + \ _{1}^{3}H \rightarrow \ _{2}^{4}He + \ _{0}^{1}n$ (17.6 MEV)
fission: $_{92}^{235}U + \ _{0}^{1}n \rightarrow$ fission products (200 MEV)

Does fusion or fission provide more energy for a fixed mass of nuclear fuel consumed? By what factor is it larger? Assume that the atomic weights of the fuel components are equal to the mass numbers.

Glossary

Abundance ratio the relative amounts in existence of the various isotopes of an element.

Alpha particle a particle composed of two protons and two neutrons, identical to the nucleus of the helium atom.

Atomic number, Z the fixed number of protons in the nucleus of an element.

Beta particle an electron of high energy and velocity, emitted by a radioactive nucleus.

Biological concentration biological magnification: processes in which living organisms extract certain substances from the environment, concentrate them, and pass them in increasing amounts up the food chain.

Boiling-water reactor (BWR) the simplest design of a nuclear power-generating unit.

Chain reaction a series of repeated reactions in which one reaction produces the initiator for the next. The chain may *expand* if more than one initiator is produced per reaction.

Control rods rods containing boron or cadmium that can be adjusted to subdue the rate of reaction in a nuclear reactor by absorption of neutrons.

Critical mass that amount of fissionable material in which the chain reaction neither grows nor dies.

Curie a measure of radioactivity: 1 curie (Ci) $= 3.7 \times 10^{10}$ radioactive disintegrations per second.

Decay curve exponential decay curve: graph showing how the amount of a radioactive substance decreases exponentially with time.

Doubling time the fixed time period during which an exponentially growing quantity will double.

Fast breeder reactor breeder (fuel-producing) reactor that uses neutrons that travel much faster than do thermal neutrons.

Fertile isotopes isotopes that can be converted or "bred" to fissionable isotopes by nuclear transformations. They are U-238 and Th-232.

Fission the splitting of a very large nucleus to two intermediate-sized nuclei.

Fuel rods rods in the core of a nuclear reactor, containing U-235 fuel enriched to 2 to 4 percent.

Fusion the joining of two small nuclei to form a larger nucleus.

Gamma ray a very energetic form of electromagnetic radiation, originating in a radioactive nucleus.

Half-life the time required for half of the original nuclei of a radioactive sample to disappear.

Ignition temperature the minimum temperature that will cause enough fusion to produce a net energy release.

Isotopes atoms of the same element that have the same number of protons but different numbers of neutrons.

Lethal dose to 50 percent (LD_{50}) the amount of a toxin which, given to a certain population, causes the death of 50 percent within a short time period.

Linear theory theory stating that the risk of sickness and death from radiation is proportional to radiation dose.

Liquid metal fast breeder reactor (LMFBR) breeder reactor being developed, in which sodium is the liquid coolant.

Magnetic bottle   a circular line of magnetic force that confines charged particles to follow it in a spiral path; proposed as part of a possible fusion reactor.

Mass defect   the difference in mass between the nucleus and the sum of its separate protons and neutrons. Changes in mass defect are the source of energy released in nuclear reactors.

Mass number, A   the sum of the number of protons and neutrons in the nucleus.

Maximum permissible concentrations (MPC)   limits set by the AEC on the amount of radioactivity in air and water released from nuclear power plant sites.

Moderator   a substance surrounding the fuel rods in a reactor that slows the neutrons to thermal velocities.

Mushroom-shaped cloud   the rising fireball residue of a fission bomb and the column of air sucked up beneath it; a symbol of our nuclear age.

Nuclear disintegration series   the decay chain by which one radioactive isotope gives rise to another which in turn gives rise to a third, and so on until a stable isotope is reached.

Nuclear forces   quantum mechanical forces that hold the protons and neutrons together in the nucleus.

Plasma   a gas composed of charged particles at very high temperatures.

Pressurized-water reactor   design of a nuclear power unit, more complicated than the boiling-water reactor and also releasing less radioactive material to the atmosphere.

Rad   a measure of energy absorbed from radioactive fragments by tissue or other material; 1 rad = transfer of 0.0024 calories per kg of material.

Radioactive   describes an isotope that emits high-energy particles or rays.

Radioactive fallout   vaporized radioactive fission products of a fission bomb, which, upon cooling, attach to dust particles or water drops and fall to earth.

Radioactivity   the emission of nuclear radiation.

Radioisotope   isotopes that are unstable, or radioactive.

Relative biological equivalent (RBE)   an adjustment factor used in conversion of rads to rems: accounts for differences in biological effect of particles having the same energy.

Rem   a measure of biological damage; 1 rem = 1 rad × RBE (see Relative biological equivalent).

Sigmoid curve   an S-shaped curve that is thought by many to describe the relation of biological damage to radiation dosage.

Stable   describes an isotope that does not tend toward radioactive decay.

Subcritical   describes a mass of fissionable material in which an explosive chain reaction cannot occur because too many neutrons are lost before they can induce fission; less than the critical mass.

Thermal neutrons   neutrons traveling at thermal velocity: slower than "fast" neutrons emitted from radioactive nuclei.

Thermal pollution   discharge of excess heat into waterways.

Thermonuclear reaction   name given to fusion reactions because of the very high ignition temperatures required.

Threshold theory   states that below a certain dosage, no injury is caused by radiation exposure.

Toroidal   doughnut shaped; describes the path of a nucleus in a magnetic bottle.

Yield   the number of long-lived radioactive nuclei produced for each 100 fission processes of a given radioisotope.

Additional Reading

Many excellent articles on nuclear-age topics can be found in *Bulletin of the Atomic Scientists*. Additional articles can be found in the magazines *Environment* and *Science*. The books listed below also contain much valuable information. Starred selections are most suitable for the nontechnical reader.

Casarett, Alison P., *Radiation Biology*. Englewood Cliffs, N.J.: Prentice-Hall, 1968.

*Curtis, Richard, and Elizabeth Hogan, *Perils of the Peaceful Atom*. New York: Ballantine, 1970.

Feld, B. T., T. Greenwood, G. W. Rathjens, and S. Weinberg, eds., *Impact of New Technologies on the Arms Race*. Cambridge, Mass.: The MIT Press, 1971.

Foreman, Harry, ed., *Nuclear Power and the Public*. Minneapolis: University of Minnesota Press, 1970.

*Lapp, Ralph, *The Weapons Culture*. New York: Norton, 1968.

Lewis, Richard S., and Jane Wilson, eds., *Alamogordo plus Twenty-Five Years*. New York: Viking, 1971.

*Nuclear Explosives in Peacetime: An SIPI Workbook. New York: Scientists' Institute for Public Information, 1970.

*Plate, Thomas G., *Understanding Doomsday*. New York: Simon & Schuster, 1971.

Upton, Arthur C., *Radiation Injury*. Chicago: University of Chicago Press, 1969.

U.S. Congress, *Environmental Effects of Producing Electric Power: Hearings before the Joint Committee on Atomic Energy*. Part 1, 1969; Part 2 (Vols. I and II), 1970.

INDEX

Italic numbers indicate illustrations. **Boldface** numbers indicate glossary entries.

Physical change, 153–55
Phytoplankton, 3, 379
Pica, 351, **357**
Picocurie, 407
Planck's constant, 54n
Plant debris, degradation of, 298–307
Plants, green, 323, *324*
Plasma, 454, 465
Plastics, 91, 343
Platinum, 157, 336
Plowshare, 457–58
Plutonium, 79, 403, 429–30, 435, 436
Plutonium-239, 392, 424, 426, 427, 431, 434, 445
Poisoning: arsenic, 354; cadmium, 353–54; lead, 349–51; mercury, 343–49. *See also* Toxicity
Polar bonds, 99–101, **107**, 112, 285, 286, 287, 291, 298–99, 314
Pollutants: amount of, 170–76; concentration of, 28; elements as, 77–79; masses of, 167–70; mixing and diffusion of, 11, 14; primary and secondary, 253, **275**
Pollution: acid, 321; agricultural, 298, 316, 384; automobile, 78, 234, 264–70, 351–53; biochemical effects of, 134–44; and chain reactions, 170–76; chemical, 322; chemical analysis of, 21; elements in, 77–79; environmental, 3–5, 49, 84; exhaust, 78, 234, 264–70, 352–53; food, 29; heavy metal, 331–57, 338–43; industrial, 10, 78, 172–74, 204, 206, 243, 257–63, 290; insecticide, 370–84; measurement of, 24, 26, 28, 29; and mixing, 9, 11; organic sources of, 126; particulate, 242–47; and periodic table, 77–79; by pesticides, 359–87; pH, 320–21; photochemical, 247, 250; radiation, 440–41, 442–46; and separation, 19–22; of stratosphere, 212; and synthetic chemicals, 23–24. *See also* Air pollution, Health, Lead, Mercury, Ocean, Radioactive pollution, Thermal pollution, Water pollution
Pollution control, 22–23, 174, 353; economics of, 228–30, 233–34; in coal-burning plants, 259–63; of oxides of nitrogen, 250–52; particulate, 246–47; of photochemical oxidants, 256–57; of sulfur compounds, 235–41; water, 307–11. *See also* Wastewater treatment
Pollution episodes, 226, 227–28, **274**
Polonium, 46, 399
Polonium-214, 407
Polyatomic ions, 121–26, **148**
Polyfunctional molecules, 134
Polymerization, 139, 140, **148**, 253
Polymers, 138–39, 141, **148**
Polynuclear aromatics, 131, **148**, 255
Population, 10, 163, 235, 305
Positive crankcase ventilation, 269, **275**
Positive exponential numbers, 31–32
Positive polyatomic ions, 125–26
Potassium, 77, 86, 332, 418
Potassium-40, 421–22
Potassium formate, 241
Power plants, 78, 240, 242, 257–64, *261*, 397, 461; coal-burning, 240, 242, 257, 461; fusion, 453–57; nuclear, 313–14, 392, 415–16, 429, 436–46
Precipitation, *20*, 21, **43**, 212
Precipitator, electrostatic, 247, 261

Pressure: in gasoline engine, 265; vapor, 153–54, **183**
Pressure (hydrostatic) force, 21
Pressurized-water reactor, 437, 465
Protein chains, 140–41
Proteins, 6, 136, 138–41, **148**, 157, 336, 337, 337–38, 340. *See also* Enzymes
Proton-neutron balance, 426–27
Protons, 45, 48, 65, **81**, 168–69, 335, 389–441
Pure substance, 7, 8, **43**
Purification, 19–20, 22, 278, 307–11. *See also* Separation, Wastewater treatment
Purines, 413
Pyrethrum, 360
Pyrimidines, 413

Quality factor, 408
Quantum behavior of electrons, 57–60
Quantum mechanics, 52, 58–69, **81**
Quaternary structure, proteins, 140, **148**

Radar, 54
Radiation, 206; electromagnetic, 53–57, **81**, 159, 200–202, 390, 400; exposure to, 421–22; infrared, 54, 200–202; low-frequency, 54; nuclear, 389–466. *See also* Alpha particles, Beta particles, Gamma rays, Ultraviolet radiation, X rays
Radiation damage, 404, 407–408, 409–19, 427, 442–46
Radical hydrocarbons, 133–34, 138–39, **148**, 301
Radioactive carbon, 434
Radioactive decay, 46
Radioactive fallout, 433–34, 458–59
Radioactive gases, 440–42
Radioactive pollution, 78, 79, 415–16, 422, 440–41
Radioactive wastes, 440–41
Radioactivity, 23, 24, 48, 144, 226, 290, 322, 389–465, 419–22
Radioisotopes, 391–441
Radio waves, 54
Radium, 399, 418
Radium-226, 407, 419
Radon, 415
Radon-222, *417*, 419, 422
Rads, 408, 412, **465**
Rain, 205, 234, 237, 281–82
Randomness, 161
Randomness, Criterion of, for Natural Change, 14–15, 16, 18
Rankine engine, 265, 270, **275**
Rankine scale, 152n
Rare earth metals, 70
Rasool, 207, 208
Reactions, chemical, 155, 157, 158, 160–63, 170–76, **182**, 187–90, 209–10. *See also* Endothermic reactions, Exothermic reactions, Hydrolysis reactions, Oxidation reactions, Photochemical reactions, Reduction reactions
Reactive hydrocarbons, 253, 256
Reactivity, chemical, 49, 51, 133–34, 158–83. *See also* Reactions, chemical
Reactors. *See* Breeder reactors, Fusion reactors, Nuclear reactors
Recycling, 14, 331n, 284
Reduction reactions, 159, **182**, 292, 368. *See also* Photosynthesis
Relative biological equivalent, 408, **465**
Rems, 408, **465**

Resonance, 95–97, 98, **107**, 113, 115–16, 118, 119, 125, 162, 333
Resources, 170–76
Respiration, 159, 191–93, *197*, **218**, 292
Respiratory diseases, 228, 229, 256
Reverse osmosis, 20, 22, **43**, 310
Ribose, 144
Ring compounds, 130–32, 137
Rivers, 3, 10, 291, 295–96. *See also* Water pollution
RNA (ribonucleic acid), 143–44
Rodenticides, 361
Roentgens, 408, 409
Rotary engine, 270, **275**
Rotational motion, 151, *152*, **183**
Rubidium, 66, 86
Rubidium-90, 427

Sabadilla, 360
Salinity, 314–17, **328**
Salt, *7*, 8, *9*, 18, 19, 85. *See also* Sodium chloride
Salt-water animals, 332
Sand filtration, 322
Saponification, 300, **328**
Saturation, 138, **148**, 291, 292
Scandium, 66
Scrubbers, 246
Second Law of Thermodynamics, 11–18
Second-order reactions, 157n
Sedimentation, 18, *20*, 22, **43**, 192, 194, 258, 287–88, 289, 322
Sedimentation tanks, 307, **328**
Selective vaporization, 21
Selenium, 66, 322
Semianthracite, 259
Semimetals, 70
Separation, 18–22, **43**
Settling chamber, 24–26
Sevin, 364
Sewage, 297–98, 305, 307–11. *See also* Wastewater treatment
Shared electron-pair bonds. *See* Covalent bonds
Sheet structure, 140, **148**
Shell model, atomic, 60–69
Shells, electron, 60–69, **81**, 84, 110, 127, 335, 398
Shielding, 72, 73, **81**
Sigmoid curve, **465**
Significant figures, 38–40, **43**
Silica, 259
Silicon, 65, 72, 332, 398
Silt, 287–88, 289, **328**
Silver, 336, 338
Sludge tanks, activated, 308, **327**
Smelters, 78, 173–74, 338
Smog, 24. *See also* Photochemical smog
Smoke, 206, *224*
Soap, 299–301
Sodium, 65, 72, 77, 86, 332
Sodium carbonate, 302n
Sodium chloride, 85, 88, 90, 109, 314–15, 316, 332. *See also* Salt
Sodium fluoride, 23
Sodium sulfite scrubbing, 241
Sodium tripolyphosphate, 301, **328**, 334
Soil, 231, 303
Solid-waste disposal, 243
Solubility, **43**, 84, 102–103, 286, 287, 291
Sound waves, 54
Spectrometer, mass, 169, 395
Spectroscopy, 56, **81**
Spectrum, wavelength, 54, *55*
Spider mites, 381

The Periodic Table of Elements